Darwin to Einstein

Historical Studies on Science and Belief

Darwin to Einstein

Historical Studies on Science and Belief

**Edited by Colin Chant and John Fauvel
at the Open University**

Published by Longman
in association with
The Open University Press

Longman Group Limited

Longman House
Burnt Mill, Harlow, Essex, UK

Published in the United States of America
by Longman Inc., New York

Selection and editorial material
copyright © The Open University 1980

First published 1980

British Library Cataloguing in Publication Data
Darwin to Einstein.
　1. Belief and doubt – Addresses, essays, lectures
　2. Science – Philosophy – History – Addresses –
essays, lectures
　I. Chant, Colin　　II. Fauvel, John
　121'.6'01　　BD215　　80-40579

　ISBN 0-582-49156-8
　ISBN 0-582-49157-6 Pbk

Set in 10/11pt Comp/Set Times Roman
Printed in Great Britain by McCorquodale (Newton) Ltd., Lancashire

Contents

Acknowledgements

We are grateful to the following for permission to reproduce copyright material:

Cambridge University Press for extracts from 'Natural Theology, Victorian Periodicals and the Fragmentation of A Common Context' by Robert M. Young. Paper presented to the King's College in Research Centre Seminar on Science and History 1969, to appear in *Darwin's Metaphor and Other Studies of Nature's Place in Victorian Society*, 1981; Clinical Psychology Publishing Co. Inc. for extracts from pp. 316–34 'The Harvard Pareto Circle' by Barbara S. Heyl in *Journal of the History of the Behavioural Sciences* Vol. 4, No. 4, October 1968; Daedalus, Journal of the American Academy of Arts and Sciences for extracts from pp. 636–73 'Mach, Einstein and the Search for Reality' by Gerald Holton in *Daedalus* Spring 1968, *Historical Population Studies*; The Johns Hopkins University Press for extracts from Chapter 1 'Philosophic Movements in the Nineteenth Century' by M. Mandelbaum from *History, Man and Reason: A Study of Nineteenth Century Thought*, 1971; pp. 161–98 'Science Politics and Spontaneous Generation in Nineteenth Century France: the Pasteur Pouchet Debate' by John Farley and Gerald L. Geison in *Bulletin of the History of Medicine*, 48, No. 2, 1974; pp. 1–116 'Weimar Culture, Causality, and Quantum Theory 1918–1927: adaptation by German Physicists and Mathematicians to a Hostile Intellectual Environment' by Paul Forman, *Historical Studies in the Physical Sciences* edited by McCormmach, 3, 1971; Natural History and the author, Stephen Jay Gould for extracts from 'Piltdown Revisited' ('Smith Woodward's Folly') in *Natural History* March 1979. Copyright The American Museum of Natural History 1979; Pergamon Press Ltd for extracts from pp. 213–27 'Maxwell's Demon' by Edward E. Daub in *Studies in History and Philosophy of Science* Vol. 1, No. 401, 1970; Science History Publications Ltd and the author, Dr John Hendry for extracts from 'Weimar Culture and Quantum Causality' to appear in *History of Science*, 1980; Springer Publishing Company for extracts from pp. 180–204 'Nineteenth Century Evolutionary Debate' by Robert M. Young from *Historical Conceptions of Psychology* edited by Mary Henle, Julian Jaynes and John S. Sullivan. Copyright © 1973 by

Springer Publishing Company Inc. Used by permission; Neale Watson Academic Publications Inc. for extracts from 'Kelvin and the Age of the Earth' by Joe D. Burchfield in *Lord Kelvin and the Age of the Earth* Chapter 3, 1975; Yale University Press for extracts from 'Victorian Scientific Naturalism' by Frank Miller Turner from *Between Science and Religion: The Reaction to Scientific Naturalism in Late Victorian England* Chapter 2, 1974.

Whilst every effort has been made, we are unable to trace the copyright owner of 'The Sexual Politics of Victorian Social Anthropology' by Elizabeth Fee and would appreciate any information which would enable us to do so.

Introduction

This anthology is part of an Open University third-level course, *Science and Belief: from Darwin to Einstein*. It consists of studies by historians of science published between 1968 and 1980, and aims to give as representative a selection of recent historical work in the area of science and belief as is compatible with the requirements of the course and the size of the volume. An anthology which tried to be even reasonably comprehensive would have to have been many times larger, but some indication of the diversity of modern historical approaches is evident here even from glancing down the contents list.

It will be evident, too, that the concepts of 'science' and 'belief' implicit in the choice of these articles are broader than their immediate associations may suggest. The earliest paper here, Barbara Heyl's, is concerned with the way scientific analogies were used in the 1930s to express beliefs about how society functions; the latest is a study by John Hendry of whether popular beliefs within the Weimar Republic substantially influenced theoretical physics.

So these studies may be read, fairly translucently, as a collection of informed views on Western belief structures over the last century or more. But there is another level on which the book may be used, and to which the Open University course addresses itself. Historians of science and beliefs are themselves people in history, with their own beliefs. To what extent these beliefs are apparent in, appear to influence, or can be divorced from, their historical studies is an important set of questions no more easy to answer than the equivalent questions about scientists and *their* beliefs. This book provides no answers, but through its range of approaches will serve as a collection of primary sources for such an inquiry.

It is for this reason that the subtitle's reference to 'recent historical studies' does not precisely mirror that of this book's complementary volume, *Darwin to Einstein: Primary Sources on Science and Belief* (eds Noel Coley and Vance Hall, Longman 1980). Besides, the term 'secondary source' seems less universally understood. Since, in our experience, some historians take 'secondary' to mean 'auxiliary' or 'inessential', and students sometimes understand 'second-rate' or 'undesirable', we felt the term had better lapse.

The ordering and division of items – more fully described in the chapter introductions – has been decided by subject-matter, and so as to produce a compromise between moving forward in time and moving from the 'softer' to the 'harder' sciences, both of which (not altogether compatible) progressions are implicit in the title *Darwin to Einstein*.

Colin Chant and John Fauvel
The Open University
February 1980

Chapter 1

1.0 Introduction

The history of science can hardly be studied from the starting-point of the beliefs of scientists and interpreters of science, without a decision being made about what beliefs are to be admitted, or what are to be admitted as beliefs. In the period before Darwin, it is the relations of science and religious belief, be they fractious, harmonious or even symbiotic, which have largely captured the attention of historians. In the debates which surrounded the publication of Darwin's *Origin of Species*, many of the participants moved towards a clear demarcation of the domains of science and religion. Accordingly, many modern historians find misconceived the once common view that the two were in conflict in those times, even though scientists and clerics no longer pursued common endeavours and goals within a framework of natural theology.

This chapter rests on the premise that those upheavals in humanity's perception of itself and the world, associated in particular with developments in physics and the biological and human sciences after Darwin, make it imperative that historians look not only to religion, but also to philosophical categories in their exploration of the genesis and reception of scientific theories. It is primarily with that thought in mind that the following extract from Maurice Mandelbaum's *History, Man and Reason* was chosen. In it, he defines and illustrates what he considers the two dominant movements in nineteenth-century philosophy: metaphysical idealism and positivism. Materialism is discussed in similar detail, though accompanied by a forthright rejection of the common characterization of the later nineteenth century as dominated by materialistic beliefs. In his exposition of unorthodox religious views, Mandelbaum dismisses the conflic model of science and religion, and identifies as the most influential strand of nineteenth-century thought that which sought to divorce science and religion, while maintaining the value of each. It is instructive for our purposes that science and scientists crop up so often in what is offered as a preliminary sketch of the history of nineteenth-century philosophy in a wide-ranging study of nineteenth-century thought. Note especially Mandelbaum's location of 'philo-

sophically inclined scientists' at the forefront of the movement he terms
critical positivism, and his argument that the later development of that
movement into a critique of science led certain scientists and
philosophers of science to idealist metaphysics, in apparent contra-
vention of positivism's historical starting-point.

1.1 Maurice Mandelbaum, Philosophic Movements in the Nineteenth Century

1. Metaphysical idealism

When taken as referring to a form of metaphysical doctrine, 'idealism'
may be defined in a variety of ways, depending to some extent upon the
context in which the definition is to be used. For the sake of offering a
rather inclusive, non-technical characterization I would suggest the
following: *metaphysical idealism holds that within natural human
experience one can find the clue to an understanding of the ultimate nature
of reality, and this clue is revealed through those traits which distinguish
man as a spiritual being.* Implicit within this characterization is the fact
that idealism, like every other traditional form of metaphysics, would
regard it as both meaningful and important to speak of 'the ultimate
nature of reality', thus drawing a distinction between reality and
appearance. In using the singular form, 'the ultimate nature of reality',
this characterization is also intended to suggest that idealists assume
some form of oneness in the world, in contrast to doctrines which are
ineradicably dualistic or pluralistic. That which is common to all things
must of course be present in man; however, the foregoing characterization stresses the idealist contention that we come into closest contact
with the nature of ultimate reality through entering most deeply into our
own natures, rather than seeking reality in external and apparently alien
aspects of the world. In attempting to discover that which is most ultimate, we need rely only upon natural human experience: according to
the traditions of metaphysical idealism, we are not dependent upon a
revelation to apprehend the truth. Thus, in sum, the metaphysics of
idealism finds man's own spiritual nature to be the fullest expression of
that which is to be taken as basic in reality.

According to this characterization, an idealist metaphysics would not
only be opposed to materialism, but would be distinct from, and
opposed to, a variety of other metaphysical systems, such as those of
Aquinas or of Descartes. Thus, the above characterization, whatever its
inadequacies, should help us to anticipate some of the more specific
philosophic and religious convictions which are to be found in
nineteenth-century thought. Further, it should help to clarify not only
the similarities which exist among various forms of idealism, but the
point at which their differentiating characteristics are to be expected. If
the preceding characterization is taken as being substantially correct, the

most important differences among idealist systems would lie in their conceptions of what constitutes the essential nature of man as a spiritual being, and how we are to apprehend that nature. Consequently, issues concerning metaphysical idealism would, in part at least, be separated from epistemological arguments concerning the independent existence and ultimate nature of material objects, with which they have too frequently become confused.

If we now turn from these generalities and examine the period with which we are to be concerned, it immediately becomes evident that the critical philosophy of Kant does not, in any particular, conform to the characterization of idealism which we have proposed.[1] And this is as it should be. This does not of course imply that the Kantian system did not influence – both positively and negatively – the development of the classic forms of German idealism. It does imply that the sources of that idealism are not to be found in Kant alone. To be sure, it is easy to trace the development of German idealism from Fichte through Schelling to Hegel in terms of technical problems within the Kantian system; these problems did of course have a decisive formative influence on the thought of all of Kant's immediate successors, idealists and non-idealists alike. However, the German idealist movement as it developed at the close of the eighteenth and the beginning of the nineteenth century cannot be adequately interpreted unless it is seen as part of a more general rebellion against the conceptions of man and of nature which characterized the Enlightenment. It was out of new convictions concerning the inner spiritual forces in the individual, in natural objects, and in cultures, and out of a conviction that there was a unity in all of these forces, that German idealism arose. The forms of argumentation within idealism may have originally been parasitic on the Kantian system, but it was not that system, and its difficulties, which can be said to have engendered what was new in this movement.

Once one acknowledges that the classic period of German idealism is not to be construed primarily as an attempt to resolve tensions within the critical philosophy of Kant, one can see that the idealist tradition in Germany did not end with the Hegelian system, and that Schopenhauer – in spite of his avowed relationship to Kant – is to be regarded as an important link in the continuing influence of idealism. To identify ultimate reality with the Will, and not with Reason, is not to give up metaphysical idealism with its distinction between appearance and reality, its emphasis upon the one all-embracing totality, and its faith that this reality is to be interpreted in terms of that which constitutes the inner, spiritual nature of man. For Schopenhauer, this inner nature was poles apart from what Hegel conceived it to be, but it would be a mistake to hold that an identification of man's spiritual nature with his reason (which is in fact Platonic rather than Hegelian), or with his historical and cultural achievements (which is closer to Hegel's own view), is a necessary attribute of idealism. Nor need idealism be pervaded by optimism. When these points are noted, it should be obvious that

Schopenhauer stands squarely within the tradition of metaphysical idealism in a sense in which his master Kant did not.

It was not only in Schopenhauer that the tradition of metaphysical idealism persisted in Germany among those who rejected Hegel. To be sure, much of the intellectual life of the period turned away from all metaphysical issues; nonetheless, during the middle years of the century, two philosophers trained in the practice of the empirical sciences, Lotze and Fechner, formulated new and influential systems of metaphysical idealism.[2] It was their aim to provide a means by which idealism could do justice to the mechanical view of nature which other idealists, following Hegel, were ready to condemn. Meanwhile, Schopenhauer's influence continued to spread, and in 1869 Hartmann published his *Philosophy of the Unconscious*, which had an immediate and spectacular success.[3] This success was doubtless in part due to the very great influence of Schopenhauer (or to whatever in the times made voluntaristic pessimism congenial to a wide audience), but it should not be overlooked that Hartmann shared the concern of Lotze and of Fechner, endeavouring to find a means by which the concrete findings of science could be incorporated within the framework of an idealist metaphysics. The task of relating philosophy and the sciences to one another can in fact be said to have been the dominant task of German philosophy in the latter half of the century. As we shall see, the short-lived materialism of Büchner and others had this as its aim; so too did the far more widespread and influential semi-positivistic forms of Kantianism which grew up among Helmholtz and other scientists of the period. At this point it is only important to note that these movements did not in the least obliterate the traces of metaphysical idealism; because of the influence of Lotze in particular, that tradition maintained itself. For example, it is of interest to notice – as one can in the case of Wundt – that when the major scientific figures in Germany in the latter part of the century departed from Kantianism and from positivism, and sought to undergird their thought with metaphysical notions, it was to idealism and not to materialism or any form of metaphysical dualism that they turned.

If we now shift our attention from the German scene to those who represented academic philosophy in France and in England, we do not find that idealism flourished in the earlier decades of the century. In France, Eclecticism was entrenched, and in England the influence of the Scottish school and of Hamilton were especially marked. However, by mid-century – at precisely the time at which it is usually assumed that 'scientism' was triumphing in philosophy – we find that, in both countries, metaphysical idealism was for the first time coming into its own. To be sure, the new idealist interests and systems did not wholly dominate the scene; in France the influence of positivism was strong, and in England John Stuart Mill was advocating Utilitarian moral and social philosophy and developing a form of critical positivism in his theory of knowledge. However, by mid-century one of the truly dominant forces in the philosophy of both France and England is to be found in the birth of a strong idealist tradition.

For example, in France, if we trace the course of the Eclectic School against which both the positivists and the new idealists reacted, we can see that it was tending away from its earlier reliance upon the Scottish philosophy toward an idealism which had strong affinities with Schelling.[4] The influence of Maine de Biran, which the Eclectics did much to promote, led French philosophers to be concerned with those problems of the self which eventuated in their characteristic personalistic idealism, emphasizing volitional activity and freedom. For example, it was the adoption of a volitional psychology that led Ravaisson, one of the seminal figures of the time, to reject monistic idealism and embrace spiritualistic personalism.[5] One can also note the extent to which he emphasized the importance of Leibniz' thought, and used Leibnizian doctrines in his evaluations of the thought of other philosophers. A similar tendency toward spiritualistic personalism is to be found in both Renouvier and Lachelier. Each of these philosophers started from a combination of Kantian epistemology and a fundamental concern for concrete human freedom; Renouvier ended his work at the turn of the century with an explicit metaphysics of pluralistic idealism which he himself was willing to term a monadology, and Lachelier's idealism, which contained a far-reaching critique of the ultimacy of causal explanations in our understanding of nature, prepared the ground for much that was to follow in French metaphysical thought.

To what extent the new pluralistic idealism in France was also influenced by the emphasis upon contingency found in Cournot's philoscphy of nature and of method, I am not in a position to estimate.[6] However, the same theme of contingency is elaborately expounded by Lachelier's pupil Boutroux in connection with his defense of pluralistic idealism.[7] Within Boutroux's idealism, as he later developed it, there was also contained an interpretation of the nature of scientific thought which merged with developments which we shall trace within the history of positivism. This interpretation of science came to be characteristic of French thought at the turn of the century: it was basic to the thought of Poincaré (Boutroux's brother-in-law), of Milhaud (who was originally a Comtean), of the Catholic Pierre Duhem, and of Bergson. But it is not only important to note the widespread acceptance of this now familiar interpretation of science, it is also important to notice its compatibility with a full-fledged metaphysical idealism, and the affirmation of individual creativity and freedom which was one of the main themes of nineteenth-century French idealism.

We shall later have occasion to refer to Bergson's system in another context. Here it is only necessary to remark that with the advent of Bergson,[8] the course of French philosophy for a time became fixed, his influence augmenting through the first decades of this century. Thus, from the middle years of the nineteenth century well into this century, metaphysical idealism formed a continuing strand in French thought. Whether this idealism was most influenced by Schelling or by Leibniz, or whether indeed a compromise was sought between it and Catholic theism, it had a strong voluntaristic tone which submerged the last vestiges of Hegelianism and tended toward those personalistic forms of idealism which could place emphasis on human freedom.[9]

In England, there was also a continuing strand of metaphysical idealism. The influence of German metaphysical idealism on Coleridge and on Carlyle is well known, and James Martineau's influence, although less frequently noted, was also considerable.[10] When one recalls that it was in 1850 that Mill could regard Coleridge as one of the two recent thinkers (the other being Bentham) whose thought had left the greatest impress on the age, and that it was in 1865 that Stirling published his *Secret of Hegel* and in 1866 that Edward Caird took up his post in Glasgow, one can recognize that in spite of the dominance of the Utilitarians and the furors of Darwinism, idealism took root within British philosophy in the middle years of the nineteenth century. To be sure, the development of this idealism was slightly different from that which was characteristic of Germany and France, for in Britain, during the course of the century, idealism tended to move from the dominant influences of Kant and Schelling and Fichte toward Hegelianism, while in Germany and France the Hegelian philosophy was left almost wholly behind.[11] In spite of this historically conditioned difference, the aim and the upshot of the idealist movement was in all three cases the same: it constituted a reaction against the philosophy of positivism, asserting the claim that ultimate reality was accessible to man; however, it was claimed to be accessible only if man abandoned the equation of scientific knowledge with truth, looking inward rather than outward for the clues to that which lay behind the realm of nature with which science was destined to deal.

These remarks on German, French, and British philosophy should be sufficient to show to what extent idealism constituted one of the two major strands in nineteenth-century thought.[12] We shall now consider positivism.

2. Positivism

It can be taken for granted that the systems of Comte and of Spencer were of very great importance in the history of nineteenth-century thought; yet if one were to define positivism with special reference to these systems, one would probably reach the conclusion that it was a much less widespread movement than metaphysical idealism had been. A similar conclusion would be forced upon one if positivism were defined solely with respect to the philosophy of science which, late in the century, came to be associated with thinkers such as Ernst Mach. However, there are many respects in which these two forms of positivism do in fact converge. While their programs were marked by sharp differences, the presuppositions which distinguished them from other philosophic positions were in large measure the same.

In order to characterize the positivist position in a manner that will include the systematic positivism of both Comte and Spencer, and will also include the forms of critical positivism represented by Huxley and by Mach, one may proceed as follows. First, positivism rejects metaphysics on the ground that the questions with which metaphysics is concerned presuppose a mistaken belief that we can discover principles of explanation or interpretation which are more ultimate than those which are directly derived from observation and from generalizations concerning observations. For positivists, any attempt to pass from the realm of 'phenomena' to a more ultimate reality is a hopeless and unjustifiable enterprise, no matter how deeply rooted the urge to do so may be. However, various positivists have adopted differing positions with respect to the traditional philosophic distinction between the phenomenal and the noumenal. Some have believed it necessary to assume an unknowable reality lying outside of all possible experience; others have denied that there is any such sphere. Still others, with more consistency, have swept aside all discussion of the question since it cannot be formulated in terms which (even in principle) are verifiable within experience. Thus, in so far as its first basic thesis is concerned, positivism can be seen to have connections with some traditional epistemological discussions: in some cases it has been closely related to a Kantian form of phenomenalism, in others to a philosophy of pure experience, and in still others to a position which resembles that of Hume.

This first thesis, on the basis of which positivists reject metaphysics, does not provide a sufficient characterization of what they affirm. What distinguishes positivists from others who may accept a philosophy of pure experience, or who accept some form of Kantian or Humean phenomenalism, is a second thesis: that the adequacy of our knowledge increases as it approximates the forms of explanation which have been achieved by the most advanced sciences. At this point, of course, other opponents of metaphysics frequently diverge from positivism. However, to complete the characterization of positivism one further step must be taken, and that is to note what constitutes an advanced science according to positivism. Since positivism confines all human knowledge to what has been experienced or can be experienced, it claims that a science which has freed itself from metaphysical preconceptions will restrict itself to discovering reliable correlations within experience; it is on the basis of such observed and repeated correlations that future events can be predicted, and it is on the same basis that past events are explained. According to this view, a scientific explanation does not involve appeal to any immanent forces nor to any transcendent entities: to explain a phenomenon is to be able to subsume it under one or more laws of which it is an instance. A law, in its turn, is simply a well-authenticated general descriptive statement of uniformities which have been observed to occur in the past. Any alternative interpretation of the nature and the aims of the sciences is rejected by positivism as involving a mistaken metaphysical attempt to transcend the limits of experience. To summarize, then, positivism may be said to be characterized by three interlocking theses: first, a rejection of metaphysics; second, the contention that science constitutes the ideal form of knowledge; third, a particular interpretation of the nature and the limits of scientific explanation.[13]

Taken in this sense, the positivist position was one which was widely espoused in the nineteenth century. And it is worth noting that its interpretation of science – which in 1865 Mill could quite properly regard as 'the general property of the age'[14] – even came to be absorbed into the idealist tradition. In tracing the history of the movement, as we shall now do, it will not be necessary to deal with it in terms of national compartments; it will, however, be necessary to distinguish between its two branches, which are to be termed the systematic and the critical forms of positivism.

Systematic positivism, which in its first formulation is to be identified with Comte, was a distinctively new movement in the history of thought. Other systems, to be sure, had rejected the belief that metaphysical questions were capable of solution, and other systems had also hailed the discoveries of science as illustrating the manner in which truth was progressively being attained. However, there was genuine novelty in the conception of philosophy which systematic positivists avowed, and this novelty was recognized. One looks in vain for any prior modern attempt to transform philosophy into a synthesis of the sciences. For Comte, Spencer, and others of the school, the task of philosophy became 'the organization into a harmonious Doctrine of all the highest generalities of Science'[15]; its method was to examine the empirical results attained by all sciences, seeking out the most general laws in each, and bringing these laws together into an integrated pattern of knowledge which was more general than that attainable by any single science. To be sure, neither Comte nor Spencer was able to confine himself to discussions which conformed to this definition of the aim and the method of the new philosophy. However, this fact need not blind us to the originality which they claimed for themselves, and which many were willing to grant them. Whatever their inconsistencies, the conception of philosophy which they represented was a new conception, and they were convinced that the whole of the future belonged to it.[16]

The claim to have discovered a new method and a new aim for philosophy does not of itself explain the impact of systematic positivism on nineteenth-century thought. In addition there was, of course, the prestige which science possessed, and this prestige grew appreciably during the century. Moreover, the thought of both Comte and Spencer was dominated by the view that there had taken place, and was taking place, a progressive development of man and society. This conception, the history of which we shall trace, was deeply rooted in nineteenth-century thought; the fact that both Comte and Spencer placed extraordinary emphasis upon it, helped to ensure that their systems would attain widespread popular influence. If I am not mistaken, there is one further point which must be noted in order to account for the appeal of Comtean positivism in particular. In the eighteenth century it had been an important article of faith that intellectual enlightenment provided a basis through which societies might be transformed. This heritage of the Enlightenment was widely shared in England and in France during the nineteenth century, and was not the property of any one school of thought; however, no other philosophy of history made claims as bold as Comte's concerning the role which intellectual life played in organizing all aspects of society. To those who tended to link intellectual enlightenment and social reform, and who also placed the empirical sciences in the forefront of knowledge, the Comtean system had tremendous appeal. As a consequence, its spread in intellectual circles was out of all proportion to its acceptance and defense as a viable philosophic system.

In noting this connection between philosophy and social reform, one should not overlook the fact that, in general, nineteenth-century positivism was no less closely connected with moral and political issues than was the idealist movement. Looking back upon the nineteenth century in terms of those aspects of positivism which have had the most marked influence on our own thought, it is easy to underestimate the extent to which its theory of knowledge was linked with ideals of social change. However, it is not possible to study Comte's work at first hand without seeing that his *Cours de philosophie positive* is not an isolated treatise on the sciences, their methods and their classification, but is a part of a larger systematic whole the aim of which is moral and social. It must also be noted that a socially directed motivation was intimately connected with the systematic positivism of others, both in France and in England. In England, for example, one finds this motivation in Spencer, and in such lesser representatives of the movement as Lewes and Harrison. It is also worth reminding ourselves that in an account of his reasons for writing his *System of Logic*, Mill laid stress on his hope that it would be of use in combatting false and injurious social doctrines.[17]

Although the systematic form of positivism was frequently accepted at its own evaluation of itself, many philosophers and scientists who themselves shared the basic presuppositions of positivism subjected it to severe criticism. The primary objections were directed against the attempts of Comte and Spencer to provide a rigid codification of the results of the sciences. In the light of the new discoveries which were being made in physics, chemistry, experimental biology, and psychophysics, the attempts of the systematic positivists to find a single all-inclusive system in which all empirical discoveries would fit seemed less and less plausible. Thus, although the views of Comte and of Spencer continued to exert a considerable influence throughout the nineteenth century, particularly in sociology, they did not attract important new adherents from the ranks of scientists or scientifically oriented philosophers.

However, this did not mean that the general movement of positivist thought failed to spread. On the contrary, even as systematic positivism spent its force, there was a remarkable growth in the acceptance of positivist theses regarding the nature of scientific method and an increasing stringency in the interpretation of these theses.[18] On the part of scientists, this tendency was not always connected with a general philosophic position; it sometimes remained a methodological principle only. However, among philosophers and among many philosophically inclined scientists, it did develop into an important philosophic movement which is to be designated as critical positivism. The aims of critical positivism were twofold: on the one hand to analyze the foundations of scientific knowledge, on the other to examine the true sources and meanings of all concepts which tended to be used in an uncritical, metaphysically charged manner, whether those concepts were employed by laymen or by scientists.

This new mode of positivistic thought had an earlier major exponent in John Stuart Mill. In addition to his psychological interest in analyzing the sources of our knowledge in experience and in the association of ideas, Mill was intent upon bringing to the test any unjustifiable common-sense moral or social or metaphysical doctrine. Unlike later critical positivists, he saw no special need to analyze the sources of specifically scientific concepts; he took such concepts to be adequately grounded in experience and adequately justified by the universality of the testimony which experience yielded. Unlike Kant and Hume, Mill did not stress the limitations of science, and it was perhaps for this reason that he was not looked upon by later critical positivists as being a major forerunner of their views. It was, rather, to Kant and to Hume that they tended to trace their lineage.

What led to the new interpretation of science was, I believe, a combination of factors, each of which deserves attention; I shall therefore deal with the history of critical positivism in some detail. The factors which I regard as having been of primary importance in its development were three. First, there arose among practicing scientists a conviction that their investigations needed to be liberated from preconceptions which were metaphysical either in origin or import. They held that so long as their colleagues were interested in upholding or in combatting particular metaphysical positions, scientific inquiry could not be free; nor would it be free so long as there was an uncritical acceptance of concepts to which there clung associations derived from earlier, metaphysical forms of thought. Thus, a positivist interpretation of science was brought forward by scientists themselves in the interest of advancing scientific inquiry. A second source of critical positivism was the rise of interest in what may broadly be termed the psychophysical problem, which was occupying some of the foremost physiologists and physicists of the day. These investigations led many to hold that a positivistic phenomenalism was the only epistemological position in accord with the methods and results of the sciences. Third, there arose in connection with the development of evolutionary theory what may be called a pragmatic interpretation of the human mind, and that interpretation was then applied to scientific thought itself. To illustrate these three factors I shall briefly consider some important representative figures who contributed to the growth of critical positivism among scientists in the latter half of the nineteenth century.

With respect to the first of these factors let us consider Rankine and Robert Mayer, who were not themselves positivists, but represent those scientists who felt it necessary to adopt a methodological positivism in order to free science from constrictions placed upon it by preceding modes of thought. In their attempts to liberate observation and generalization from the influence of hypothetical, mechanical constructs, their influence lay on the side of a growingly positivistic interpretation of the sciences. Rankine stated his position with respect to a contrast between two types of procedure:

According to the ABSTRACTIVE method, a class of objects or phenomena is defined by describing, or otherwise making to be understood, and signing a name or symbol to, that assemblage of properties which is common to all the objects or phenomena composing the class, as perceived by the senses, without introducing anything hypothetical. According to the HYPOTHETICAL method, a class of objects or phenomena is defined, according to a conjectural conception of their nature, as being constituted, in a manner not apparent to the senses, by a modification of some other class of objects or phenomena whose laws are already known.[19]

To be sure, Rankine did not wholly reject the hypothetical method, as a positivist would have done; and, as Clerk Maxwell pointed out, he did not abandon a realistic interpretation of science, as contrasted with the phenomenalistic positions adopted by critical positivists.[20] In these respects Mayer's views were similar to those of Rankine.[21] Nevertheless, the insistence of both on liberating empirical inquiry from constrictions placed upon it by hypothetical constructs had an important influence on the development of positivism among later figures in the history of thermodynamics, where positivism had one of its chief centers.[22]

Turning now to a scientist who went beyond Rankine and Mayer, using a full-fledged positivism in defending his objectives and methods of work, we may cite the case of Claude Bernard. In his *Introduction à l'étude de la médecine expérimentale*, published in 1865, Bernard set out to argue that the experimental method applies to living beings no less than to inorganic things, and that determinism characterizes all phenomena associated with life. To support this position it was necessary to discredit vitalism, which constituted the chief opposition to his view. The form in which he cast his attack was to argue that only a critical positivism was consonant with the methods to be followed in scientific inquiry; and this of course precluded in principle all possibility of appealing to vital forces as a means of explaining the functioning of living things. Such an identification of a particular interpretation of science with the possibility of pursuing free and original scientific inquiry, unhampered by past prejudices, was an extremely effective argument at the time. It was on the basis of it, rather than on the basis of philosophical argument, that many scientists, including Bernard, came to accept a critical positivism.[23]

Thomas Henry Huxley presents a second example of the same tendency. While it was undoubtedly true that Huxley drew much from Hume and from Berkeley, and a good deal from Kant (as he interpreted him), it was not primarily on the basis of philosophic considerations that he either put forward or defended a positivistic position. Running throughout his essays one finds emphasis on the lessons which are to be learned from science. The primary lesson was the necessity for acknowledging limitations in human knowledge: a second lesson was the fact that if men scrupulously adhere to the discipline which the practice of science is able to instill, they are capable of pushing back the boundaries of ignorance. Thus, science trains us to recognize what we can and cannot know; its training serves to teach us the necessary rudiments of epistemology. This aspect of Huxley's thought represents a very widespread tendency to interpret critical positivism as the natural outgrowth of accepting and using the methods of scientific inquiry. For those who adopted this view, philosophy was not to be regarded as a means by which the competence or the implications of science could be assessed; on the contrary, science was itself capable of showing the range of problems which men can solve. However, science could exercise this function only so long as it remained free of metaphysical assumptions, and of the vestiges of those assumptions which lay concealed in earlier modes of scientific thought. Thus, the scientists of the period stressed the need for methodological purity, that is, for remaining within the boundaries of those forms of explanation which positivism held to be the only forms which it was legitimate for scientists to use.[24]

I turn now to the second factor which was important to the rise of critical positivism: the development of psychophysics. To be sure, psychophysics does not in itself demand the adoption of any form of positivism: however, at the time, it was widely held to have important implications with respect to the boundaries of human knowledge, and implications for the interpretation of science itself. As an example of a scientist for whom critical positivism was linked to a concern with psychophysics, Helmholtz is especially noteworthy; his own vast contribution to the development of the subject, and his stature as a scientist, made his position particularly significant in Germany. We shall later examine his views in more detail: here it is only necessary to suggest their general import.

Unlike the other philosophically inclined scientists with whom we have just been dealing, Helmholtz did not regard critical positivism as being merely an extension of the method and habits of thought followed by scientists who had freed themselves from a metaphysical heritage; for him, it was a specifically philosophical position which he took to be affiliated to the position of Kant.[25] However, one can scarcely imagine a version of Kantianism more antithetical to the presuppositions of Kant's own thought, for the universal and necessary forms of experience were interpreted by Helmholtz as consequences of the nature of our sensing organs. Thus, his form of phenomenalism rested directly upon the limits of the human organism, not on a priori categories of the mind. Helmholtz developed this supposedly Kantian rejection of metaphysics in an early lecture, on the occasion of his receiving his professorship at Konigsberg. In that lecture he offered an interpretation of experience in which sensations are construed as symbols for unknown relationships; according to him, the specific nature of what we experience is not to be identified with anything existing independently of us any more than the written name of a man is to be identified with that man himself.[26] However, the fact that we can note an orderly connection among our sensations gives us the possibility of knowledge, which is the discovery of patterns of relationship among sensory elements. As Helmholtz recognized, this view of human knowledge ruled out all traditionally metaphysical questions, including all philosophic issues concerning epistemological realism and idealism.[27] Thus, as his student Heinrich Hertz later pointed out, Helmholtz assumed that psychophysics was able to establish an epistemological position.[28] This position, which was Helmholtz's version of Kant's Copernican revolution, had a considerable influence throughout the latter part of the century. As we shall later see, it was basic to the claims of DuBois-Reymond regarding the necessary limitations of knowledge; and Lange's influential *History of Materialism*, which was published in 1865, represents a similar physiologically oriented Kantianism. By the mid-seventies an acceptance of this type of position was said by Wundt to have almost been taken for granted in the German scientific community.[29]

The factors which I have already discussed were sufficient to establish a very solid base for critical positivism, but its widespread extension was associated with still another movement in late nineteenth-century thought: this was a new interpretation of science which grew out of a pragmatic, or economical, view of the human mind. According to this view, we have a tendency to order our experience in whatever ways serve to make it most assimilable to our needs, interests, and expectations. To appreciate the radical transformation which this assumption introduced into the interpretation of science, one need only recall the positivism of Bernard or of Helmholtz. For both, there was determinism in nature, and the order of our experience depended upon that determinism. Thus, although we could not know things-in-themselves, it was assumed that the relationships which we found within experience were not attributable to us, but to nature itself. This, for them, was the basis on which the very possibility of science was founded, and they did not inquire on what grounds they, as positivists, had a right to make such an assumption.[30] On a pragmatic-economical view of the human mind, however, even the order which science establishes in experience, and which we are inclined to look upon as an order fixed by nature itself, may be a product of our own tendencies to arrange and summarize experience in a manner useful to us. This conception of the human mind, and its implications for an interpretation of science was presumably accepted by Mach and by others as early as the 1860s, but it was not until the 1880s that one finds the first high water mark of its development.[31] The chief factors which contributed to that development were three.

In the first place, the Darwinian theory of evolution was interpreted as having proved that all aspects of living creatures must have an adaptive function in the struggle for survival.[32] As we see in Darwin himself, this assumption was applied to the evolutionary development of human impulses and human intelligence, no less than to the evolutionary development of bodily structures. Having interpreted the development of man's intelligence in terms of its usefulness in satisfying needs, it was but a small step to the further claim that science itself was to be regarded as a form of adaptation. This connection between Darwin's theory and an acceptance of a pragmatic-economical interpretation of science, as well as of all other knowledge, can be seen in one of the first essays in which Mach offered a systematic statement of the latter view; and in Laas, among others, there is an equally strong emphasis on the ways in which the ordering of experience into a system of knowledge reflects life-serving needs. It should be obvious that this conception of science helped to extend the arguments previously advanced by critical positivists, for its acceptance entailed that even science could not be assumed to have access to relationships existing outside of the domain of experience.

There was a second factor which influenced the development of a pragmatic-economical interpretation of science, adding its weight to evolutionary interpretations of mind. This factor is to be found in the type of analysis of experience which dominated the psychological theories of the period, for example, the views of the associationists, of Helmholtz, and to some extent the earlier views of Wundt. In line with the heritage of traditional British empiricism, it was assumed by associationists and by Helmholtz that all of our everyday concepts, such as our concepts of ordinary material objects, develop on the basis of a series of simple, recurring sensations, some of which ordinarily recur in groups. Past recurrences create tacit expectations regarding future sensations, and when apparently stable groupings are formed we fix them by giving them a name. The fullest application of this general view, extending it to the differentiation between objects and the self, is, of course, to be found in the philosophy of pure experience of Avenarius, but it had also been developed in Mach's influential *Contributions to the Analysis of Sensations* of 1886. The phenomenalistic implications of this type of analysis of perception were not of course new: they had been wholly apparent to Hume and to John Stuart Mill, to name only two. And in Hume, at least, there is recognition of a connection between this analysis and a conception of the human mind which (so long as we leave abstract reason out of account) closely resembles what came to be the pragmatic-economical view. However, it was not central to Hume's concerns to trace the implications of his views of experience for the procedures which should be followed – or should be avoided – by scientists, although there are passages in which he did discuss that issue. In the case of Mach, however, it was precisely here that his major interest lay. One of his aims – which was in line with the whole development of positivism, both systematic and critical – was to purge the sciences of all concepts in which traces of metaphysical assumptions could be found. Another was to show that it was theoretically legitimate to pass back and forth among the data of physics, physiology, and psychology in exactly the ways that Mach's own research, and the research of others, had shown that it was feasible to do. Mach was able to achieve both these aims by emphasizing the view of the human mind which had, in general, only been implicit in earlier, traditional empiricist analyses of perception: he made that view explicit, and applied it with rigor to the manner in which science itself proceeds. Just as perceptible objects are, in the last analysis, merely the sensory elements which we find often appearing together and to which we attach a name, so scientific concepts, properly conceived, only represent bundles of the elements of experience. This view allowed Mach to argue that concepts such as 'force' and 'atoms' had no proper place within science; it also permitted him to hold that different sciences are distinguished from one another solely with respect to the manner in which they find it useful to group the elements of experience. According to these assumptions, science is confined to ordering the flow of experiences according to whatever patterns permit us to predict future experiences.[33] This interpretation of science, which was wholly consonant with the dominant psychology of the day, made the pragmatic implications of evolutionary theory all the more plausible, and thus assisted the spread of critical positivism among the large number of scientists who were already predisposed to accept some form of a positivistic view.

This self-criticism and self-limitation on the part of scientists was soon to have consequences unfavourable to the general aims of positivism, for it opened the way to a reintroduction of metaphysics. Those who had previously been attacked by positivists for pursuing metaphysical questions were in a position to point out that scientists themselves no longer interpreted science as a presuppositionless system of knowledge, but as a function of our own practical need to organize experience in some readily manageable form. Thus, the pragmatic-economical interpretation of science deprived positivists of the possibility of offering any objective reasons for their belief that scientific procedures should serve as a model for philosophy or, indeed, as a model for any other form of thought. It was precisely at this point that a strong reaction against positivism, and against other forms of 'scientism', set in. However, before sketching the forms which that reaction took, it will be necessary to cite a third factor which helped to show that science was relative to the presuppositions with which it operated, thus undermining the earlier positivist view that science offered an all-inclusive and necessary way of organizing experience. This extension of a pragmatical-economic interpretation arose out of the heavy emphasis newly placed on the roles of hypothesis and theory-construction in scientific method.

Until almost the very end of the century, those philosophers and scientists who subscribed to positivism tended to interpret all forms of scientific theory in terms of generalizations drawn directly from repeated observations. While not all shared Mill's view that inductive inference must proceed from particular to particular, they did tend to regard repeated observations as the source from which any hypothesis useful to the sciences could be derived. However, after the middle of the century, developments in both geometry and physics began to undermine this emphasis on observation, and began to pose very grave problems for those who accepted a strictly positivist view of the sources of scientific knowledge in sense experience. For example, these developments seemed to be wholly at odds with the empiricist interpretations of mathematics given by Mill and by Helmholtz[34]; they also seemed to be at odds with Mach's view, according to which all so-called axioms are the products of instinctive psychological functions which have operated uniformly in the past and have thereby become firmly rooted in our experience.[35] While Mach attempted in *Erkenntnis und Irrtum*, and in other later writings, to defend his epistemological views against all difficulties which had arisen in connection with new, non-observational developments within the sciences, it was not generally conceded that he had been successful in that attempt. Views such as those espoused by Poincaré at the turn of the century became highly influential, and what was then emphasized was not the role of experience in science, but the creative, constructive aspects of scientific imagination in the formulation of hypotheses and of intellectual models.

The philosophic upshot of the new emphasis on constructive imagination which one finds in Hertz and in Poincaré, among others, was not in all respects different from Mach's own view of the general relations between science and experience. Although the assumptions of the two schools concerning the nature of the mind were extremely different, both in fact held that the role of the mind in scientific theorizing was that of selectively ordering experiences in such a way that further experiences could be predicted. Neither held it to be within the scope of science (nor within the power of man) to say that one set of constructs more nearly approximated the characteristics of nature than did another: the test of the adequacy of a scientific theory lay wholly within the results which could be obtained by ordering past and future experiences in terms of that theory. The upshot of such a pragmatic-economical interpretation of science buttressed the view that the human mind, so long as it followed the path of scientific inquiry, could not pretend to a knowledge of the ultimate nature of reality; that, on the contrary, it was only an instrument which selectively ordered its experience according to its own interest in simplicity, coherence, and predictability.

It is precisely at this point that the critical positivist movement became able to meet and merge with the idealist movement. To be sure there were persons in the positivist tradition who remained wholly committed to the general theses of positivism and who, after accepting a pragmatic-economical interpretation of science, refused to acknowledge the legitimacy of attempting to transcend science. But there were others who, at the turn of the century, were equally willing to make this attempt. Some among them sought to re-establish a metaphysics, now that science had not only been shown to be neutral on all metaphysical issues, but had, through its self-criticism, been shown to be necessarily confined to a world of appearances. Among those who followed this path, Bergson was the most influential. However, it was far more frequent to find that once this self-criticism was accepted, philosophers sought to show that since science was limited in its methods, and therefore in the results which it was able to attain, other methods had an equally valid claim to consideration. Thus, in James the pragmatic-economical interpretation of science was closely connected with his willingness to accept forms of pragmatic justification which lay wholly outside of what was traditionally viewed as knowledge. Even more frequently, however, attempts were made to distinguish the method of the natural sciences from other methods of knowledge which had an equal right to be considered as fundamental forms of human understanding. This position did not stem primarily from the pragmatic-economical interpretation of science, but it was able to converge with it. In its origins, it stemmed chiefly from the Neo-Kantian revival and from the growth of interest in humanistic historical studies. These studies, it seemed, demanded a theoretical justification which would show them to be of an equal importance with science. The beginnings of this movement can perhaps be traced as far back as Zeller, in whom we find both the attempt to revive Kant and the attempt to do justice to the place of historical studies in the economy of learning[36]; certainly we find the latter motive strongly represented as early as 1883 in Dilthey's *Einleitung in die Geisteswissenschaften*. However, it was not until almost the turn of the century that the movement reached its height.[37] From then until our own time it has not been unusual to find science interpreted as one of two distinct and equally legitimate means of ordering experiences, although we may here note that the differences between these forms of knowledge have not always been fixed with the precision which one might think was demanded by the importance of the question.

Thus, at the end of the nineteenth century, positivism had turned into a self-criticism of science, largely at the hands of practicing scientists, and the earlier systematic form of positivism had to all intents and purposes lost its hold upon the major streams of thought. What had once seemed to be the philosophic import of the physical sciences no longer carried the same conviction, and the way was open for twentieth-century philosophy to interpret these results in the most diverse fashions, the chief of which were that of setting up other modes of knowing side-by-side with science, or that of restricting the import of science in order to leave room for faith.[38]

3. Materialism

In the two preceding sections we have been at pains to trace the development of the dominant strands in nineteenth-century philosophy. In these discussions of metaphysical idealism and of positivism little mention was made of the doctrine of materialism. However, because it is so frequently claimed that the nineteenth century was a period in which the dominant modes of thought were oriented toward materialism, it will now be necessary to examine the materialist movement. Before attempting to analyze its nature and fix the range and the time-span of its influence, let us first be clear as to some of the causes which have led to the misconception that materialism represents the fullest expression of the dominant modes of thought in the nineteenth century.

There are doubtless many such causes, but one of them is surely the fact that during the nineteenth century there was a tremendous growth in material goods and an enthusiasm for material progress. Since the word 'materialism' is sometimes used to connote a concern for material goods as well as being used to designate a particular metaphysical position, and since it has often been believed that the latter position must lead to the former, the two distinct meanings of 'materialism' have tended to coalesce. However, there is no necessary relation between these two meanings. Furthermore, if the term 'materialism' is to be used to connote an overweening concern for material goods to the exclusion of what might be designated as 'moral idealism', then the nineteenth century cannot by any stretch of the imagination be called a materialistic age: it was characteristic of all schools of thought in the nineteenth century – and characteristic of the materialists no less than of the positivists and metaphysical idealists – that they were imbued with 'moral idealism'. Each school in fact sought to show that its philosophic doctrines were the only sure foundation for moral progress.

It is not with 'materialism' in this loose sense of the term that we are to be concerned. 'Materialism' has also been said to have characterized the nineteenth century for a specifically philosophic reason: during that century orthodox Christian theism was not held in high repute by a majority of the most influential thinkers. Yet this provides no adequate reason for characterizing the period as an age of materialism. In fact, many nineteenth-century idealists explicitly rejected a theistic position. In general, it was only among those who reacted against Hegelianism that theism came to be included as part of the idealistic position, and even among them this theism was frequently unorthodox. Thus, it is obvious that a rejection of the traditional Christian theist position is by no means equivalent to materialism.

A third, and better, reason for considering the nineteenth century to have been an age of philosophic materialism lies in the fact that during that period there was an increasing number of persons who challenged the traditional dualistic view of the mind-body relationship. As Mill noted in one of his letters,[39] the conventional definition of materialism which was current equated the materialist doctrine with the doctrine that all mental impressions resulted from the activities of the bodily organs. Such a definition of materialism is still sometimes held. However, those who acknowledge the dependence of all mental processes on bodily states need not be materialists, as the examples of Lotze and of other idealists clearly show.[40] And Huxley, too, took such a position, asserting that an acceptance of what he called automatism, and which might have been termed epiphenomenalism, did not entail materialism, and indeed had no specific metaphysical implications. Whether so extreme a position is or is not justified, it is best to avoid using this psychophysiological thesis as a means of defining materialism as a philosophic position. Instead, I shall propose a definition which can serve to make clear exactly where materialism differs from both idealism and positivism.

Unlike positivism, materialism is itself a metaphysical position. Materialists, like idealists, seek to state what constitutes the ultimate nature of reality, and are willing to distinguish between 'appearance' and that which is self-existent and underlies appearance. Taken in its broadest sense, materialism is only committed to holding that the nature of that which is self-existent is material in character, there being no entities which exist independently of matter. Thus, in this sense, we would class as a materialist anyone who accepts all of the following propositions: that there is an independently existing world; that human beings, like all other objects, are material entities; that the human mind does not exist as an entity distinct from the human body; and that there is no God (nor any other non-human being) whose mode of existence is not that of material entities.

While this type of characterization of materialism is one which would be very generally accepted, there is a stricter and more precise sense in which that term may be used. In the stricter sense materialists not only deny that there are entities which are not material; they also hold that whatever properties or forms of behaviour particular material objects exhibit are ultimately explicable by means of general laws which apply equally to all of the manifestations of matter. It should be obvious that this stricter definition of materialism excludes most forms of a naturalistic metaphysics which accept the doctrine of emergence. For example, what is generally termed 'emergent naturalism' would hold that while all entities are material in character, the varying forms of organization which matter may possess give rise to diverse properties and diverse modes of behaviour, neither of which can be adequately explained, even in principle, by an appeal to any single set of laws.[41] Strict materialists, on the other hand, put forward the claim that if one had a knowledge of the relevant conditions concerning their forms of organization, all of the diverse forms of organized matter would be explicable in terms of one set of basic laws. Thus, a strict materialism is a reductionist philosophy in a two-fold sense: it not only claims that all entities, however immaterial they may appear to be, have a material basis by means of which they are to be explained, but it also claims that whatever properties these entities reveal are explicable in an identical set of terms, regardless of their apparent disparities. To be sure, even the strictest materialist need not claim, and presumably will not claim, that at any one date the physical sciences have accurately and fully understood the properties of matter. For example, he need not hold (and today could not hold) that the atomism of Boyle and of Newton provided an ultimate explanation of all entities. However, the second element in his reductionism does demand a commitment to the view that there should be one all-embracing and basic science of nature which would, in principle, be capable of explaining all aspects of the behaviour of material entities by means of a single basic set of properties, regardless of how these entities are organized. In *traditional* materialist systems, the relevant set of basic properties has usually been identified with the properties which it is the goal of physics to discover, and physical laws (or physical-chemical laws) have been regarded as the basic laws of matter; however, in *dialectical* materialism it is not physics but the method of dialectical explanation itself which is held to serve as a basis for interpreting all manifestations of matter. Thus, dialectical materialism is not wholly reductionistic (in one usual sense of that term), since a full-fledged doctrine of emergence is compatible with it. However, dialectical materialism does go beyond what one usually identifies with the position of emergent naturalism, since it holds that one fundamental set of laws (though these are not identified with the specific laws of physics) is regarded as providing the basic explanation of change at every level of existence.

These distinctions will help to clarify the extent to which there may be said to have been widespread acceptance of materialism in the nineteenth century. If one were to identify materialism with a rejection of mind–body dualism, then materialism was indeed not uncommon: many philosophers and philosophically oriented scientists gave up all attempts to support the view that the human mind was an entity, and that it could function independently of the human nervous system. The positions which were most often adopted with respect to this problem were either a neutral monism or a psycho-physical parallelism, both of which were compatible with positivism, and each of which could also be interpreted as compatible with idealism, as is proved by the examples of Clifford and Wundt. However, if 'materialism' is construed in a more usual sense, and is taken as a position which is an alternative to idealism and to other forms of metaphysics on the one hand and to positivism on the other, then there were relatively few materialists in the nineteenth century. One looks in vain for any in France after Saint-Simon, and in England Tyndall stands out as an almost unique example.[42] Only in Germany does one find that materialism represented an important philosophic position.

The reason why this fact has so often been overlooked is that commentators have too frequently accepted the views of those opponents of positivism who refused to acknowledge that there was any difference between positivism and materialism. In the light of the repeated, explicit disavowals of materialism on the part of Comte and Spencer, and on the part of Bernard, Huxley, and Mach, among others, one might wonder how any such confusion was possible.[43] The answer lies in the fact that positivism and materialism had two elements in common: both were opposed to all traditional theologies, and (except for Feuerbach) both held that the sciences represented the most reliable knowledge attainable. It was on the basis of these similarities that those who opposed what was later to be called 'scientism' felt justified in identifying positivism with materialism, helping to give currency to the myth that materialism dominated philosophy in the nineteenth century.

Another source of an overemphasis on the scope of philosophic materialism in the nineteenth century lies in a failure to perceive the very marked differences between two originally distinct philosophic movements which existed in Germany. On the other hand, there was a significant materialist controversy which originally centered in the problem of vitalism, erupting in the early 1850s, and giving rise to a series of semi-popular expositions of materialism over a period of some twenty or twenty-five years. On the other hand, there already existed a position which identified itself with materialism, but which had in large measure arisen as a reaction against Hegelian idealism on the part of Feuerbach. In its original form, this position had little to do with most of the metaphysical issues which had always been central to materialism. Thus, in Germany during the period with which we are here concerned, there were two forms of materialism which must be taken into account. Their aims and even their conceptions of philosophy were so disparate that they might never have become related had it not been for the polemical writings of Engels. Since these writings served to interlace the two movements, it will be useful to view nineteenth-century materialism from the perspective afforded by Marxism.[44] However, it should not be forgotten that the two movements were distinct; nor should one overlook the fact that it would be anachronistic to attribute to Marxism in the nineteenth century a philosophic influence comparable to that which it has had in the twentieth.

To understand even the definition of the concept 'materialism' which is given by both Marx and Engels, and to understand aspects of their epistemology and of their views of man, one must go back to Feuerbach. The manner in which Feuerbach expressed what he found most objectionable in Hegel was the relation which the latter conceived to exist between thought and being. As Feuerbach never tired of emphasizing, existence precedes thought, and thought arises out of the problems posed by existence. This conviction, which was shared by Marx and by Engels, constituted the primary positive thesis which linked their thought to his. From this it followed that the thought with which they were concerned was the concrete thought of existing individuals, not thought in the abstract, and not thought removed from the problems of existing individuals. That from which they started was, then, the concrete, living, breathing human being, whose thought arises in the course of his struggles with nature, and in his relationships with his fellows. To adopt this starting point, to view thought as arising out of the problems of an organic being's existence, was for them equivalent to being a materialist. This equivalence could, of course, only be maintained so long as they rejected a dualistic view of man's nature, and held that as a part of nature man was in essence a material being whose capacity for thought and for action was a function of the nature of his bodily organs. On this point all three insisted.[45] However, this does not constitute their definition of materialism. Throughout, they define materialism in terms of their opposition to Hegel's system, that is, in terms of the relations between thought and existence.

Now, taken strictly, this is not, as we have seen, an adequate characterization of materialism. In fact, unless one interprets materialism in a very loose sense, Feuerbach cannot be classified as a materialist at all. While he did not explicitly reject both idealism and theism, and while he insisted upon the organic basis of thought, his own interpretation of the relation between thought and existence was such as to make him regard all genuine philosophy as being 'anthropological' in character, and not metaphysical. The problems of philosophy were for him solely the problems of the nature of man, and these problems were not to be solved by any attempt to push knowledge beyond that knowledge which we can reach through our own specifically human experience.

However one may interpret Marx's own thought, it is clear that Engels goes beyond Feuerbach's starting point and develops a materialistic system of metaphysics. What divides his Marxism from Feuerbach's views was not only the fact that he and Marx developed a sociological interpretation of man's nature, that they rejected Feuerbach's ethics of love, and that they had a hostility to religious modes of thought; more important was the fact that Engels constantly looked to the sciences for a knowledge of nature, and placed emphasis on the need for an understanding of nature in general if we are to attain an understanding of the nature of man. Genuine philosophy was not therefore to be confined to the self-knowledge of man through his personal and historical experience, but was to yield an all-embracing knowledge of existence. For both Marx and Engels the tools for such knowledge were to be found in the empirical sciences: to understand himself man could not merely reflect upon his experience, but was forced to employ categories of science. These categories, they claimed, were dialectical categories. Thus Engels held that the materials out of which a comprehensive philosophy could arise were being provided, and would continue to be provided, by the sciences. The tools with which these materials were to be worked into a philosophic system were simply the tools of formal and dialectical logic.

It is at this point that the thought of Engels made contact with that other form of materialism to which we have already alluded: the materialism of Moleschott, Vogt, and Büchner which arose in the early 1850s, and which represented a traditional form of materialism in which all processes were to be explained in terms of physical laws.[46] The position of Engels approached that of these 'itinerant preachers of materialism' only with respect to the fact that both claimed to rest their materialistic positions upon science: they diverged in their views of the methods and the conclusions of the sciences. With respect to these conclusions, Engels repeatedly insisted that the most fundamental new discoveries of science – the principle of the transformation of energy, the cell structure of living matter, and the theory of evolution – rendered obsolete a view of nature based upon the science of mechanics. He therefore regarded the materialism of Moleschott, Vogt, and Büchner as residual examples of an eighteenth-century mechanicalism which could not be squared with these new developments. However, the real basis of his charge does not lie in the conflict between an up-to-date and an outmoded knowledge of science, as Engels would have one believe, but in a difference between the two schools with respect to their views of the goal of scientific explanation, and therefore with respect to their views regarding the methods which it was appropriate for science to follow.

Insofar as Marx and Engels viewed all of reality as a developing process, the goal of scientific understanding was for them the elucidation of the place of any phenomenon within this process, that is, the ability to show how it was related to other phenomena out of which it arose, and how it was related to those phenomena to which it in turn would give rise. The method of science was therefore conceived as a method in which the concrete and total nature of one type of entity was understood with reference to its relations to the concrete and total nature of other types of entity with which it was developmentally connected. In Moleschott, Vogt, and Büchner, on the other hand, what might be denominated as the classic method of scientific explanation was assumed to be the proper method for science to follow: a scientific explanation consisted in analyzing the concrete events to be explained into a set of specific characteristics, attempting to establish constant relationships among various of these characteristics, and from these relationships deducing by ordinary logic what consequences would follow in any given case. Explanations of this type involve a piecemeal consideration of concrete entities, abstracting specific characteristics from the contexts in which they are embedded and examining them one by one. It also involves disregarding the temporal context in which such entities appear, on the assumption that what holds good of an entity at one particular time will hold good of it, or of any similar entity, at any time. Having tacitly accepted this model of scientific explanation, the non-Marxist materialists then proceeded to generalize on the basis of the results which had been (or which presumably would be) established by science. Then generalizations led to their specifically philosophic view that the total system of nature was one whole in which the place and function of each part was strictly determined by the fundamental laws of matter and energy as formulated by physics and chemistry. It was against this form of reduction that Engels protested. But while he protested on the basis of the claim that their very knowledge of the sciences was inadequate – that they held, in short, to a 'mechanical' view of nature – the true basis of his criticism, like that of Feuerbach's criticism of Moleschott, lay in the fact that they had taken the classic model of scientific explanation for their own, while he wished to found a specifically *modern* materialism: that is, a materialism grounded in the belief that nature is a dialectically developing, evolutionary process.[47]

That this furnished the underlying motive for Engels's slighting references to Moleschott, Vogt, and Büchner, is also evident from the nature of his attack on Dühring's metaphysics. Dühring was incomparably more competent as a philosopher than were Engels's other opponents, and he scorned their dilettantism, with its easy reliance upon specific principles which had been utilized in physics, and its failure to be concerned with the problem of knowledge.[48] Yet he, too, shared the traditional materialist position, in which the basic understanding of all nature was to be found in the physical sciences; and he therefore looked upon himself as the continuer of the eighteenth-century materialist tradition, to which he, unlike the dilettantes, would give a fundamental philosophic justification. His language, like the language of Moleschott, Vogt, and Büchner, often suggested a purely 'mechanical' view of nature, yet he was thoroughly cognizant of the inadequacies of traditional atomistic mechanism. What he sought to accomplish was to deal with the problem of the differences between the various gradations of the material world in terms which would make it possible to comprehend all of them through one system of categories.[49] But, to do so, he felt himself obliged to add to the types of concepts which physics employed, and to hold that in physical phenomena themselves there were, for example, self-fulfilling impulses, and that there was a law of opposition and contrast (*Differenz*) which applied equally to inorganic, organic, and mental states. By such means he sought to assimilate the nature of living things, and their evolution, to the nature of matter itself. Since he was not succesful in establishing the validity of using such concepts in physics, it sometimes appears as if he, no less than Marx and Engels, was relying upon the doctrine of emergence. Yet this was not his aim. His aim was to show that all of reality, including human and social existence, could be adequately understood in terms of the underlying nature of universal matter, and that this nature could be adequately grasped by human thought if the basic philosophic categories necessary for the physical sciences were uncovered. To this reductionism Engels of course objected, and it is in his attack upon Dühring that his own insistence upon the meaning of modern, or dialectical, materialism becomes most clear. At almost every point in this attack, the genuinely philosophic argument (as distinguished from mere invective) turns upon the fact that for Engels the manifestations of reality cannot be understood in terms of any universal category except that of a dialectical development, and this category makes it impossible to view thought as grasping any eternal structure of things or to regard things as manifesting the same specific properties throughout all of their developments.

In summary of the contrast between the opposed schools of materialism in Germany we may then make the following statements. Feuerbach, Marx and Engels, in their opposition to Hegel, found the essential basis of materialism to be the relation of thought to existence. If one could accept such a definition of materialism, all three could be unambiguously classified as materialists. However, the definition is inadequate, and the self-styled materialism of Feuerbach can only be construed as a form of materialism if one takes the latter term in its broadest sense, that is, as holding that all phenomena have a material basis, and that minds as separate from body – or God as separate from the world of nature – are non-existent. In a stricter sense, Feuerbach cannot be counted a materialist. His main effort was in the first place 'anthropological' rather than metaphysical; in the second place, even his metaphysical commitments were not in line with an interpretation of all phenomena in terms of the categories which were adequate to deal with the physical world. In his oft-cited phrase, the type of materialism represented by Moleschott was true of the foundations of human nature and human knowledge, but not of its superstructure: 'Rückwärts stimme ich den Materialisten vollkommen bei, aber nicht vorwärts.'[50]

The Marxism of Engels, however, represents a materialism in the strict sense of that term. It was his contention that all events were manifestations of the fundamental nature of matter, and that there was one fundamental science which could explain all of these manifestations by means of its grasp of the nature of the material world. This science, however, was not physics, but was the dialectical interpretation of nature and man. In identifying the fundamental science with dialectics, rather than with physics, he departed from traditional materialism, and could espouse the doctrine of emergence: the emergence of new and irreducible properties in nature was taken to be a manifestation of the fundamental dialectical self-transformations of matter, and an acceptance of these properties did not therefore collide with materialism.

In Moleschott, Vogt, and Büchner, a more traditional form of materialism was upheld. For them the definition of what constituted 'materialism' was not the relation of thought to existence, but the problem of whether all aspects of reality could be analyzed in terms of the categories which the sciences of physics and chemistry were successfully applying to all inorganic things. They were firmly convinced that this was not only a possibility for the future, but that it was gradually beginning to be achieved in their time. It was this faith that lay at the foundation of their materialism. And it was therefore the growth of critical positivism, with its interpretation of the necessary limitations of science, that served to undermine materialism among those who followed the new movements of thought in the last quarter of the century.[51]

Dühring's brand of materialism had an initial and striking success,[52] but partly due to the vagueness of his fundamental concepts and partly due to the unfortunate polemical style which grew out of his personal afflictions, he shortly lost the influence which he had first exerted, and he is now chiefly known as the butt of Engels's most savage attack.

Looking back, then, on the materialist movement in the nineteenth century, one can see that it was of relatively short duration. Even were one to include Feuerbach among its exponents, it would have first been newly affirmed in the 1840s, and the last of its original as well as influential statements was probably Dühring's *Cursus der Philosophie*, published in 1875. Were it not for Haeckel's subsequent popularization of his evolutionary monism, and were it not for the influence exerted by the political and sociological views of this school, the materialist movement of the middle years would have been of only antiquarian interest by the turn of the century: all of the specifically philosophic reasons which have tended to operate against it in the last decades were by then already manifest, and were in fact definitive of the new positivism and of the new idealism. It is for this reason that it is necessary to reject the common view that the nineteenth century was an age in which materialism flourished. It was almost totally absent in both England and France. And even in Germany it was not a movement which had the continuity and the pervasiveness of either idealism or critical positivism; nor was it fortunate enough to possess any figures of the same philosophic ability as were represented in both of the other movements.

4. Variant views of religion

The conventional view of the place of religion in the thought of the nineteenth century holds that science and religion were ranged in open hostility, and that unremitting warfare was conducted between them. The source of this belief is to be found in the very obvious fact that the Darwinian theory of evolution had widespread repercussions on the religious thought of the times, and was combatted in varying ways, and to different degrees, by a number of theologians. And the thesis can be made even more plausible by identifying the rise of the historical criticism of the Bible with the method of scientific thought – an identification which the historically minded times would not have denied.

This stereotyped interpretation of the place of religion in nineteenth-century thought, and of its relations with science, is one which cannot be accepted. Perhaps the real relations can best be illustrated by a passage from Tyndall who, it will be recalled, probably was the clearest proponent of materialism in nineteenth-century England. In his celebrated Belfast address, Tyndall expressly stated that 'the facts of religious feeling are to me as certain as the facts of physics',[53] and he held that in spite of the fact that many religions both past and present, are 'grotesque in relation to scientific culture', yet they are 'forms of a force, mischievous if permitted to intrude on the region of *knowledge* . . . but being capable of being guided to noble issues in the region of *emotion*, which is its proper and elevated sphere',[54] In the same connection he admitted that 'without moral force to whip it into action, the achievements of the intellect would be poor indeed'.

What led to the stereotyped conviction that the nineteenth century represented an age in which science and religion stood in open hostility was the undoubted fact that science and the historical criticism of the Bible came into conflict with widely held theological doctrines, and that there was a continuing battle concerning the proper interpretation to be placed upon specific teachings of the organized churches. This does not signify that those who were combatting various forms of theological doctrine felt themselves to be combatting religion. In their minds (though not of course in the minds of their opponents) they were combatting certain theological doctrines in the interests of religion itself. At the very outset of the century, Schleiermacher had insisted on separating religion from theology, and this insistence was no less characteristic of most of the theologians of the century than it was of, say, Carlyle, Matthew Arnold, Huxley, and Clifford.

The criticism of theological doctrines on scientific and historical grounds was not, of course, new in the nineteenth century; it was an inheritance from the eighteenth century and was not challenged even by those who were in other respects most opposed to the tendencies of eighteenth-century thought. All of the leading figures in religious thought at the outset of the nineteenth century were fully prepared to distinguish between religious belief and an acceptance of traditional interpretations of the doctrines of the Christian churches. One can see this most clearly in the profound influence which pantheism exerted in the early years of the century. The struggles evoked by this tendency were theological struggles in which those who most vigorously attacked orthodox Christian positions did so on behalf of religion itself. A similar situation arose in 1835. In that year David Friedrich Strauss published his celebrated *Das Leben Jesu*, and from it one may conveniently date nineteenth-century conflicts engendered by the historical criticism of the Bible. However, Strauss was utterly convinced that his criticism of the historical authenticity of the New Testament account of Jesus' life in no way undermined religious faith in the essential truth of Christianity.[55] The same position persisted among others throughout the struggles which historical criticism evoked: for example, it was clearly the position of those who in the 1860s were condemned for unorthodoxy in the two most celebrated cases in English theology of the century, *viz.*, the heresy trials of the seven authors of *Essays and Reviews*, and of Bishop Colenso of Natal. Furthermore, one can see precisely the same point in the struggles between geology and theology, and especially in the controversies which followed the Darwinian theory of evolution. While the orthodox felt that the Darwinian theory was one which undermined all genuine religious belief, those who used it in order to attack currently accepted theological doctrines did not feel that they were attacking or in any way undermining what was most significant in religion. Huxley was typical of this strain of thought, for while he insisted that one must attempt to break up theological dogma, he wished to do so in order to enable man to start 'cherishing the noblest and most human of man's emotions, by worship "for the most part of the silent sort" at the altar of the Unknown'.[56] Thus, in all three of the most notable cases in which nineteenth-century thought came into conflict with the established churches – in the pantheism struggle, in the growth of the critical historical treatment of theological documents, and in controversies concerning evolution – the position of the unorthodox was one in which theological dogma was being attacked not for the sake of undercutting religious faith, but as a means of freeing that faith for what were regarded as being nobler and more adequate forms in which it could find expression.

In spite of the fact that there was this common feature in all of the major theological struggles of the century, there was a significant transformation from the opening of the century to its close. It had been characteristic of German idealism to stress the unity of art, religion, and philosophy through insisting upon the identity of that which they revealed. Even though natural science was not taken to be the highest expression of this unity, it was claimed to be comprehended within it. Yet, writing just after the close of the century, Boutroux found a striking contrast between earlier views of the relations between science and religion and those which had come to be widely accepted:

To sum up, the relation between Religion and Science which had established itself in the course of the nineteenth century was a radical dualism. Science and Religion were no longer two expressions (analogous in spite of their unequal value) of one and the same object, viz. Divine Reason, as they were formerly in Greek philosophy; they were no longer two given truths between which the agreement was demonstrable, as with the Schoolmen; Science and Religion had no longer, as with the modern rationalists, a common surety – reason: each of them absolute in its own way, they were distinct at every point, as were distinct, according to the reigning psychology, the two faculties of the soul, intellect and feeling, to which respectively, they corresponded. Thanks to this mutual independence, they could find themselves in one and the same consciousness; they existed there, the one beside the other, like two material, impenetrable atoms side by side in space. They had come to an understanding, explicitly or tacitly, in order to abstain from scrutinizing one another's principles. Mutual respect for the positions achieved, and on that very account, for each, security and liberty – such was the device of the period.[57]

One need not agree with Boutroux's assessment of the relations between science and religion in other ages, nor with the suggestion that the relationship which evolved by the end of the nineteenth century was unique in Christian thought. Yet, his general assessment of the trend of the nineteenth-century struggles regarding the relations of science and religion can scarcely be challenged. It is to the task of briefly tracing some of the major factors contributing to this trend that I shall now turn.

So far as the problem of religion was concerned, the heritage which Kant bequeathed his immediate successors, the German idealists, was twofold: in the first place, it involved a critical and largely negative view of traditional interpretations of theological concepts; in the second place, in Kant's system there was a complete cleavage between the realm of pure, or scientific reason, and the realm of the moral and religious. The first of these aspects of Kant's view was willingly accepted by the idealists, but the second was philosophically intolerable for them. To understand their position, it is not sufficient to recall that there were grave technical difficulties within the Kantian system, which they believed that only a new monistic metaphysics could overcome; one must also take into account the appeal which the doctrine of divine immanence exerted upon German thought at the time. One finds that doctrine in Lessing, Herder, Goethe, and Novalis, as well as in Fichte, Schelling, and Hegel. However, it would be impossible to espouse such a view and yet remain within the framework of the Kantian system. According to the latter, the realm of nature as present within direct experience, and as known by science, cannot be considered as a manifestation of the divine: the formative power of the human mind in moulding the alien materials of sensation yields an orderly world, but not a world which manifests within itself a unity which is independent of us and to which we also belong. Nor, according to Kant's view, is there any concrete form of experience through which, within ourselves, we find a unity between the phenomenal and the noumenal; nor are there any concrete circumstances which elicit from us a total response in which all aspects of our nature – and not merely the noetic, or the moral, or the aesthetic – are fully incorporated and fully expressed. In order to see that there could not really be such experiences, according to Kant's system, one need merely think of his attempt to split sensibility from reason, or of his separation of inclination from awareness of duty. In short, Kant's system made it impossible to find any form of ultimate unity within experience, either between man and nature or within man himself.

It was against such a view that Kant's idealist successors revolted. Behind all of their variant technical arguments, each in his way sought that higher unity which was part of the metaphysical pathos of the times, and each sought it in an idealist form of the doctrine of divine immanence. One can see this in Fichte's attempt to make the moral nature of man the clue to the whole of the world of nature; in Schelling's attempt to overthrow Newtonian views (upon which Kant had so heavily relied), in favour of a conception through which man's alienation from nature would be overcome; and, above all, one sees it in Hegel's interpretation of the ultimate reality as that which is mediated in all things, but reaches its highest expression in the self-conscious objectification achieved within art, religion, and philosophy. Thus, for all three, man's supreme achievements place him in harmony with the totality of being, and in these achievements he is able to find the clue to reality. For those who identified a doctrine of divine immanence with true religion, this form of metaphysical idealism was itself a religious position. That it was not necessarily an orthodox position did not, of course, trouble them. On their view, that through which man could establish his unity with reality, that through which he could experience this reality in both its magnitude and its inner significance was identical with the truly religious. Thus, the sphere of the religious was enlarged beyond any confines of orthodoxy, and in fact merged not only with philosophy, but with all awareness of whatever was taken to be true or beautiful or good.

As a consequence of this doctrine, not only was there a broadening of the conception of religion, but there was a demand to reinterpret Christian doctrine as being symbolically rather than literally true. The way had already been opened for such an interpretation by Hamann's teachings, and even by Kant's, and it was also made necessary by the sympathetic interest vouchsafed to cultures in which Christian doctrine had no place. Since religious belief was not viewed as a separate compartment of man's nature, the belief in the truth and beauty inherent in the products of these cultures made it impossible to restrict authentic religion to the doctrinal teachings of Christianity. Thus, while Christianity could still be regarded as the highest or most adequate form of religious belief, the specific formulations of Christian theology had to be interpreted as symbols, not as literal transcriptions of matters of fact.

A belief in the idealistic theory of divine immanence, a sympathy for the varieties of religious experience in all cultures, and a willingness to interpret theological doctrine in symbolic terms, was no less characteristic of Schleiermacher than it was of his philosophic and literary contemporaries. However, Schleiermacher radically altered the stream of theological thought by his separation of the religious aspect of experience from both the intellectual and the moral. He sought to define for his contemporaries what constituted the basic phenomenon of religious experience, an experience which he felt to be no less binding because it was autonomous with respect to the intellectual. This basic phenomenon was, of course, to be found in the realm of feeling: the Christian faith is not a body of doctrine, but is a condition of man. Therefore religion is not to be confused with theology. According to Schleiermacher's view, the latter presupposes the religious experience, but religion itself does not need doctrinal expression: it is only because of the needs of the intellect that we must attempt to formulate in theological terms that which is immediately given within religious experience itself.

One can see that such a position could lead to a number of different attitudes with respect to the relations between knowledge and faith. One might, for example, attempt to construct a theology which would interpret Christian doctrine in ways that were satisfactory to the intellect, without seeking to appraise the experience of Christian faith itself. Schleiermacher's own theology may be looked upon as setting itself this goal. However, as Schleiermacher recognized, such an attempt demanded that theology be separated from philosophy: the intellect was not to be called upon to pass judgement upon the truth of Christian faith, but only to interpret that faith. Others, however, were not willing to accept so radical a separation of philosophy and theology, and they sought to offer a philosophical basis for the acceptance of that which came through faith. This constituted an alternative to Schleiermacher's position, and also to the tradition of natural theology; it consisted in developing a philosophical position in which the realm of feeling could lay claim to a truth higher than that which arises through the intellect. One could then defend Christian theology as being guaranteed by, or as being most consonant with, the spiritual insights to which religious feeling gave rise. It was in the latter fashion that Jacobi had already proceeded, and which Coleridge, Hare, and Francis Newman were to follow.

However, there are two other forms which religious thought may take when religious feeling is held to be prior to and independent of all the propositions which theology expounds. On the one hand, it can be maintained that all of the beliefs which are identified with theology are merely 'projections' of the fundamental nature of religious feeling. This, in general, was the path followed by those who, like Strauss and Feuerbach, adopted a mythical or a psychological interpretation of the historic doctrines of religion. On the other hand, it was possible to view these doctrines as the reflections of the knowledge and experiences of certain peoples at certain times in the world's history, and to hold that the importance of religion was to be found, first, within the realm of immediate feeling and, second, in the fruits which this feeling bore. On the latter view, theological beliefs are not direct projections of feeling: they arise from sources outside of religion and they undergo change as these external factors change.

One can see that each of the last two views contains precisely that duality between science and religion which Boutroux regarded as characteristic of the development of religious thought in the nineteenth century. But one can also note that the two views differ with respect to the degree to which they are potentially hostile to religion. According to the first view, all of the content of religious belief is to be regarded as error so long as it is interpreted as being anything more than a projection of individual and social feeling. According to the second view, however, the content of religious belief is a reflection of the state of knowledge and experience of those who hold to the belief; and in so far as there is a change in the state of knowledge it is possible to reform religious belief, to make it no less adequate as an expression of feeling, but more adequate as an expression of what is known to be true of the world. Thus, on the second view, one could believe in progress within the domain of religious belief, and one could seek to reform religion to meet the needs and the knowledge of succeeding generations.

In the history of nineteenth-century thought one finds that it was the first of these views, rather than the second, which developed earlier. In Germany it was chiefly exemplified by Strauss and by Feuerbach, in France by Comte.[58] All three of these thinkers represented challenges to current Christian orthodoxy, to eighteenth-century conceptions of natural religion, and to the religious positions generally characteristic of Romanticism and absolute idealism. Each of the latter views – though they were opposed in other ways – had involved a claim that there was a noetic aspect of religion, and that questions of truth and falsity were relevant to religious commitment. However, Strauss, Feuerbach, and Comte rejected this claim. They did point out that the feelings which were associated with traditional forms of religion were capable of generating beliefs about the world, but they insisted that if these beliefs were regarded as statements of matters of fact they were erroneous, and were to be fought. So long as no such cognitive claims were associated with the basic forms of feeling which are present in all religions, these feelings were considered by Strauss and Feuerbach and Comte to be of supreme human importance.

In Comte, for example, a sharp distinction is drawn between the theological, fictive mode of thought, which was to be wholly rejected, and genuinely religious feeling, without which the unity of a good social order cannot be achieved. Similarly, Strauss never abandoned his faith in the value of religious feeling: his devastating criticism of orthodoxy was originally accompanied by his belief in the 'eternal truth' of what the myths of orthodoxy symbolized; and even at the end of his life, when he had abandoned Christianity, he, like Comte, attempted to substitute a new faith for the old. And Feuerbach, who was if anything more insistent than Comte and Strauss that the doctrines of Christianity were false, yet regarded religious feeling as that which gives man his worth. In Feuerbach, in fact, one finds an intense religious feeling, and a most passionate affirmation of the value of this feeling, combined with as radical an expression of the theory of the psychological, 'protective' origin of religious belief as the nineteenth century has to offer. In his outspoken denial that there is any object outside of man himself which corresponds to the object of religious emotion, Feuerbach's position remains unparalleled until, at the turn of the century, one comes to Durkheim's theory, or, later in this century, to the theory of Freud.

However, the tendency represented by Comte, Strauss, and Feuerbach did not in fact come to dominate nineteenth-century religious thought. It was, rather, the less radical interpretation of the relation between religious feeling and religious belief which exerted the greatest influence. This interpretation, it will be recalled, regarded religious belief as having arisen out of a natural human capacity for religious feeling, operating upon the knowledge and experience available to men at different times and in different places. Thus, religious beliefs were mutable, but each gave expression to that which was taken to be true. Criticism of religious beliefs was necessary in order to bring them into line with current knowledge, but knowledge alone was not sufficient for man: religious feeling was a natural capacity which demanded satisfaction, and was justified by the fruits which it bore.

This attitude was so prevalent among the liberal theologians in England, and among English literary men, that its existence and importance need not be documented through individual discussions of the figures concerned. What may however escape the reader's attention is the intimate connection which existed between this mode of thought and a factor which we have already noted to have been characteristic of the German Romantic and idealist philosophers of religion: the espousal of a belief in divine immanence. Such a belief (as the term is here used) denies that the object of religious worship transcends the world; instead, it finds the object of religious feeling within the totality of nature, of which man is a part. This one-worldly religious conception characterized liberal theologians such as Francis Newman, Seeley, and most, if not all, of the authors of *Essays and Reviews*; it characterized Carlyle, Matthew Arnold, and Tennyson; and it was no less characteristic of philosophers as diverse as Spencer and Bosanquet. For those who held that the object of religious worship did not reside outside the world, but at its heart, the theory of evolution posed no obstacles; indeed, it was possible for some who held this view to identify true religion with a worship of that immanent power which was at work in nature, evolving higher forms of existence, bringing mankind out of crudity, ignorance and selfishness into altruism, knowledge, and culture. This created a climate of opinion in which the boundaries of what was recognized as religion became greatly enlarged. One finds, for example, that positions characteristic of twentieth-century religious humanism are already explicitly stated when, according to Matthew Arnold, to be religious meant to worship 'the Eternal', 'the stream of tendency by which all things seek 'o fulfil the law of their being', 'the enduring power not ourselves which makes for righteousness',[59] and when, in T. H. Green's phrase, 'God is our possible or ideal self'.[60]

According to such definitions of religion, the age was not an age of irreligion. To be sure, there were those who did not take the noetic claims of religion so lightly, and who, if they were not orthodox, or were unable to struggle to an acceptable compromise with orthodoxy, found it necessary to adopt a position of agnosticism. This, for example, came to be Darwin's view,[61] and it was also the view of Leslie Stephen. However, agnosticism was not widespread, and an openly avowed atheism was even less common. As we have noted in the case of Huxley (who coined the term 'agnosticism') and in the case of Tyndall (who was a materialist), religion could be redefined in such a way that it made no noetic claims whatsoever, and the issue of agnosticism or of atheism would thus be by-passed. Such a position seems to have provided a welcome *modus vivendi* between the allegiance felt toward science and the allegiance felt toward Christian belief. This compromise was only possible because religion had antecedently been defined exclusively in terms of feeling. Once this had been accepted, science could be held to yield our most certain knowledge; at the same time, any acknowledgment of the limitations of science (such as one found among critical positivists) would open an adjoining door through which access was given to another domain of existence, in which the endless questionings of the intellect had no place. Thus, poetry could be interpreted as depicting the truths of feeling, and could be apotheosized. And thus, also, the age which saw its chief moral problem as the problem of extending the bounds of sympathetic, altruistic action, could abandon the view that faith in God was either the source or the enemy of the social good, and could instead view God and the social good as synonymous.

Such was the predominant view, during the latter part of the century, among those who had held that the essence of religion was to be found in feeling. There was, however, another tendency in nineteenth-century thought which also served to disengage religion from scientific and philosophic controversy. It consisted in interpreting religion as a manifestation not of feeling but of ideal morality. Although this tendency undoubtedly had other sources as well, it was fed by the strong Kantian revival during the latter part of the century; and just as Schleiermacher may be viewed as the fountainhead of one of these theological tendencies, so Albrecht Ritschl may be seen as the source of the other.

Ritschl, like his predecessor Kant, sought to divorce metaphysics and theology. He attempted to justify his view by holding that since metaphysics had the task of dealing with all forms of being, its categories would have to be applied to the natural world as well as to God. Yet, since he found an absolute cleft between the realms of spirit and nature, and since religion, and all theology, dealt with things of the spirit, the categories of metaphysical thought were inappropriate within the province of theology. However, he also insisted that theology was secondary to, and an adjunct of, the immediate religious experience of man. This experience he found in a self-commitment to the moral ideal. The function of theology became that of interpreting accepted Christian doctrine in terms of the moral faith of a believing Christian. Thus,

Ritschl bears a close similarity to Schleiermacher, in spite of the differences between their interpretations of the nature of religious experience. Both found the authority of religion to lie within the sphere of immediate experience; both interpreted that experience as autonomous with relation to theology, seeking through theology merely to explicate its Christian significance; and, finally, both divorced theological questions from the metaphysical questions with which, in their eyes, Christian doctrine had become burdened. Considered in the wider context of modern theology as a whole, one may say that both Schleiermacher and Ritschl stood opposed to what had formerly been regarded as the fundamental thesis of natural theology: that the truth of theism could be established through arguments based upon man's capacity to reach an explanation of the world and of his place within it. Yet neither wholly departed from the traditions of natural theology: it was within experience, and not by revelation, that man could become aware of the truth of Christianity. In this transformation of the very basis of natural theology, the influences of Schleiermacher and of Ritschl converged. Both contributed to the relaxation of demands for theological orthodoxy, since each held that the truth of religion was to be found in the inner nature of religious experience, and that what was important to religion was the spiritual fruit of this experience, not an external conformity with one or another set of conflicting theological propositions. Both also contributed to the willingness to divorce religious commitments from theoretical commitments, since both held that the object of religious worship was inwardly revealed, and could not be found through an examination of that realm of nature and of history with which the methods of empirical and rational inquiry sought to deal.

One can see how this reform of natural theology fostered the compartmentalization of religious and theoretical commitments with which we are here concerned. However, the most extreme degree of compartmentalization was only reached after the development of that form of critical positivism which interpreted theoretic knowledge as symbolic, as a construction of our experience made in the interests of practice and of the economy of thought. Just as Spencer's systematic positivism had permitted him to view both science and religion as two ever-present ways in which man sought to approach and adapt himself to the one Unknowable, so the later forms of positivism saw in the symbols of science and the symbols of religion two different modes of interpreting differing forms of experience, each mode being guided by considerations appropriate to itself. And with the growth of the belief that the realm of nature as depicted by science was different from the realm of man's history and his moral commitments, there came to be an increased emphasis upon a doctrine of the twofold nature of truth. Now, however, the two truths were not separated by the gulf between the natural and the revealed, but were to be found within man's own natural experience: differing objects were to be differently viewed, and even the same object or experience could equally well be regarded from different, purely human, points of view. It was the growth of this spirit, which first came to full flower in James, that may be regarded as the most extreme development of the tendency which dominated the religious thought of the nineteenth century. In spite of the many conservative theologians, and the conservative revivals of the French Romantic reaction and the Oxford Movement, in spite also of the persisting tradition of metaphysical idealism, the most influential strand in nineteenth-century thought was that which attempted to divorce science and religion, and maintain the value of each. Positivism had merged with idealism to limit the domain in which science had authority; and once these limits were generally agreed upon, all schools relinquished the view that religion could intrude into that domain. On the other hand, the scope of religion had been broadened, not narrowed: in giving up its claim to the possession of any literal knowledge, religion came to be identified with whatever ranges of feeling and of moral aspiration were of most significance to man. Thus the assumptions of both natural theology and revealed religion were abandoned, and it was thought that at last a means had been found to effect a permanent reconciliation of science and religion. Set against this background in the latter part of the century, John Stuart Mill's *Theism* seems no less out of touch with his age than had been Kierkegaard's passionate search for orthodoxy in an age dominated by Hegel.

Source: Taken from Maurice Mandelbaum, *History, Man and Reason: A study in Nineteenth-Century Thought,* Baltimore, Johns Hopkins Press, 1971, Ch. 1.

Notes

1. In Kroner's valuable book, the period of German idealism is interpreted as spanning forty years, from the publication of Kant's *Critique of Pure Reason* (1781) to the publication of Hegel's *Philosophy of Right* (1821). For Kroner there was a sharp break after Hegel, and the characteristics of the period then came to a close (Richard Kroner, *Von Kant bis Hegel,* 2 vols. Tübingen, Mohr, 1921, I; 1–6). However, his characterization of idealism (I; 7–10) was aimed at defining 'the idealist school' in such a way that one could trace a direct line of descent from Kant through Hegel. Such a characterization seems to me at once too broad and too narrow: too broad in its inclusion of Kant, too narrow in its assumption that idealism is to be characterized primarily in terms of a restricted movement in early nineteenth-century German philosophy.

2. Lotze's first relevant work, *Metaphysik,* was published in 1841; Fechner's *Nanna* was published in 1848 and, more importantly, his *Zendavesta* in 1851. Both men continued to develop their systems, and each published what was probably his more important metaphysical work in 1879.

3. According to Höffding, in the five years succeeding the publication of von Hartmann's *Philosophy of the Unconscious,* fifty-eight works dealing with his philosophy were written (Harald Höffding, *History of Modern Philosophy,* 2 vols. New York [n.d.] II; 533). For Büchner's puzzled disappointment at the decline of materialism, see his *Am Sterbelager des Jahrhunderts* [1898] 2nd edn. Giessen, Roth, 1900; 9.

4. Cf Ravaisson's treatment of Eclecticism in his admirable and influential *La philosophie en France* (1868). This work was itself one of the major influences on the French idealist movement of the latter part of the century. It is also characteristic of the self-evaluation of the Eclectics that the third volume of Ferraz' *Histoire de la philosophie en France au XIXe siècle* (1887) should have been entitled *Spiritualisme et liberalisme*. It may also be noted that in 1857, in *Les philosophes français du XIXe siècle*, Taine regarded Eclecticism as a species of idealism.

5. For example, Félix Ravaisson, *La philosophie en France*. Paris, Imprimerie Impériale, 1868; 258–62, *et passim*.

6. There are of course interesting parallels between Cournot and specific doctrines of the idealists. On this point, cf. Dominique Parodi, 'Le criticisme de Cournot', *Révue de métaphysique et de morale*, xiii (1905), especially; 474–5; also, cf. Parodi, *La philosophie contemporaine en France* (Paris, Alcan, 1919) in which Renouvier and Cournot are compared with respect to their interest in the problem of the limitations of science.

7. See the conclusion of *La contingence des lois de la nature* (1874). As in the case of Ravaisson, who was also his teacher, Boutroux manifested great interest in Leibniz, whose *Monadology* he edited, contributing a long introduction.

   In the development of Boutroux's thought there is a striking difference between his earlier exposition of the factor of contingency and the exposition which one finds in his *De l'idée de loi naturelle* (1895). It is difficult to estimate whether this difference was due to the manner in which Cournot had developed his own thought in *Materialisme, vitalisme et rationalisme* (1875). There would also have to be considered the mutual influences of Boutroux and the Tannerys, and of Boutroux and Poincaré. Given the importance of Boutroux as a teacher, and his close relationship to other important figures in the philosophy of science, one would think that there should be more interest in his thought than has recently been shown.

8. It is often forgotten that Bergson published his *Essai sur les données immédiates de la conscience* (translated into English as *Time and Free-Will*) in 1889; a second edition was published in 1898, and thereafter new editions followed in rapid succession. When one considers the date of this publication, as well as its content, one sees in how close a relationship he stood to the continuous tradition of nineteenth-century French idealism. (Cf. Isaac Benrubi, *Les Sources et les courants de la philosophie contemporaine en France*, 2 vols. Paris, Alcan, 1933, II; 741 *et passim*, and Parodi, *La philosophie contemporaine*; 254, *et passim*.)

9. The denial of freedom which was felt (with some justice) to be implicit within French positivism, doubtless also contributed to the French rejection of Hegel. (One notes that Ravaisson was impressed by the manner in which Schelling's late lectures stressed human freedom; cf. *La philosophie en France*; 264.)

10. From an interesting letter from Mill to Martineau, cf. *Letters of John Stuart Mill*, Hugh Elliot (ed.), 2 vols. London, Longmans, 1910, I; 62. With respect to Martineau's earlier relations to Mill and for the development of his idealism, see Martineau's preface to the first edition of his *Types of Ethical Theory*.

11. Martineau's reaction against Hegelianism was, I believe, the nearest

approximation in England to the changes taking place within idealism in Germany at mid-century, and to the French idealist movement. Like the latter movement Martineau insisted upon the concepts of freedom and personality as ultimate, and therefore regarded Hegelianism as a threat.

In Green, of course, the same two concepts played an important role, but his system represented a merging of Kantian and Hegelian elements, rather than the development of a personalism. Apart from Martineau, we find the development of personalism most clearly exemplified in Seth's *Hegelianism and Personality* in 1887, in some of the contributions to *Personal Idealism* (1892), and in James Ward, who was greatly influenced by Lotze. McTaggart's pluralistic idealism was, in his own eyes at least, closely related to Hegel.

In Sir John Herschel, at an earlier date, we find the rudiments of a monadology, but I am not aware of any influence which his views on this subject exerted. (Cf. his essays 'On atoms' and 'On the origin of force' in his *Familiar Lectures on Scientific Subjects.*)

12. As is universally recognized, metaphysical idealism was far and away the dominant strand in nineteenth-century American thought, and was extremely widespread in Italy as well.

13. Cf. Mill's statement of what Comte took to be the essential nature of the Positive Philosophy: 'We have no knowledge of any thing but Phenomena; and our knowledge of phenomena is relative, not absolute. We know not the essence, nor the real mode of production, of any fact, but its relations to other facts in the way of succession or similitude. These relations are constant; that is, always the same in the same circumstances. The constant resemblances which link phenomena together, and the constant sequences which unite them as antecedent and consequent, are termed their laws. The laws of phenomena are all we know respecting them. Their essential nature, and their ultimate causes, either efficient or final, are unknown and inscrutable to us.' (John Stuart Mill, *The Positive Philosophy of Auguste Comte* [1865]. Boston, Spencer, 1867; 7–8.)

Those acquainted with Leszek Kolakowski's *The Alienation of Reason: A History of Positivist Thought* (Garden City, Doubleday, 1968) will note two major differences between his characterization of positivism (pp. 2–10) and that offered here. (1) While stressing the first and third features which I have discussed, Kolakowski omits the second; as a consequence, he links Locke and Berkeley (among others) with the positivist tradition, and places Hume squarely within it. (2) He regards non-cognitivism in value-theory as a defining characteristic of positivism, yet neither Comte nor Mill represents non-cognitivism as he defines it. In addition, and perhaps most importantly, we differ in the fact that he fails to distinguish between the methods and aims of systematic positivists and critical positivists.

14. *The Positive Philosophy of Comte*; 9–10.

15. G. H. Lewes, *Biographical History of Philosophy*, 2 vols, 3rd edn. London, Longmans, 1867, II; 654.

16. The followers of Comte did not necessarily attribute finality to his system. In spite of his self-evaluation, and in spite of how highly they rated his achievement, it was the new method rather than his system itself which evoked their unbridled confidence. One sees this not only in Littré (cf. Benrubi, *Les sources et les courants de la philosophie contemporaine en France*, I; 23–5), but in Lewes (*History of Philosophy*, II; 640–1).

17. Cf. J. S. Mill, *Autobiography* [1873]. New York, Columbia University Press, 1924; 157-9.
18. Cf. Antonio Aliotta: 'We must distinguish two periods in the history of positivism: of these the first is marked by a dogmatic belief in physical science, which is set up as a mode for every form of knowledge; the second, dating from around 1870, goes still farther, and subjects science itself to searching criticism in order to eliminate any traces of metaphysics which might be sheltering themselves beneath the cloak of experimental theories.' (*The Idealistic Reaction against Science*. London, Macmillan, 1914; 53.)
19. William J. M. Rankine, 'Outlines of the Science of energetics' (1855), *Miscellaneous Scientific Papers*. London, Griffin, 1881; 210. Cf. Robert Mayer, *Ueber die Erhaltung der Energie: Briefe an Wilhelm Griesinger*, ed. W. Preyer. Berlin, Paetel, 1889, especially letters X and XIII, written in 1844.
20. Cf. Tait's Memoir prefixed to Rankine's *Miscellaneous Scientific Papers*; xxix.
21. Cf. Mayer, *Ueber die Erhaltung der Energie*, especially letters I, III, and V.
22. Ernst Cassirer tended to identify the mid-century developments in physics with the overthrow of a realistic interpretation of science (cf. *The Problem of Knowledge: Philosophy, Science and History Since Hegel*. New Haven, Yale University Press, 1950, Ch. 5), but Rankine, Mayer and others of the period should really be counted as precursors of that change, not its representatives. In this connection, it is to be noted that Cassirer made his point through citing (in the main) Ostwald and Helm, who belong to a later generation.

    One can also see the fallacy of identifying the mid-century developments in physics with the later overthrow of a realistic interpretation of science if one considers the more philosophical passages in Clerk Maxwell's writings. To be sure, Maxwell not infrequently expressed impatience with metaphysics; for example, he characterized a metaphysician as 'a physicist disarmed of all his weapons – a disembodied spirit trying to measure distances in terms of his own cubit, to form a chronology in which intervals of time are measured by the number of thoughts they include, and to evolve a standard pound out of his own self-consciousness'. (Lewis Campbell and William Garnet, *The Life of James Clerk Maxwell*. London, Macmillan, 1884; 436.) It is also true that one can find statements in Maxwell which anticipate the economical view of thought which was later associated with Mach's form of critical positivism. While these statements appear to have had a certain influence on Boltzmann and others, it can scarcely be said that Maxwell consistently espoused critical positivism. He never extended such statements through to their full implications, and he was himself frequently absorbed in discussing metaphysical issues. (Cf. Campbell and Garnet, Ch. 8 and Ch. 14.) The most famous of these discussions is to be found in his essay 'Molecules' (1873), where he was led from a consideration of molecular structure to a proof for the existence of God. (Cf. *Scientific Papers*, 2 vols. Cambridge University Press, 1890, II; 361-78).
23. For illustrations of Bernard's thought with respect to the preceding points, see in particular the following sections of his *Introduction à l'étude de la médecine expérimentale*: Part I, Ch. II, Sect. IV; Part II, Ch. I, Sects. I and IX; and Part III, Ch. IV, Sect. IV.
24. Kirchoff, who was regarded by Mach as one who anticipated his own position (cf. 'On the principle of comparison in physics', in Mach's *Popular*

Scientific Lectures), seems to have subscribed to a view of the sort here discussed.

25. Cf. Helmholtz's *Treatise on Physiological Optics*, 3 vols. Rochester, Optical Society of America, 1924, III; 2 and 25–36. Also, cf. *Popular Lectures on Scientific Subjects*, 2 vols. London, Longmans, 1884, II; 230. Of course, Helmholtz explicitly rejected Kantian views regarding our perception of spiritual relations (for example, cf. *Physiological Optics*, III; 17–18, 36, *et passim*) and regarding the status of geometrical axioms (cf. *Popular Lectures*, II; 68 and *Wissenschaftliche Abhandlungen*, 3 vols. Leipzig, Barth, 1882–95, II; 640–60).

26. Cf. *Wissenschaftliche Abhandlungen*, II; 608. (For a further discussion of this view, cf; 293–4 and 297–8.)

27. Cf. *Die Thatsachen in der Wahrnehmung*. Berlin, Hirschwald, 1879; 24–40: also, *Popular Lectures*, II; 284–5.

28. Cf. Hertz's lecture on the occasion of Helmholtz's seventieth birthday (1891), in Heinrich Robert Hertz, *Miscellaneous Papers*. London, Macmillan, 1896, especially; 335–7. It is to be noted that at the end of this passage Hertz expressed a degree of scepticism with respect to the view that epistemological claims legitimately followed from psychophysics. Hertz's own views more clearly approached a true Kantianism, as one sees in his *Principles of Mechanics* [1894] (London, Macmillan, 1899) for example on; 1–3 and 296–307. (Cf. also Cassirer's discussion of Hertz and Mach in *The Problem of Knowledge*; 105–8.)

29. Cf. Wilhelm Wundt. *Ueber den Einfluss der Philosophie auf die Erfahrungs wissenschaften*. Leipzig, Engelmann, 1876; 6.

30. This assumption is perfectly explicit throughout Bernard's *Introduction à l'étude de la médecine expérimentale* (1865), and is equally clearly expressed in Helmholtz's lecture in 1869, 'The aim and progress of physical science' (translated and reprinted in his *Popular Lectures on Scientific Subjects*). As we shall later see, it was also characteristic of the views of DuBois-Reymond.

31. Lange, in his treatment of ideals, anticipated it in 1865, and in various places it was also suggested by Clifford; in fact, in Clifford's series of lectures entitled 'The philosophy of the pure sciences', delivered in 1873, one finds an expression of each of the three reasons which I single out as the primary reasons why a pragmatic-economical view was held. (Cf. William K. Clifford, *Lectures and Essays*, ed. Leslie Stephen and Sir Frederick Pollock, 2 vols, 3rd edn. London, Macmillan, 1901, I; 301–36).

    According to Höffding, Mach himself had actually arrived at his economical interpretation of science as early as 1863 (cf. *Modern Philosophers*; 116), and had regarded Kirchhoff and Maxwell (though perhaps with little justification, as I have indicated) as his forerunners. Though Mach may well have arrived at this view earlier than the 1880s, it was at that time that he came to develop it in a whole series of lectures and books. (For Mach's own later account of the development of this aspect of his thought, cf. *Die Leitgedanken meiner naturwissenschaft lichen Erkenntnislehre, und ihre Aufnahme durch die Zeitgenossen* [1910]. Leipzig, Barth, 1919.

32. To this day it is insufficiently recognized that this is not in fact an implication of Darwin's theory. All that his theory entails is that no factor which is markedly deleterious will persist. In the *Descent of Man* (1871) Darwin explicitly recognized the error of assuming that every new factor

must be of positive value in order to persist, and he attributed that error to the difficulty of abandoning his earlier teleological habits of thought. (Cf. Charles Darwin, *Descent of Man and Selection in Relation to Sex*, 2 vols. New York, Appleton, 1871, I; 146–47.) In the second edition Darwin rearranged his order of discussion, and this passage is to be found two pages prior to that section of Ch. II which is labeled 'Conclusion'. The original passage in the *Origin of Species*, which it was obviously Darwin's aim to correct, appears in Ch. VI shortly before the concluding summary, and contains a reference to Paley.

33. In addition to his *Contributions to the Analysis of Sensations* and his lecture 'On transformation and adaptation in scientific thought', which have already been cited, the following other works of the same period are to be especially recommended as giving insight into his philosophic thought: 'The economical nature of physical inquiry' (1882) and 'On the principle of comparison in physics' (1894), both of which are republished in his *Popular Scientific Lectures* [1864–94]. Chicago, Open Court, 1894. In addition, his classic *Science of Mechanics* (1883) is directly relevant.

34. Mill's analysis of mathematics is to be found in Part II, Ch. 5 and 6, of his *System of Logic*; for Helmholtz's discussions of geometry see his *Wissenschaftlichen Abhandlungen*, II, articles 77, 78, 79, and for his discussion of arithmetic see III, article 129, of the same collection.

35. Cf. 'On the economical nature of physical inquiry', in *Popular Scientific Lectures*; 190.

36. Cf. *Ueber die Aufgabe der Philosophie und ihre Stellung zu den uebrigen Wissenschaften*. Heidelberg, 1868.

37. Cf. Windelband, *Geschichte und Naturwissenschaft* (1894), republished in his *Praeludien*; Rickert, *Die Grenzen der naturwissenschaftlichen Begriffsbildung* (1896) and *Kulturwissenschaft und Naturwissenschaft* (1899); also Croce: 'La storia ridotta sotto il concetto generale dell'arte' (1893), reprinted in *Primi Saggi* (Bari, 1919).

38. Among other examples of this movement at the turn of the century are to be found Balfour and Boutroux. Cf. Balfour's development of his earlier criticism of 'scientific philosophy' (*A Defense of Philosophic Doubt*, 1879) in his *Foundations of Belief* (New York, Longmans, 1895) for example; 243 and 301. Cf. Boutroux's development from *De la Contingence* (1874) to *La Science et la Religion dans la Philosophie contemporaine* (1908).

39. Cf. *Letters of John Stuart Mill*, II; 286.

40. As Pollock pointed out, Clifford's views were such that they would 'in a loose and popular sense be called materialist', even though Clifford in fact accepted a form of metaphysical idealism. (Cf. Pollock's biography, prefixed to Clifford, *Lectures and Essays*, I; 50.) Although Pollock does not thus state it, the causes of this erroneous ascription of materialism to Clifford would have been the two which we have mentioned: an attack on Christian orthodoxy and an attempt to hold, in Mill's terms, that 'toute nos impressions mentales resultent du jeu de nos organes physiques'. These are in fact the two most important grounds upon which Hutton criticized Clifford. (Cf. R. H. Hutton, *Criticisms on Contemporary Thought and Thinkers*. London, Macmillan, 1894, I, essays 26 and 27.)

41. There are two points concerning the doctrine of emergence to which I should here like to call attention: first, that it can be associated with a vast variety of otherwise very different philosophic positions, and, second, that it is not a doctrine which was new in the twentieth century, as has often been

supposed by those who identify it with the thought of Samuel Alexander and C. Lloyd Morgan.

One can find idealists such as Hegel, materialists such as Marx and Engels, positivists such as Comte, non-dualists such as Alexander and R. W. Sellars, and dualists such as Lovejoy and Broad, all holding doctrines of emergence which (with the exception of Hegel's) were remarkably similar. Thus, it is a mistake to treat the concept of emergence as if it were necessarily associated with an interest in metaphysics, or with a particular position, or with a denial of mind-body dualism.

The foregoing remark should also be sufficient to suggest that an acceptance of the doctrine of emergence was prevalent in all schools of thought in the nineteenth century. It is of special interest to note that in Book III, Chapter V of his *System of Logic* Mill gave the first careful analysis of the difference between the principle of the 'Composition of Causes' and what he called 'chemical causes'. G. H. Lewes developed the first full-fledged natural philosophy which was based upon the principle of emergence in his *Problems of Life and Mind* (especially in Volume II), which appeared in 1874–75; and it is said that it is to him that we owe this use of the term 'emergence'. Among the many other instances of an acceptance in the nineteenth century one further example may be noted, for it also serves to illustrate the way in which that doctrine migrated freely from system to system. Claude Bernard accepted a doctrine of emergence, holding that the living was not reducible to the non-living, and it was this doctrine – and not his positivism, nor his arguments for an experimental science of medicine – that was of primary influence on the next generation of philosophers, most of whom used this aspect of his thought in support of some form of idealist metaphysics. (Cf. Ravaisson's treatment of Bernard in *La philosophie en France*; 125–7, *et passim*, and Bergson's 'La philosophie de Claude Bernard', in *La pensée et le mouvant*. For one of Bernard's own clearest statements of his doctrine, cf. *Leçons sur les phenomènes de la vie, communs aux animaux et aux végétaux*. Paris, Baillière, 1878, I; 50.)

42. For Tyndall's position, see his well-known lectures, 'The scope and limit of scientific materialism' (1868), and the 'Address delivered before the British Association in Belfast' (1874), which was then published with additions.

In support of the view that Tyndall's position was the chief manifestation of materialism in England, one may cite James Martineau's exchanges with him, as well as the accounts of these exchanges in Hutton's *Aspects of Religious and Scientific Thought*, essays 10 and 11, and Tulloch's 'Modern scientific materialism', in *Modern Theories in Philosophy and Religion*.

To be sure, in 1851 the *Letters on the Laws of Man's Nature and Development* by Henry George Atkinson and Harriet Martineau caused something of a scandal. The position they represented was akin to that of the German materialists of the time. However, within two years Harriet Martineau published her translated abridgement of Comte's *Positive Philosophy*, and the influence of Atkinson was no longer significant.

It is finally necessary to note that in England the working-class movement seized upon the eighteenth-century thought of Tom Paine, and Paine's influence persisted in the Secularism of Bradlaugh and Holyoake. However, this can scarcely be taken as indicating that materialism had established itself as an important philosophic movement in England.

43. In this connection, the following representative passages may be listed:
    (1) Comte, *System of Positive Policy* [1851–4], 4 vols. New York, Franklin, 1967, I; 40–1.
    (2) The concluding section in all editions of Spencer's *First Principles*.
    (3) Claude Bernard's *Introduction to the Study of Experimental Medicine* [1865]. New York, Dover, 1957; 66.
    (4) The concluding remarks in Huxley's essay 'On the physical basis of life' (1868), in *Collected Essays*, vol. 1.
    (5) Mach, *Contributions to the Analysis of Senations*, [1866] 5th edn. New York, Dover, 1959; 12.

44. In what follows concerning the doctrines of Marx and Engels I wish to avoid current controversies over whether or not one can assume that Marx subscribed to that form of dialectical materialism which one finds developed in Engels' *Anti-Dühring* (1878) and would have subscribed to Engels' *Ludwig Feuerbach und die Ausgang der klassischen deutschen Philosophie* (1888), which are the two texts of greatest relevance to what follows. While I am inclined to believe that current interpretations of Marx's own thought are too much under the influence of the *Economic and Philosophic Manuscripts of 1844*, I shall attempt to phrase my remarks in such a way as to leave it as an open question how one is to relate the philosophic position of Marx to that of Engels.

45. It is, I believe, a mistake to view Feuerbach's early attack on Dogurth (1838), in which he took a fundamentally dualistic view of the mind–body relationship, as being consistent with the views which he came to develop as soon as he had completely rejected Hegel. However, in *From Hegel to Marx*, Sidney Hook interprets Feuerbach in this way. On the basis of his other writings of this period, I should agree with the interpretation of Jodl in *Ludwig Feuerbach*, Ch. I.

46. The following are perhaps the most characteristic and significant works of these three materialists: Moleschott, *Kreislauf des Lebens* (1852); Vogt, *Köhlerglaube und Wissenschaft* (1854); Büchner, *Kraft und Stoff* (1855).

47. In his *Ludwig Feuerbach and the Outcome of Classical German Philosophy* (Marxist Library XV. New York, International Publishers, 1935), Engels says:

The old metaphysics which accepted things as finished objects arose from a natural science which investigated dead and living things as finished objects. But when this investigation had progressed so far that it became possible to take the decisive step forward of transition to the systematic investigation of the changes which these things undergo in nature itself, then the last hour of the old metaphysics sounded in the realm of philosophy, also. And in fact, while natural science up to the end of the last century was predominantly a *collecting* science, a science of finished things, in our century it is essentially a *classifying science*, a science of the processes, of the origin and development of these things and of the interconnection which binds all these natural processes into one great whole (p. 55).

Cf. also the contrast drawn by Engels between 'mechanical' and 'modern' materialism in *Herr Eugen Dühring's Revolution in Science (Anti-Dühring)* [1878]. Marxist Library XVIII. New York, International Publishers, 1939; 31–2.

48. Cf. *Cursus der Philosophie.* Leipzig, Koschny, 1875; 56ff.

49. Cf. *Das Werth des Lebens* [1865] 8th edn. Leipzig, Reisland, 1922; 102ff and the passage in *Cursus der Philosophie* (p. 104) which Engels selected as the passage through which he attacked Dühring's natural philosophy of the organic world (*Herr Eugen Dühring's Revolution*, opening of Ch. 7).

50. *Briefwechsel und Nachlass*, ed. Karl Grün, 2 vols. Leipzig, Winter, 1874, II; 308. On the relations between Feuerbach and Moleschott, cf. A. Levy, *La Philosophie de Feuerbach* (Paris, 1904).

51. As Lenin recognized, the most important philosophical enemy of dialectical materialism was critical positivism, and he therefore attacked Mach and Mach's followers in *Materialism and Empirio-Criticism* (1909).

52. In 1876 Hans Vaihinger looked upon Hartmann, Dühring, and Lange as the three outstanding newer philosophers (cf. his *Hartmann, Dühring und Lange*, Iserlohn, Baedeker, 1876; 4) and in the *History of Modern Philosophy* (1894) Höffding selected Dühring as one of the five German philosophers of the period from 1850 to 1880 to whom he devoted individual discussions, the others being Lotze, Fechner, Hartmann and Lange, all of whose names remain far more familiar to us.

53. *Address Delivered before the British Association in Belfast.* New York, Appleton, 1875; 4.

54. Ibid; 63–4. Also, cf. the preface to Hennill's *Inquiry Concerning the Origin of Christianity* (1838) in which Christianity is regarded as 'an elevated system of thought and feeling', but one which must be freed of 'fables'. By Tyndall's time 'free-thinkers' regarded Christianity as an elevated system of feeling, but not of thought.

55. Cf. the preface to *Das Leben Jesu*, and the appendix to its second volume. To be sure, Strauss's position later changed, as can be seen in *Der alte und der neue Glaube* (1865); by that time his religious position had shifted toward pantheism.

56. 'On improving natural knowledge' (1866), in *Collected Essays*, 9 vols. New York, Appleton, 1894, I; 38.

57. Emile Boutroux, *Science and Religion in Contemporary Philosophy* [1908]. London, Duckworth, 1909; 35. Boutroux, of course, sought to overcome this dualism.

58. There does not seem to have been any independent analogue to this position in England. To be sure, the works of Strauss and of Feuerbach were widely known through the translations of George Eliot; Comte, of course, had a significant number of followers.

59. Cf. Arnold, *Literature and Dogma,* Ch. 1 Sect. 4 and 5. (*The Complete Prose Works of Matthew Arnold*, R. H. Super (ed.), vols. I–[VI]. Ann Arbor, University of Michigan Press, 1960–[1967], VI; 189–201). Fiske, in *The Ideas of Good as Affected by Modern Knowledge* (1885), takes up and utilizes Arnold's phrases as defining the object of his own religious worship.

60. Cf. Otto Pffeiderer, *Development of Theology.* London, Swan, Sonnen-schein, 1893; 345.

61. Cf. my article, 'Darwin's religious views', *Journal of the History of Ideas,* XIX (1958); 363–78.

Chapter 2

2.0 Introduction

The common theme which runs through the studies in this chapter may be loosely put as science, society and religion in the nineteenth century. In the previous chapter Maurice Mandelbaum was concerned with the history of science only in so far as it bore on nineteenth-century history of philosophy and religion. Turner, Young and Farley and Geison not only reverse that emphasis, but, in their own ways, seek to relate issues of science and belief to social contexts. John Farley and Gerald L. Geison are quite explicit in their approach, and indeed having discussed the role of 'socio-political factors' in the debate over spontaneous generation between the Frenchmen Louis Pasteur and Félix Pouchet, they conclude by considering whether such 'external factors' may have influenced their own historical interpretation.

The section by Robert M. Young and the extract from Frank Miller Turner's *Between Science and Religion* are complementary in terms of subject-matter, and are both relevant to Mandelbaum's position on the supposed conflict between science and religion during the nineteenth century. It is evident from Turner's subsequent published work, as well as from the chapter here, that in his opinion the conflict was real.[1] Though much of the chapter which follows is concerned with the philosophical stances of Victorian scientific naturalists, and critiques by those of their non-Christian opponents who are the principal subjects of Turner's book, the conflict is explained not as a clash of ideas but as a contest for cultural leadership, in which the naturalists, reflecting growing professionalism in science, aimed to subvert clerical dominance of the educational system, and to secularize society. Note Turner's emphasis on the polemical value of the naturalists' dismissal of ontological questions (or in similar vein, his characterization of agnosticism as a cultural stance rather than a theory of knowledge), and his charge that their matter theory was chosen for the purposes of popularization rather than on any scientific grounds.

Young's paper deals with the other side of the conflict equation, in particular the response of theologians and theologian-scientists to scientific developments in the nineteenth century. He professes himself at the outset dissatisfied with 'internalist history of science' and 'internalist intellectual history', and aligns himself with works attempting to build bridges between history of scientific and philosophic ideas, and history of social and political ideas. In fairness, this is more a statement of future research plans than a foretaste of what is to follow in the paper, the latter being largely confined to the kind of history he wishes to marry with social, political and economic history (this is particularly true of the sections on Whewell, Baden Powell and the Metaphysical Society). The final section invokes more of a *social* context by analysing the fragmentation of a common *intellectual* context of natural theology in terms of the decline of highbrow periodicals, increasing professionalization of the sciences, and the growth of specialist societies and periodicals. His position on the conflict between religion and science is intermediate to Mandelbaum's and Turner's. He accepts the existence of conflict on one level, though recognizing that theologians made a variety of moves to overcome it. In the case of Whewell this involved a disorderly retreat, whereas Baden Powell ran headlong into the enemy's arms. In short, the price theologians paid for avoiding conflict between religion and science was the loss of natural theology's unifying role. In later published work, Young has attempted to follow through his research programme, and to build the bridges referred to in this paper. As a result, he has laid less emphasis on intellectual conflict in the debate over man's place in nature, and argued that at bottom the participants differed only about the best means of rationalizing the existing social and political order.[2]

The article by Farley and Geison is on the surface more restricted that the other two sources, being concerned with a relatively brief episode in the history of French biology. But in attempting to refute the traditional view that the outcome of the Pasteur–Pouchet debate represented the triumph of the true 'experimental' method, the authors go well beyond the debate's immediate temporal and disciplinary boundaries. A parallel is drawn with Cuvier's debate with Etienne Geoffroy St Hilaire in the 1820s and 1830s over spontaneous generation and transformism, while Pasteur's own inconsistent attitude to spontaneous generation is traced throughout the Second Empire and into the Third Republic. The authors do not undermine Pasteur's professed objectivity by asserting that all science incorporates the socio-political values of its practitioners; rather, they invert the traditional view, arguing that whereas Pouchet's adherence to heterogenesis flew in the face of his own political and religious beliefs, very similar views to the latter were instrumental in the attack on Pouchet led by Pasteur and the Académie des Sciences. It can be inferred both from the article's characterization of policies during the Second Empire, and from the two quotations given from Taine, and from Pasteur's Sorbonne lecture of 1864, that the debate took place within an intellectual *milieu* of apparent conflict. However, the polarizing of the spontaneous generation issues in that way relies on a questionable interpretation.

Notes

1. See in particular Frank M. Turner, 'The Victorian conflict between science and religion: a professional dimension', *Isis*, **69** (1978): 356–76.
2. See Robert M. Young, 'The historiographical and ideological contexts of the nineteenth-century debate on man's place in nature' in *Changing Perspectives in the History of Science: Essays in Honour of Joseph Needham*, Mikuláš Teich and Robert Young. (eds). London, Heinemann, 1973: 344–438.

2.1 Frank Miller Turner, Victorian Scientific Naturalism

The new nature and its champions

'Science touched the imagination by its tangible results', declared G. M. Young of early Victorian England.[1] Over the course of Queen Victoria's reign, those tangible results multiplied rapidly and extensively. The average Englishman came to enjoy better food, softer clothing, and a warmer home. Although his landscape might have become less lovely and the air he breathed less pure, he could live longer and dwell in greater security from the vicissitudes of nature than any man before him. The seventeenth-century Baconian vision of science come to the aid of man's estate reflected itself in so many facets of daily life that by the year of the queen's golden jubilee, T. H. Huxley could boast without exaggeration of a 'new Nature created by science' manifested in 'every mechanical artifice, every chemically pure substance, employed in manufacture, every abnormally fertile race of plants, or rapidly growing and fattening breed of animals'.[2]

Huxley and many of his contemporaries considered the quality of life so fully transformed as to require an almost complete reorientation in the thought and expectations of men and society. The New Nature itself, rather than God, heaven, or human nature, now furnished the chief points of reference for the organization and direction of life. '[T]his new Nature begotten by science upon fact', continued Huxley, had come to constitute

the foundation of our wealth and the condition of our safety from submergence by another flood of barbarous hordes; it is the bond which unites into a solid political whole, regions larger than any empire of antiquity; it secures us from the recurrence of pestilences and famines of former times; it is the source of endless comforts and conveniences, which are not mere luxuries, but conduce to physical and moral well being.[3]

Huxley and others believed the New Nature and the scientific theories associated with it sufficient for the expression, explanation, and guidance of human life. A wholly secular culture seemed altogether possible.

Nevertheless, Huxley realized that before the complete physical and moral benefits of the New Nature could be enjoyed, two tasks must be accomplished. First, the ordinary Englishman must be persuaded to look toward rational, scientific and secular ideas to solve his problems and to interpret his experiences rather than toward Christian, metaphysical, or other prescientific modes of thought. Second, scientifically trained and scientifically oriented men must supplant clergymen and Christian laymen as educators and leaders of English culture.[4] The champions of the New Nature set out to publicize the advantages of embracing scientific and secular ideas and of acknowledging the cultural preeminence of men of science. That the tomb of Newton in Westminster Abbey stands surrounded by the graves and monuments of Victorian scientists illustrates not only the achievements of the latter but also the success of the nineteenth-century publicists in establishing a favorable image of science in the public mind.

Not since the genius of the seventeenth-century virtuosi stirred learned imaginations had so many eloquent voices praised the cause of science. The leadership of this effort to educate and to persuade the public consisted of Huxley himself, professor of biology at the Royal School of Mines and chief apologist for Charles Darwin; John Tyndall, a physicist and successor to Faraday as superintendent of the Royal Institution; Herbert Spencer, the philosopher par excellence of evolution; W. K. Clifford, an outspoken mathematician at University College, London; and Sir Francis Galton, cousin of Charles Darwin, a eugenicist, a statistician, and an advocate of professionalism in science. Another coterie related to this core leadership but more closely associated with English Positivism included Frederic Harrison, a lawyer and leading English Positivist; John Morley, a freethinker and the editor of *Fortnightly Review*, in which most of these men published; and G. H. Lewes, Positivist, historian, and psychologist. Anthropologists, such as Edward Tylor and John Lubbock, extended the theories of science into the study of society. Biologist E. Ray Lankester and physician Henry Maudsley wrote and spoke on behalf of naturalistic ideas. Among essayists and men of letters who advocated the cause of science, Leslie Stephen was the outstanding author. He was joined by lesser literary figures, such as Grant Allen and Edward Clodd. The list could be greatly expanded.

These men rarely stood in complete agreement with one another. None of them embraced all the specific ideas and theories associated with the movement in the public mind. On more than one occasion they indulged in the luxury of internecine quarrels. Huxley was an avowed enemy of Positivism; Spencer was a Lamarckian. Few scientific writers were so enamored with the determinism of the nebular hypothesis as was Tyndall. The movement for scientific publicism was thus far from monolithic; it was not even coterminous with Victorian science. However, these authors knew and visited one another, enjoyed mutual friends, cited one another in their books and articles and sparred with mutual enemies. A composite of their leading ideas may risk presenting a caricature of the thought of any single figure but will nevertheless suggest the image that even the sophisticated reading public gleaned from their works. For what impressed contemporaries far more than their differences was the unity of their conviction that 'in the struggle of life with the facts of existence, Science is a bringer of aid; in the struggle of the soul with the mystery of existence, Science is a bringer of light'.[5]

Settling upon a label for a movement united more by common sentiment than by specific ideas or goals was a problem that plagued the participants as well as the historian. John Morley recalled that during the early days of the *Fortnightly Review*, 'People quarreled for a short season whether we should be labelled Comtist, Positivist, Naturalist. They were conscious of a certain concurrence in the writers, though it was not easy to define.'[6] While no single word or phrase will exactly describe the cluster of men and ideas associated with Huxley's New Nature or perhaps avoid creating another of what Lovejoy termed 'trouble-breeding and usually thought-obscuring terms', *Victorian scientific naturalism* appears more inclusive than any other phrase.[7] *Positivism* cannot serve because of Huxley's vehement antipathy to the sect and because in England Positivism was simply one part of the larger effort to advance science in the public forum. *Scientism* was coined during the period but seems to have had little vogue until the twentieth century. *Rationalism* fails to suggest the crucial role of science in addition to critical reasoning. *Free thought* is simply too imprecise. By the close of the nineteenth century, the words *naturalistic* appear to be the terms employed most frequently to denote the movement associated with contemporary scientific men and ideas. By 1902 'Naturalism' had received an entry (discarded in 1970) in the *Encyclopaedia Britannica*.[8]

Though the nature of its thought stemmed from the seventeenth-century scientific revolution and represented a further extension of the 'touch of cold philosophy', the movement was distinctly Victorian in that it arose and flourished in circumstances unique to the period 1850 to 1900. By that era the reality of the New Nature had fully emerged so as to grant plausibility to a wholly secular life. Moreover, the midcentury public-health campaign, the effort after the repeal of the Corn Laws to make farming more efficient, and later the necessity of meeting the threat of German industrial competition intimately related science to the affairs of business and everyday life. Furthermore, during the second half of the century numerous men of science, partly because of the earlier debates over geology and partly because of the necessity of finding employment, relinquished, if they did not openly renounce, the time-honoured belief that the scientist's occupation complemented that of the clergy by discovering the glory of God in nature.[9] From the 1850s onward scientists consciously moved toward greater professionalism involving social and intellectual emancipation from theology and financial independence from aristocratic patronage. Finally, this period marked what would appear to be the last era when the essential theories of science could be understood by the layman without training in advanced mathematics. Such a comprehension on the part of laymen was essential for the popularization of new scientific theories around which they were encouraged to organize their lives.

The movement was scientifically naturalistic in that it derived its repudiation of supernaturalism and its new interpretations of man, nature, and society from the theories, methods, and categories of empirical science rather than from rational analysis.[10] Beatrice Webb described the 'religion of science', which for a time attracted her, as 'an implicit faith that by the methods of physical science, and by these methods alone, could be solved all the problems arising out of the relation of man to man and of man towards the universe'.[11] In this regard Victorian scientific naturalism represented the English version of a general cult of science that swept across Europe during the second half of the century and that was associated with the names of Renan, Taine, Bernard, Buchner, and Haeckel as well as with various forms of scientific materialism and scientific socialism.[12]

The intimate relation of Victorian scientific naturalism to contemporary science renders the movement not only datable but also distinctly dated. Its exponents aligned themselves with the specific physical theories of the third quarter of the century and uncritically accepted the current concept of scientific law. The new physics of Rutherford and Einstein rendered their picture of nature obsolete. Twentieth-century philosophy of science redefined the nature of scientific laws. Consequently, the specific concepts of Victorian scientific naturalism became largely outmoded and no longer proved a source of present or enduring wisdom. As one historian has commented, 'None of the nineteenth-century scientific publicists is of any great importance as a philosopher.'[13]

Nevertheless, between 1850 and 1900 the advocates of scientific naturalism exercised a considerable influence in Britain. Any thinking man or woman found himself compelled to consider what they said. Beatrice Webb, remembering her youthful friendships with Spencer, Huxley, Tyndall, and Galton, wrote:

[W]ho will deny that the men of science were the leading British intellectuals of that period; that it was they who stood out as men of genius with international reputations; that it was they who were the self-confident militants of the period; that it was they who were routing the theologians, confounding the mystics, imposing their theories on philosophers, their inventions on capitalists, and their discoveries on medical men; whilst they were at the same time snubbing the artists, ignoring the poets, and even casting doubts on the capacity of the politicians?[14]

As the dissolution of the Christian faith led many intelligent people to seek alternative means of organizing their lives, the achievements of science and the surface plausibility of the views of the scientific publicists rendered confidence in the methods and results of science a convincing substitute for the discarded religion.

However naïve the mid- and late-Victorian optimism about science may now appear it then seemed productive of entirely constructive ends.

As Charles Kingsley once reminded an audience, '[S]cience has as yet done nothing but good. Will any one tell me what harm it has ever done?[15] Moreover, scientific naturalism could flourish because the most vocal intellectual opposition stemmed from writers who critized it while defending Christian or scriptural positions weakened by the higher criticism or scientific theory.[16] Effective criticism had to originate in dissatisfaction with scientific naturalism as an alternative non-Christian and self-sufficient world view.

James Ward, a Cambridge psychologist, philosopher, and former Christian, contributed the most significant and systematic discussion of scientific naturalism as a would-be complete interpretation of man and nature. In the Gifford Lectures, delivered at Aberdeen between 1896 and 1898, Ward explained:

This naturalistic philosophy consists in the union of three fundamental theories: (1) the theory that nature is ultimately resolvable into a single vast mechanism; (2) the theory of evolution as the working of this mechanism; and (3) the theory of psychophysical parallelism or conscious automatism, according to which theory mental phenomena occasionally accompany but never determine the movements and interactions of the material world.

Naturalistic writers employed these three theories to formulate what they considered a complete *Weltanschauung* that 'separates Nature from God, subordinates Spirit to Matter, and sets up unchangeable law as supreme'.[17]

In 1902 in the *Encyclopaedia Britannica* article on 'Naturalism', Ward distinguished between 'objective' and 'subjective' naturalism. The latter consisted of a mechanical explanation of mind; the former, a mechanical explanation of physical nature. Since the time of Hume and Hartley, subjective naturalism had merged with objective naturalism to produce associationist psychology. This illicit combination formed the keystone of current naturalism. Ward declared, 'As long as association of ideas (or sensory residue) is held to explain judgment and conscience, so long may naturalism stand.'[18] Once spontaneity had been purged from the mind, there was no difficulty in extending the check to spontaneity throughout all nature.

Though in both the lectures and the article Ward set mechanistic naturalism in opposition to spiritual or idealistic interpretations of the universe, he did not regard naturalism as materialistic. In this regard he displayed considerably more perception than did the Christian critics of naturalism. Ward regarded the chief characteristic of naturalistic metaphysics to be a neutral monism that *tended* toward materialism when employed to deal with any specific problem.

[W]hen the question arises, how best to systematize experience as a whole, it is contended we must begin from the physical side. Here we have precise conceptions, quantitative exactness and thoroughgoing continuity; every thought that has ever stirred the hearts of men, not less than every breeze that has ever rippled the face of the deep, has meant a perfectly definite redistribution of matter and motion. To the mechanical principles of this redistribution an ultimate analysis brings us down; and – beginning from these – the nebular hypothesis and the theory of natural selection will enable us to explain all subsequent synthesis. Life and mind now clearly take a secondary place; the cosmical mechanism determines *them*, while they are powerless to modify it. The spiritual becomes the 'epiphenomenal', a merely incidental phosphorescence, so to say, that regularly accompanies physical processes of a certain type and complexity.

This neutrally monistic, mechanical interpretation of the universe had produced 'cosmological naturalism'. He concluded that the psychological naturalism of associationism and the cosmological naturalism of mechanism had over the course of the nineteenth century merged to form 'absolute naturalism', which totally denigrated life and mind.[19]

Ward's definitive treatment of scientific naturalism pointed up two important features that permitted a non-Christian critique of the whole system. First, Ward very carefully refused to equate scientific naturalism with science itself. Rather, he considered naturalism to be 'a philosophy of . . . being which is especially plausible to, and hence is widely prevalent among, scientific men'.[20] Ward was quite correct in drawing this distinction. The naturalistic publicists sought to expand the influence of scientific ideas for the purpose of secularizing society rather than for the goal of advancing science internally. Secularization was their goal; science, their weapon. Ward realized that one of the chief solvents of the naturalistic philosophy might well come from the internal development of science itself. New discoveries and new theories might lead men of science away from the narrow reductionism that had characterized the popular theories of the third quarter of the century. If this were the case, contemporary scientific naturalism might very well be rejected on empirical and scientific grounds.

Second, Ward recognized that naturalism represented an alternative to Christianity rather than an attack upon it. The scientific publicists viewed contemporary scientific methods and ideas as capable of gratifying what G. H. Lewes termed 'the great desire of this age' for 'a Doctrine which may serve to condense our knowledge, guide our researches, and shape our lives, so that Conduct may really be the consequence of Belief'.[21] Ward and others argued that naturalism would fulfill that desire only at the cost of life itself. If it could be demonstrated that naturalism could guide men's lives only by ignoring whole areas of human experience, by dismissing inescapable human questions, and by applying inapplicable scientific categories to other areas and questions, naturalism might be rejected as incapable of meeting the stated goals of its advocates. Such a rejection would in no manner necessarily constitute a defense of Christianity. However, during the 1860s and 1870s, the halcyon years of scientific naturalism, such criticism had been most difficult because the advocates of naturalism defined the nature of truth and the means to its discovery so as to exclude the validity and reality of any experience or question not amenable to their own methods.

The truth according to scientific naturalism
In 1866 Huxley delivered to a group of workingmen a lecture entitled
'On Improving Natural Knowledge'. In his usual manner he outlined the
most recent scientific triumphs and enthusiastically pointed to the
benefits derived from scientific knowledge by all men in all stations of
life. In a crescendo of exuberance, he told his audience:

If these ideas be destined, as I believe they are, to be more and more firmly
established as the world grows older: if that spirit be fated, as I believe it is, *to
extend itself into all departments of human thought, and to become co-extensive
with the range of knowledge*; if, as our race approaches its maturity, it discovers,
as I believe it will, that *there is but one kind of knowledge and but one method of
acquiring it*; then we who are still children, may justly feel it our highest duty to
recognise the advisableness of improving natural knowledge, and so to aid
ourselves and our successors in our course toward the noble goal which lies
before mankind.[22]

In this address Huxley voiced the general convictions of adherents to
scientific naturalism. He appealed to the traditional view of science as
playing a normative role in ascertaining truth and to the widespread
belief in the unity of truth.[23] However, Huxley broke with both
traditions in the breadth of validity he ascribed to science and in his
assertion that the scientific method was the only means of acquiring true
knowledge.

What Huxley, Spencer, Clifford, and other naturalistic writers said
about the nature of truth was not original. The roots of their opinions
lay in the empirical philosophy of the English Enlightenment and in the
works of Comte and J. S. Mill. Epistemologically the spokesmen of
scientific naturalism were positivists. To the jesting Pilate's question,
'What is Truth?', they responded that truth is always description of the
phenomena of the external world and of the mind and description of the
laws of succession and coexistences of those phenomena. With this
concept of truth they sought to sustain their own scientific positions and
simultaneously to preempt those of potential opponents. Naturalistic
truth included no information about the nature of the underlying reality
(if there were such a reality) of phenomena. Naturalistic writers usually
classed speculations upon this later question as metaphysics that
pertained to a prescientific pattern of thought.

The publicists of scientific naturalism claimed to possess neither old
truth nor new truth, neither absolute truth nor relative truth, but rather
the only truth accessible to men and the only truth sufficient and
necessary for men dwelling in the New Nature. G. H. Lewes spoke for
most of his fellow champions of the New Nature when he defined this
truth as 'the correspondence between the order of ideas and the order of
phenomena, so that the one becomes a reflection of the other – the
movement of Thought following the movement of Things'.[24] Only when
internal human thought came to reflect the external movement of things
or phenomena could human actions approximately correspond to the
actual nature of the perceivable world and become genuinely

meaningful and effective. In this process the human mind contributed little more than its capacity to receive external impressions from the senses and to arrange them according to the laws of association.

Lewes argued that there were essentially two modes of approaching empirical evidence so as to achieve a proper correspondence between thought and things. The first was the subjective method; the second, the objective method. According to the former the direction of truth was determined by thought; according to the latter it was controlled by things. Both methods employed inference, but the subjective method employed unverifiable inference.

Lewes associated the subjective method with religion, intuitionism, metaphysics, and prescientific thought. Each of these committed the fundamental error of attempting 'to explain the scheme of the visible from the invisible, [to] deduce the knowable from the unknowable'. The subjective method could not lead to truth because it lacked the chief virtue of the objective method – 'vigilant verification'. The absence of verification meant that the subjective method confounded 'concepts with precepts, ideas with objects, conjectures with realities' and tried to make things conform to thought.[25] Experience proved that this method encouraged men to reach for chimeras. No evangelical Christian feared the snare of the inner man more than did the adherents of naturalistic truth.

The objective method, on the other hand, accepted no inference that could not be verified by observable empirical facts and correspondence with the known laws of nature. This method prevented an investigator from confusing ideas with objects or his own hypothetical concepts with reality. The objective method assured the proper coordination between the materials of experience and the ideas of the mind. It assured that the latter would always correspond to the former. Such knowledge permitted men to act with certainty and prevented them from pursuing phantoms that arose from subjective illusions.

In subscribing to this particular standpoint, the advocates of scientific naturalism chose among alternative views of the scientific method. Various studies of the last twenty-five years have revealed one crucial fact about Victorian science – namely, that there was little agreement among the scientists themselves as to what exactly constituted the method of science. For example, much of the serious objection to natural selection concentrated on whether Darwin's method had been genuinely 'scientific'. Consequently, the adherents of naturalism never spoke for the entire scientific community. In describing their own view of truth, they had chosen empiricism over idealism, objectivity over any mode of subjectivity, and most significantly the logic of Mill over that of Whewell.[26]

It is questionable how much attention the naturalistic writers paid to the more technical aspects of Mill's laws of induction. It is even less sure that the scientists among them ever consciously employed Mill's *System of Logic* to guide their own research. Mill's system could explain scientific laws after they had been discovered, but it could not account for the mode of discovery itself. The naturalistic publicists, having embraced an essentially descriptive view of science, encountered considerable difficulty in coming to grips with the role of hypothesis in scientific investigation. For this reason in 1900 Sidgwick could speak of 'the actual impossibility of finding a satisfactory scientific explanation of the development of scientific knowledge'.[27]

Nevertheless, two of Mill's theories were of utmost importance for the naturalistic view of knowledge. First, the spokesmen for naturalism accepted Mill's idea that man could have knowledge of phenomena alone. Second, they accepted Mill's definition of causation as being the law of the invariable succession of phenomena. The significance of these two ideas for scientific naturalism can hardly be overestimated. They relieved naturalistic authors from having to deal with any formal ontology. In actual practice most of them oscillated between a near Berkelian idealism and a naïve realism. However, by formally ignoring the ontological issue or, like Spencer, by subsuming it under the Unknowable, naturalistic thinkers could apply scientific concepts to any area of intellectual endeavor merely by reducing the subject matter to phenomena. It made little difference whether these natural phenomena related to matters of physical nature, mind, art, society, or morals.

The dismissal of ontological considerations and the restricted view of knowledge were polemically (rather than personally) important as a means of permitting naturalistic authors to ignore questions, issues, and experiences inimical to their secular vision. The naturalistic theory of truth thus led directly to a self-serving agnosticism. The employment of agnosticism by Huxley and Leslie Stephen against the pretensions of religious writers is familiar. Any religious belief that could not be sustained by verifiable empirical facts became open game for the agnostic challenge.[28]

Agnosticism also served scientific naturalism in two other capacities. First, agnosticism was an instrument for clearing away certain metaphysical remnants in practical scientific research. This was especially true in psychology, where metaphysical terminology, such as 'soul', plagued those men who wanted to turn the discipline into positive science. Henry Maudsley, a London physician and physiological psychologist raised the question, 'Is metaphysics anything more than supernaturalism writ fine?' Maudsley had commenced in the 1860s to rid applied psychological theory of metaphysical concepts. He argued that the latter had no reality in thought, fact, or consequences.

The very terms of metaphysical psychology have, instead of helping, oppressed and hindered him [the psychologist] to an extent which it is impossible to measure: they have been hobgoblins to frighten him from entering on his path of inquiry, phantoms to lead him astray at every turn after he has entered upon it, deceivers lurking to betray him under the guise of seeming friends tendering help.[29]

Only after the imperceivable metaphysical fictions had been vanquished could meaningful and genuinely answerable questions be posed. The latter questions alone held the promise of health and healing.

Second, and for this study most important, agnosticism more nearly represented a cultural stance than a theory of knowledge per se. By asserting that men lacked sufficient knowledge to decide whether the universe was material or spiritual or whether it was ruled by a deity, the agnostics rejected a culture and cultural values that depended upon answers to such questions. They rejected a culture founded upon improbable answers to what they considered invalid or unanswerable questions. These questions and answers had no place in the New Nature, where only problems and experiences subject to the 'vigilant verification' of the objective method might be considered. The naturalistic theory of knowledge and the agnostic stance toward issues inaccessible to such knowledge constituted the primary intellectual apology for both the necessity and the adequacy of an entirely secular culture.

In their zeal for scientific enlightenment, the advocates of naturalism often tended to overstate their case in order to stir an audience or to provoke further public discussion. On the lecture platform or in the pages of the liberal journals, their tone was arrogant, their confidence unbounded, and their metaphysics reductionist. They always gave themselves room to retreat without surrendering the major issue at hand. This was their public side. In the privacy of letters and conversation, they were less fulsome, more introspective, and even self-doubting. No one who has read their letters and the anecdotes of friends can believe that they were grim reductionists unaware of or insensitive to the emotional requirements of human nature. However, these private selves were familiar only to intimate friends and a few students. The world at large had to await the turn-of-the-century biographies to discover this side of their characters. What contemporaries witnessed were impatient, hard-headed polemicists reluctant to yield or to compromise. To this public character reply had to be directed and reaction measured.

Christians could and did object to both the manner and substance of the naturalistic theory of truth for ignoring the Scriptures and the destiny of the human soul. However, non-Christian writers, such as Henry Sidgwick. Samuel Butler, and James Ward, raised more impressive and pointed criticisms. As earlier suggested, Sidgwick considered empiricism an inadequate philosophy of scientific investigation. Butler thought the objective method was rooted in illusion. Ward argued that empirical examination of mental experience suggested that the activity of the mind determined part of the perception of phenomena. All of them believed there were certain unavoidable human questions that required nonempirical and nonobjectively verifiable concepts for rational answers.

In addition, non-Christians believed that the naturalistic view of truth illicitly excluded consideration of subjective empirical experiences that had practical consequences in the lives of human beings. The a priori exclusion of subjective experience meant that secular life must include only a part of the normal existence of normal human beings. In this regard, Sidgwick, Wallace, Myers, Butler, Romanes, and Ward firmly agreed with an argument William James once advanced to a group of his English friends.

The personal and romantic view of life has other roots beside wanton exuberance of imagination and perversity of heart. It is perennially fed by *facts of experience,* whatever the ulterior interpretation of those facts may prove to be. . . . These experiences have three characters in common: they are capricious, discontinuous, and not easily controlled; they require peculiar persons for their production; their significance seems to be wholly for personal life. Those who preferentially attend to them, and still more those who are individually subject to them, not only easily *may* find but are logically bound to find in them valid arguments for their romantic and personal conception of the world's course.[30]

A view of truth and knowledge that systematically excluded the consideration of those experiences could never allow men to discover the whole truth or to understand the full dimensions of personality and existence. Moreover, a cosmology that did not take into account such experience could never adequately portray the conditions and circumstances of human life.

The cosmology of scientific naturalism

W. K. Clifford once remarked, 'The character of the emotion with which men contemplate the world, the temper in which they stand in the presence of the immensities and the eternities, must depend first of all on what they think the world is.'[31] The exponents of scientific naturalism endeavored to persuade their fellow Englishmen that nature was essentially a mass of uniformly evolving atoms and energy. Since careful examination of these atoms and energy revealed no supernatural beings, nature should be contemplated in a wholly secular temper.

The advocates of naturalism constructed their model of external nature from three seminal theories of nineteenth-century science, which in their hands converged into what Ward termed 'absolute naturalism'. These were Dalton's atomic theory, the law of the conservation of energy, and evolution. Of this triad Huxley observed in 1887:

The peculiar merit of our epoch is that it has shown how these hypotheses connect a vast number of seemingly independent partial generalizations; that it has given them that precision of expression which is necessary for their exact verification; and that it has practically proved their value as guides to the discovery of new truth. All three doctrines are intimately connected, and each is applicable to the whole physical cosmos.

Huxley and his fellow naturalistic writers, such as Spencer, employed the three theories 'to interpret the detailed phenomena of Life, and Mind, and Society, in terms of Matter, Motion, and Force'.[32] Reduced to the common denominators of evolving matter and energy, all natural phenomena could be explained mechanically and interpreted without reference to God, supernatural agencies, or independent mind.

Undergirding the model of nature derived from these theories was the concept of continuity, described by George Eliot as 'the great conception of universal regular sequence, without partiality and without caprice'. In 1866 W. R. Grove reminded the British Association for the Advancement of Science that continuity was 'no new word, and used in no new sense, but perhaps applied more generally than it has hitherto been'. This was very much an understatement. The seventeenth-century concept of uniform natural activity had by the 1860s been extended to chemistry, physics, biology, psychology, medicine, and the emerging sciences of society. Although the concept was a nonempirical assumption and not subject to 'vigilant Verification', it was the essential postulate that allowed the scientists and naturalistic writers to interpret nature rationally. As Huxley once explained, the value of the extension of this principle was 'to narrow the range and loosen the force of men's belief in spontaneity, or in changes other than such as arise out of that definite order [of nature] itself'.[33] Explanations of natural phenomena that did not reduce the latter to such uniformity stood condemned as unscientific. By definition what was not subject to uniformity had no existence save in illusion.

The atomic theory of matter furnished the exponents of scientific naturalism with the primary example of the continuity of physical nature. However, the manner in which they popularized the theory illustrates why the movement could not be equated with the advance of science, why some opponents of naturalism could claim to be more scientifically advanced than the leading spokesmen of naturalism, and how the latter unnecessarily tied themselves to theories that became scientifically obsolete by the turn of the century.

Tyndall, Huxley, Spencer, and the psychologist Alexander Bain advocated Dalton's atomic theory. From the chemists' standpoint this theory allowed for a mechanical and mathematical concept of matter, the development of the law of definite proportions and the completion of the periodic chart of the elements.[34] On the popular level, this variety of atomism permitted an easily visualized image of matter as composed of small, round, solid, indestructible particles.[35] All the material manifestations of the physical world stemmed from particular arrangements of such atoms and the molecules formed by them. However, in choosing to popularize the solid-ball theory of atoms, Huxley, Tyndall, and Spencer consciously chose among competing theories. For example, Spencer specifically dismissed Boscovich's theory. Huxley refused to accept Maxwell's views that the similarity of atoms bespoke a planned manufacture and that theoretically more than one atom might occupy the same space. Moreover, all of these writers paid scant attention to William Thomson's vortex theory, which interpreted the internal composition of atoms in terms of a flowing fluid.[36] A desire for conceptual convenience in popular explication rather than experimental or theoretical considerations primarily determined the publicists' choice of theory. They thus tied themselves and the philosophic view that they expounded to a concept of matter that became increasingly untenable in the latter part of the century when subatomic particles were discovered. By that time the 'scientific' idea of matter associated with scientific naturalism was no longer associated with the scientific community.[37]

The atomic theory served to portray nature as one vast mechanical structure. The law of the conservation of energy explained the operation of the machine and established the limits to what was scientifically and naturally possible within the realm of nature. It held that the quantity of energy in the universe remained fixed throughout all the varied transformations of energy and atoms. The mechanism of nature remained closed to all external interference.[38] As Tyndall described energy in 1861:

The proteus changes, but he is ever the same; and his changes in nature, supposing no miracle to supervene, are the expression, not of spontaneity, but of physical necessity. A perpetual motion, then, is deemed impossible, because it demands the creation of energy, whereas the principle of Conservation is – no creation but infinite conversion.

The contemporary significance of this law was immense and probably more destructive to a supernatural interpretation of nature than was evolution by natural selection. In 1868 the Archbishop of York declared, 'The doctrine most in favour with physical philosophers at this moment is that of conservation of force'.[39] The theory struck down any religious explanation entailing miracles, spirits, or God by suggesting such modes of interference would require new infusions of external energy into the closed mechanism of nature.[40]

Evolution completed the naturalistic cosmology. Despite the numerous discussions of this topic, considerable confusion still surrounds the contemporary meaning of the term. There were two primary definitions interpreted and associated respectively with physics and biology. In physics, evolution was linked with the nebular hypothesis. Spencer and Tyndall were its major advocates. In its simplest form, it explained the state of the physical universe at any particular moment as the result of the continuously developing arrangement of matter and energy from all previous time. In *First Principles* Spencer set forth the following definition:

Evolution is an integration of matter and concomitant dissipation of motion: during which the matter passes from an indefinite, incoherent homogeneity to a definite, coherent heterogeneity; and during which the retained motion undergoes a parallel transformation.[41]

This variety of physical evolution in conjunction with continuity, atomism, and the conservation of energy allowed naturalistic writers to suggest that the human mind itself might be little more than an as-yet-unexplained manifestation of atoms and energy.

Among biologists, evolution applied to the species question. Prior to Darwin and Wallace, numerous biologists had been convinced that evolution or transformation of species occurred. However, until the formulation of the theory of natural selection, biologists possessed no thoroughly mechanistic concept to account for those modifications. Natural selection, in the words of Huxley, 'was a hypothesis respecting the origin of known organic forms which assumed the operation of no causes but such as could be proved to be actually at work'.[42] Natural selection brought organic forms under a theory of change analogous to that under which the nebular hypothesis brought physical forms. It reduced modifications in organic structures to rearrangements of matter and energy requiring no supernatural agencies. In this regard, it is important to recognize that evolution by natural selection represented only the final element of the broader naturalistic synthesis of man and nature that had arisen over the course of the century.[43]

The exponents of naturalism derived their idea of man directly from their view of physical nature. As Henry Maudsley explained, 'There are not two worlds – a world of nature and a world of human nature – standing over against one another in a sort of antagonism, but one world of nature, in the orderly evolution of which human nature has its subordinate part.'[44] The cosmology of scientific naturalism, like its epistemology, simply excluded all those facets of human nature that did not fit into the preconceived pattern of physical nature.[45] Naturalistic authors could and did answer the questions about man that Huxley propounded in *Man's Place in Nature*.

The question of questions for mankind – the problem which underlies all others, and is more deeply interesting than any other – is the ascertainment of the place which Man occupies in nature and of his relations to the universe of things. Whence our race has come; what are the limits of our power over nature, and of nature's power over us; to what goal are we tending; are the problems which present themselves anew and with undiminished interest to every man borne into the world'[46]

So long as the discussion was limited to the 'universe of things', naturalistic writers could discuss man. He was an animal descended from more brutish animals. Science gave him power over physical nature so that he might progress toward his end, which was the civilization of the New Nature. But of man considered as a problem to himself and within himself or of man confronting existential dilemmas, such as death, naturalistic writers could not and would not speak in public. Those issues were not relevant to the New Nature.

The Christian censure of the evolutionary and naturalistic view of man and nature is well known. However, naturalistic evolution could be considered incompatible not only with Christian doctrines and the scriptures but also with the secular doctrine and ideal of moral progress associated with rationalist and enlightenment thought. Writers such as Wallace, Myers, Butler, and Ward desired a concept of evolution that would, at least in the case of man, provide for genuinely qualitative change in the universe rather than for rearrangement of qualitatively unchanging atoms and energy. The realization of the non-Christian ideal of moral improvement required a universe of 'becoming' rather than one of 'being-in-process'. Long before Bergson, they recognized that Spencer 'had promised to trace out a genesis, but he had done something quite different; his doctrine is an evolutionism only in name'.[47] They sensed a radical incommensurability between mankind's collective and individual moral awareness and the nature described in the cosmology of scientific naturalism. Like the scientific publicists, they considered man an integral part of nature. However, they contended that if man were such a part of nature, his own moral character and inner experiences must be acknowledged when describing nature as a whole. To at least some degree the macrocosm must be interpreted through the microcosm. This was exactly what the advocates of naturalism refused to do.

Moreover, the figures in this study thought man had evolved a mental nature and displayed intellectual sensibilities for which the scientific cosmology could neither account nor provide future development. If bound by that reductionist cosmology, they could only conclude with Thomas Hardy, 'We have reached a degree of intelligence which Nature never contemplated when framing her laws, and for which she consequently has provided no adequate satisfactions.'[48] Sidgwick, Wallace, Myers, Romanes, Butler, and Ward were not prepared to accept that predicament. They believed that nature must be commensurate with man's intelligence and ideals. In various ways, they set out to prove that nature included realities and experiences that would satisfactorily account for such intellect and permit the realization of human ideals.

The challenge to religious culture
Already reeling under the attacks on the scriptures from rationalism and the higher criticism, Christianity now faced from scientific naturalism a challenge to the actual validity of religion itself.[49] Frederic Myers succinctly described the new turn of mind.

The essential spirituality of the universe, in short, is the basis of religion, and it is precisely this basis which is now assailed. . . . The most effective assailants of Christianity no longer take the trouble to attack, as Voltaire did, the Bible miracles in detail. They strike at the root, and begin by denying – outright or virtually – that a spiritual world, a world beyond the conceivable reach of mathematical formulae, exists for us at all. They say with Clifford that 'no intelligences except those of men and animals have been at work in the solar system': or, implying that the physical Cosmos is all, and massing together all possible spiritual entities under the name which most suggests superstition, they affirm that the world 'is made of ether and atoms, and there is no room for ghosts'.[50]

Most naturalistic thinkers drew two conclusions from those assertions. First, as Myers suggested, they considered all existing religion a form of superstition, primitive survival, or illusion. Second, and in general departure from previous critics of religion who called for a purified faith, they argued that traditional religion, especially Christianity, and men in religious occupations were culturally dysfunctional. They called for the replacement of a religiously directed culture with a scientifically directed culture dominated by scientifically oriented men.

Among the naturalistic coterie the rising sociologists proved to be the most insidious enemies of religious culture. Tylor, Spencer, McLennon, and Lubbock, though disagreeing on particulars, were of one mind in arguing that religion had originated naturally with no divine aid or revelation.[51] They interpreted its origins in such a manner that religion should naturally and necessarily give way to advancing science.

Tylor considered 'a minimum definition of Religion' to be 'the belief in Spiritual Beings'. This belief and the religion it nurtured had originated in the attempts of primitive men to understand the operation of nature by ascribing natural processes to the activity of spirits. All religion might in one manner or another be reduced to a survival of this primitive animism.

Animism characterizes tribes very low in the scale of humanity, and thence ascends, deeply modified in its transmission, but from first to last preserving an unbroken continuity, into the midst of high modern culture. . . . Animism is, in fact, the groundwork of the Philosophy of Religion, from that of savages up to that of civilized men.

Tylor admitted that some later religions had added ethical systems to animism, but this addition did not change the nature of religion itself. Ethics was not intrinsically related to religion. Moreover, according to John Lubbock, the higher religions were essentially no better or no purer than the lower. He remarked, 'The higher faiths . . . merely superimposed themselves on, and did not eradicate, the lower superstitions.'[52] Contemporary religion consequently remained stigmatized by the original sin of its primitive founders.

This sociological analysis of religion permitted the spokesmen of scientific naturalism to contend that redemption of the primitive sin lay near at hand. The doctrines and theories of science could now correct that original misunderstanding of nature. Scientific ideas, therefore, did not rob men of their religion but rather fulfilled those very yearnings of human curiosity that had for centuries been misdirected into superstitious behavior. Tyndall explained, 'The same impulse, inherited and intensified, is the spur of scientific action today.'[53] This assertion constituted the foundation for the claim of naturalistic writers to replace the clergy as new leaders of English culture.

Certain naturalistic authors did not rest content with this sociological examination of religion. It sufficed only to exorcise religious doctrine and social justification for religious institutions. The chief stronghold of nineteenth-century religion had been and continued to be the subjective religious experience in the form of conversion, ecstasy, or some other antinomian manifestation. Henry Maudsley argued that such experiences were invalid for two reasons. First, being subjective they could not be investigated by the objective scientific method. No idea, knowledge, or intuition so received could be considered genuinely true because it could not be communicated as part of the universal experience of sensible men. Such experiences could not reveal truth.

The authority of direct personal intuition is the authority of the lunatic's direct intuition that he is the Messiah; the vagaries of whose mad thoughts notoriously cannot be rectified until he can be got to abandon his isolating self-sufficiency and to place confidence in the assurances and acts of others.

Second, the religious experience could be naturally explained and dismissed as hallucination or illusion. Maudsley perceived no difference in kind between the vision of a saint and the delusion of a madman. 'Whatever its inner essence, the spiritual ecstasy in which a person is carried *out of himself* by divine action, has all the outward and visible character of the ecstasy in which he is *beside himself* through morbid action.'[54] With such analysis Maudsley summarily dismissed the religious experiences of mankind from Saul on the road to Damascus to the converts of nineteenth-century revivals.

Through the epistemological limits ascribed to human knowledge, the portrayal of nature without supernatural entities, and the reduction of religious practices to a primitive survival and of religious experience to mental disorder, the advocates of scientific naturalism sought to displace the existing clerical and literary intellectual elite. A culture based on science must replace one founded on religion. This was what Huxley intended to indicate near the close of his life when he wrote that 'the future of our civilization . . . depends on the result of the contest between Science and Ecclesiasticism which is now afoot'.[55] The real issue at stake in the contest was whether men could lead secular lives in which their occupations, institutions, values, and aspirations would be determined through scientific method with little or no reference to matters beyond the range of naturalistic epistemology and outside the framework of the naturalistic cosmology or whether their lives should be guided by knowledge drawn from science as supplemented by revelation, intuition, or creative reason.

The emergence of the New Nature had opened a vast new realm for human life and achievement, the cultural values and leadership of which had yet to be determined. R. H. Hutton, the liberal Christian editor of the *Spectator*, described the novel situation.

A very great part of the best thought of the best men is occupied in very large degree with interests which have all the largeness and catholicity, as one may say, of something quasi-spiritual, and yet no vestige of the true spiritual world in them, no vestige in them of the great conflict between darkness and light, between evil and good, between temptation and grace. The area of perfectly disinterested and perfectly innocent and wholesome interests which are not in the least moral or spiritual interests has grown vastly in the modern world, and the effect of this is that a much larger portion of the permanent mind of good men is usually eagerly at work in tracking out clues which have neither the taint of moral danger about them on the one side, nor the inspiration of spiritual help on the other. A great part of the minds of good men is thus invested in secular interests which are not in the bad sense worldly, and which are indeed in a very real sense unworldly, though they cannot be called moral or spiritual, nay, which, far from calling upon the vision of an unseen world, only tend to give a deeper intellectual fascination to the spectacle of the seen world.[56]

Over the cultural domination of this emergent secular society and secular achievement the combatants of science and religion fought. Here Francis Galton hoped there might arise 'a scientific priesthood' that would tend to the health and welfare of the entire nation. New men as well as new ideas must direct the life of the New Nature.[57]

Source: Taken from Frank Miller Turner, *Between Science and Religion: The Reaction to Scientific Naturalism in Late Victorian England.* New Haven, Yale University Press, 1974, Ch. 2.

Notes

1. G. M. Young, *Victorian England: Portrait of an Age,* 2nd edn. New York, Oxford University Press, 1964: 7. See also Walter Houghton. *The Victorian Frame of Mind, 1830–1870.* New Haven, Yale University Press, 1957: 33–45.
2. T. H. Huxley, *Collected Essays.* New York, Appleton, 1894. I: 51.
3. Ibid.
4. John Tyndall, *Fragments of Science,* 6th edn. New York, Appleton, 1892, II: 1–8, 40–5; Francis Galton, *English Men of Science: Their Nature and Nurture.* New York, Appleton, 1875; John Morley. *The Struggle for National Education.* London, Chapman & Hall, 1873; and the anonymous article 'The national importance of research'. *Westminster Review,* **99** (1873): 343-66.
5. G. H. Lewes 'On the dread and dislike of science', *Fortnightly Review,* **29** (1878): 805. See also J. Vernon Jensen. 'The X club: fraternity of Victorian scientists', *British Journal for the History of Science* **5** (1970–71): 63–72.
6. John Viscount Morley, *Works of Lord Morley.* London, Macmillan, 1921, I: 81.
7. Arthur O. Lovejoy, *The Great Chain of Being: A Study in the History of an Idea.* New York, Harper Torchbooks, 1960: 6.
8. See Sydney Eisen, 'Huxley and the Positivists', *Victorian Studies,* **7** (1964): 337–58; 'Scientism', *Oxford English Dictionary.* Oxford, The Clarendon Press, 1933, 9: 223: A. W. Benn, *A History of English Rationalism in the Nineteenth Century.* London, Longmans, 1906. I: 1–58; J. M. Robertson, *A History of Freethought.* New York, G. P. Putnam's Sons, 1930; A. J. Balfour, *The Foundations of Belief.* London, Longmans, 1895; H. W. Blunt, 'Philosophy and naturalism', *Proceedings of the Aristotelian Society,* 3, no. 2 (1896): 43–51; C. Lloyd Morgan, 'Naturalism', *Monist,* **6** (1895): 76–90; Andrew Seth, "Naturalism" in recent discussion', *The Philosophical Review,*

5 (1896): 576–84; W. R. Sorley, *On the Ethics of Naturalism*. Edinburgh and London, Blackwood, 1885: 16–21, 277; James Ward, *Naturalism and Agnosticism*, 2 vols. London, A. and C. Black, 1899; and *Encyclopaedia Britannica*, 11th edn, s.v. 'Naturalism'. Balfour's book and the responses it evoked were primarily responsible for giving currency to *naturalism* as the term under which scientific writing and philosophy became subsumed.

9. See C. C. Gillispie, *Genesis and Geology: A Study in the Relations of Scientific Thought, Natural Theology and Social Opinion in Great Britain, 1790–1850*. New York, Harper Torchbooks, 1959; Robert Young, 'The impact of Darwin on conventional thought', in *The Victorian Crisis of Faith*, A. Symondson (ed.), London, Society for Promoting Christian Knowledge, 1970: 13–35; George Basalla, William Coleman, and Robert H. Kargon, (eds). *Victorian Science: A Self-portrait from the Presidential Addresses to the British Association for the Advancement of Science*. Garden City, New York, Doubleday, 1970.

10. Noel Annan, 'The strands of unbelief', and J. Bronowski, 'Unbelief and science', in *Ideas and Beliefs of the Victorians*, H. Grisewood (ed.). New York, E. P. Dutton, 1966: 150–6. 164–72.

11. Beatrice Webb, *My Apprenticeship*. London, Longmans, 1926: 83. Mrs Webb's chapter on the religious turmoil of her late adolescence is exceptionally revealing since much of it is taken from her diary. Her rejection of both Christianity and the religion of science closely resembles the figures examined in this study.

12. D. G. Charlton, *Positivist Thought in France during the Second Empire*. Oxford, The Clarendon Press, 1959; Friedrich Albert Lange. *The History of Materialism and Criticism of Its Present Importance*, trans. Ernest Chester Thomas. London, Routledge & Kegan Paul, 1957; Maurice Mandelbaum, *History, Man, & Reason: A Study in Nineteenth-Century Thought*. Baltimore, Johns Hopkins Press, 1971: 10–28; George Mosse. *The Culture of Western Europe: The Nineteenth and Twentieth Centuries*. New York, Rand McNally, 1961: 197–212; John Herman Randall, Jr., *The Making of the Modern Mind*, rev. ed. New York, Houghton Mifflin, 1954: 458–576.

13. John Passmore, *A Hundred Years of Philosophy*. London, Penguin Books, 1968: 48. For a series of discussions of the modifications in scientific theory that undermined the nineteenth-century scientific world view, see Herbert Dingle (ed.), *A Century of Science, 1851–1951* (London, Hutchinson's Scientific and Technical Publications, 1951) and J. Bronowski, *The Common Sense of Science* (London, Heinemann, 1951): 97–119. For a more broad-ranging discussion of the problem of epistemology in light of twentieth-century science, see Stephen Toulmin, *Human Understanding*. Oxford, The Clarendon Press, 1972.

14. Webb, *My Apprenticeship:* 130–31. See also Henry Holland, *Recollections of Past Life*, 2nd ed. (London, Longmans, 1872): 311, and George Bernard Shaw, Preface t⸍ 'Back to Methuselah' in *Complete Plays with Prefaces*. New York, Dodd, Mead, 1963, 2: ix–xc.

15. Charles Kingsley, *Health and Education*. London, Ibister, 1874: 292.

16. For a series of essays by two sensible and intelligent Christians who opposed scientific naturalism, see John Tulloch, *Modern Theories in Philosophy and Religion* (Edinburgh and London, Blackwood, 1884), and R. H. Hutton, *Aspects of Religious and Scientific Thought*, ed. Elizabeth Roscoe (London, Macmillan 1899). Two enlightening essays on the varied

Christian response to science are Noel Annan, 'Science religion, and the critical mind: introduction', and Basil Willey, 'Darwin and clerical orthodoxy', in *1859: Entering an Age of Crisis*, P. Appleman, W. A. Madden, and M. Wolff (eds). Bloomington, Ind, Indiana University Press, 1959: 31–62.

17. Ward, *Naturalism and Agnosticism*, 1: 186; ibid.
18. *Encyclopaedia Britannica*, 11th edn, s.v. 'Naturalism' by Ward.
19. Ibid. Ward, Sorley, and Blunt stood in firm agreement that the subordination of the active mind to mechanical nature was the heart of naturalism.
20. Ward, *Naturalism and Agnosticism*, 1: 39.
21. G. H. Lewes, *Problems of Life and Mind, First Series*. London, Trubner 1874, 1: 2.
22. Huxley, *Collected Essays*, 1: 41 (FMT's emphasis).
23. Walter Cannon, 'The normative role of science in early Victorian thought', *Journal of the History of Ideas*, **25** (1964): 487–502, and 'Scientists and broad churchmen: an early Victorian intellectual network', *Journal of British Studies*, **4** (1964): 65–88.
24. G. H. Lewes, *History of Philosophy from Thales to Comte*, 4th edn. London, Longmans, 1871, 1: xxxi. See also Passmore, *A Hundred Years of Philosophy*: 13–47.
25. Lewes, *History of Philosophy*, 2: 212; ibid., 1: xxxix; ibid., 1: xl. See also Lewes, *Problems of Life and Mind, First Series*, 1: 1–87.
26. Alvar Ellegard, 'The Darwinian theory and the nineteenth-century philosophies of science', *Journal of the History of Ideas*, **18** (1957): 362–91; C. J. Ducasse, 'Whewell's philosophy of scientific discovery', *Philosophical Review*, **60** (1951): 59–69, 213–34; E. W. Strong, 'William Whewell and John Stuart Mill: their controversy about scientific knowledge', *Journal of the History of Ideas*, **16** (1955): 209–31. On Mill's influence during the 1860s see David Masson, *Recent British Philosophy*. London, Macmillan, 1865: 11–13; Lewes, *Biographical History of Philosophy*, rev. and enl. New York, Appleton, 1866: xxi, note; Henry Sidgwick, 'Philosophy at Cambridge', *Mind*, **1** (1876): 235.
27. Arthur J. Balfour (ed.), *Papers Read before the Synthetic Society, 1896–1908*. London: Spottiswoode, 1909: 276. In regard to the role of hypothesis in naturalistic writers, see Lewes, *Problems of Life and Mind, First Series*, 1: 314–41; Tyndall, *Fragments of Science*, 2: 101–34.
28. W. Irvine, *Apes, Angels, and Victorians*. Cleveland and New York, World Publishing Co., 1964: 247–63, 311–30; Noel Annan, *Sir Leslie Stephen: His Thought and Character in Relation to His Time*. London, MacGibbon & Kee, 1951: 172–95.
29. Henry Maudsley, *Natural Causes and Supernatural Seemings*. London, Kegan Paul, Trench, 1886: 104; idem, *Body and Mind: An Enquiry into Their Connection and Mutual Influence*, rev. and enl. New York, Appleton, 1875: vii.
30. William James, 'Presidential address to the Society for Psychical Research, January 31, 1896', in *Presidential Addresses to the Society for Psychical Research, 1882–1911*. Glasgow, The Society for Psychical Research, 1912: 84.
31. W. K. Clifford, *Lectures and Essays*, L. Stephen and F. Pollock (eds). London, Macmillan, 1901, 2: 259. See also Herbert Dingle, 'The Scientific

Outlook in 1851 and 1951' in *European Intellectual History since Darwin and Marx*, W. Warren Wagar (ed.). New York, Harper Torchbooks, 1966: 159–83.

32. *Encyclopaedia Britannica*, 11th ed., s.v. 'Naturalism'; Huxley, *Collected Essays*, 1: 66; Herbert Spencer, *First Principles*, 4th edn. New York, Collier, 1901: 468.

33. George Eliot, 'The influence of rationalism: Lecky's history', *Fortnightly Review* (1865), reprinted in *Works of George Eliot*. New York, Collier, n.d., 12: 138; W. R. Grove, 'Presidential address', *Report of the British Association for the Advancement of Science: 36th Meeting*. London, John Murray, 1867: 56; Huxley, *Collected Essays*, 1: 39. All the naturalistic writers admitted that uniformity was an assumption which, they contended, was justified by, though not drawn from, experience.

34. D. S. L. Cardwell (ed.), *John Dalton and the Progress of Science*. New York, Barnes & Noble, 1968; Frank Greenway, *Dalton and the Atom*. Ithaca, New York, Cornell University Press, 1966; H. L. Sharlin, *The Convergent Century: The Unification of Science in the Nineteenth Century*. New York, Abelard–Schuman, 1966: 52–8.

35. A. Bain 'The atomic theory', *Westminster Review*, **59** (1853): 125–96; Herbert Spencer, *First Principles*: 43–45, 243–45, and Preface to *Epitome of the Synthetic Philosophy*, by F. H. Collins. New York; Appleton, 1889: viii–xi; Tyndall, *Fragments of Science*, 1: 39, 2: 51–74.

36. For competing theories and Huxley's, Tyndall's, and Spencer's rejection of them, see George M. Fleck, 'Atomism in late nineteenth-century physical chemistry', *Journal of the History of Ideas*, **24** (1963): 106–14; W. H. Brock and D. M. Knight, 'The atomic debates: "Memorable and Interesting Evenings in the Life of the Chemical Society"', *Isis*, **56** (1965): 5–25; and D. M. Knight, *Atoms and Elements*. London, Hutchinson, 1967; Spencer, *First Principles:* 43–45; Huxley, *Collected Essays*, 1: 60–61; W. D. Niven (ed.), *The Scientific Papers of James Clerk Maxwell*. New York, Dover, 1965, 2: 376–7, 445–84; Robert H. Silliman, 'William Thomson: smoke rings and nineteenth century atomism', *Isis*, **54** (1963): 461–74.

37. T. T. Flint 'Particle physics', and W. Wilson, 'The structure of the atom', in *A Century of Science*, ed. Dingle: 39–69.

38. C. C. Gillispie, *The Edge of Objectivity*. Princeton, Princeton University Press, 1960: 370–405; Sharlin, *The Convergent Century:* 30–7; C. Singer, *A Short History of Scientific Ideas*. Oxford, Oxford University Press, 1959: 375–78.

39. Tyndall, *Fragments of Science*, 2: 4; William Thomson, *The Limits of Philosophical Inquiry: An address*. Edinburgh, Edmonston & Douglas, 1868: 10.

40. Naturalistic writers were often as selective in their dealings with thermodynamics as they were with atomism. During the 1860s Huxley and other supporters of Darwin tried to ignore the second law of thermodynamics, which suggested a universal dissipation of energy whereby all motion and thus all life would come to a frozen halt. This law was believed to mean there had not been sufficient time for geological and organic evolution. Huxley, Spencer, and other naturalistic writers later seem to have accepted the second law although with little enthusiasm, for the morbid future implied by the law was incompatible with the optimism associated with the New Nature. See Loren Eiseley, *Darwin's Century:*

Evolution and the Men Who Discovered It. Garden City, New York. Doubleday, 1961: 238–41; E. N. Hiebert, 'The use and abuse of thermodynamics in religion', *Daedalus*, **95** (1966): 1046–80; Stephen G. Brush. 'Science and culture in the nineteenth century', *The Graduate Journal,* 7 (1967): 477–565.

41. Spencer, *First Principles:* 334. See also Spencer, *Essays.* New York, Appleton, 1868, 1: 1–60. R. M. Young, 'The development of Herbert Spencer's concept of evolution', in *Acts du XIe congrès international d'histoire des sciences.* Warsaw, 1967, 2: 273–78; Tyndall, *Fragments of Science*, 2: 170–99.

42. Leonard Huxley, *Life and Letters of Thomas Henry Huxley.* New York, Appleton, 1902, 1: 182.

43. R. Young, 'The impact of Darwin on conventional thought', in *The Victorian Crisis of Faith,* A Symondson (ed.): 13–35; Noel Annan, 'Science, religion, and the critical mind; introduction', in *1859: Entering an Age of Crisis,* Appleman, Madden, and Wolff (eds): 31–51.

44. Henry Maudsley, 'Hallucinations of the senses', *Fortnightly Review,* **30** (1878): 386.

45. See John C. Greene, 'Darwin and religion', in *European Intellectual History since Darwin and Marx,* ed. W. W. Wagar (ed.): 31–4. Herbert Dingle once observed, 'The reason why the Victorian world [of science] contained nothing corresponding to religious experience is . . . because religious experience had not been taken into account in building it up,' Herbert Dingle, *The Sources of Eddington's Philosophy.* Cambridge, Cambridge University Press, 1954: 26.

46. T. H. Huxley, *Man's Place in Nature and Other Essays.* London, Dent 1910: 52.

47. Henri Bergson quoted by A. O. Lovejoy in 'Schopenhauer as evolutionist', in *Forerunners of Darwin 1745–1859,* B. Glass, O. Temkin, and W. L. Strauss, Jr., (eds). Baltimore; Johns Hopkins Press, 1969: 433.

48. Quoted in F. E. Hardy, *The Early Life of Thomas Hardy, 1840–1891.* New York, Macmillan, 1928: 213.

49. F. L. Baumer, *Religion and the Rise of Scepticism.* New York, Harcourt, Brace, 1960: 128–86; Howard R. Murphy, 'The ethical revolt against Christian orthodoxy in early Victorian England', *American Historical Review,* **60** (1955): 800–17; Herbert G. Wood, *Belief and Unbelief since 1850.* Cambridge University Press, 1955.

50 F. W. H. Myers, *Science and a Future Life with Other Essays.* London, Macmillan, 1893: 131–2. Even the opponents of Christianity argued among themselves on the nature of religion and what, if any, religion men should adopt. See Sidney Eisen, 'Frederic Harrison and Herbert Spencer: embattled unbelievers', *Victorian Studies,* 12 (1968–69): 33–56.

51. E. Tylor, *Primitive Culture.* London, John Murray, 1871; J. F. McLennon, 'The worship of animals and plants', *Fortnightly Review,* **12** (1869): 407–27 and **13** (1870): 194–216; J. Lubbock, *The Origin of Civilization and the Primitive Condition of Man.* London, Longmans, 1870: H. Spencer, 'The origin of animal worship', *Fortnightly Review,* **13** (1870): 535–50.

52. Tylor, *Primitive Culture,* 1: 383 (see J. W. Burrow, *Evolution and Society: A Study in Victorian Social Theory,* Cambridge, Cambridge University Press, 1966: 234–59); ibid.: 385. Lubbock, *The Origin of Civilization*: 255.

53. Tyndall, *Fragments of Science*, 2: 135. See also Herbert Spencer. *The Principles of Sociology*, New York, Appleton, 1897, 2: 247–60.

54. Henry Maudsley, *Body and Will*. London, Kegan Paul, Trench, 1883: 44; idem. *Natural Causes and Supernatural Seemings*: 271.

55. T. H. Huxley, 'Mr Balfour's attack on agnosticism', *The Nineteenth Century*, **37** (1895): 530.

56. Hutton. *Aspects of Religious and Scientific Thought*: 27–8.

57. Galton, *English Men of Science*: 195.

2.2 Robert M. Young, Natural Theology, Victorian Periodicals and the Fragmentation of a Common Context

For some time I have been pondering an hypothesis about the nature of a common intellectual context in the Victorian periodical literature, the role of natural theology in that milieu, and what happened when the relatively integrated common culture of issues and publications broke up. In this frankly exploratory essay I want to ruminate some of the problems which arise when one tries to write convincingly about such large-scale relationships in history.

My general line of research is concerned with the problem of doing science about people, with particular emphasis on the influence of Cartesian mind – body dualism in nineteenth and early twentieth-century biological, behaviourial and social science.[1] My first approach to the problem was to investigate theories of the relationship between mind and brain, centred on attempts to localize functions in the brain. As this work developed I became increasingly interested in the intellectual context within which various concepts of the functions of the brain were conceived.[2] This led away from relatively internalist preoccupation with the histories of psychology, psychiatry, neurology and neurophysiology to the development of evolutionary theory in Britain. This, in turn, led to a much wider enquiry into the nineteenth-century debate on man's place in nature. When I began to look for the response to developments in psychology and the study of the nervous system in the debate, I found very little. Instead, the monographs, periodicals, and lives and letters were much more concerned with geology, Malthusianism, that is, with the role of non-material causes in nature and with the relationships among God, man and nature. Thus, instead of detailed debates on psychology and brain physiology, there was a much more general debate on the uniformity of nature, miracles, free will and the status of man in nature. The arenas for discussing man's place in nature were not, it turned out, the sciences which seemed to me to be the most relevant (from my vantage point inside current disciplinary boundaries). Instead, geological and biological theory attracted the attention of the main participants in the debate.

However, two features of the debate soon became very clear. First, although the particular sciences of psychology and neurophysiology were not explicitly at the centre of controversy, it was almost impossible to make any demarcations between disciplines over a wide range of subjects as twentieth-century scholars would define them.[3] Second, at least for the first five or six decades of the century, natural theology provided the general context for the debate.[4] That is, new and controverted scientific findings and theories were taken up in the general culture as bearing on the demonstration and attributes of God as drawn from the book of nature (as distinct from revelation). In the course of the period, the *structure* of the debate changed progressively from a relatively undifferentiated one in which specialist studies were seen as parts of that very general theistic context in the early decades of the century, to one in which disciplines were increasingly demarcated and the relationship between science and theology became – at least at one level – one of conflict. At another level that relationship became one of identifying the new claims for the laws of nature with God's governance of the universe at a level of abstraction that was virtually meaningless.

If I am right about these very general features of the debate, the problem becomes one of relating three issues: (1) the common intellectual context; (2) natural theology; and (3) the body of writings in which the debate occurred. In attempting to understand the development of the debate I have found myself employing a number of related hypotheses, and this essay is an effort toward making them explicit and beginning to test them:

1. There was a common intellectual context (one could put that anachronistically as a rich interdisciplinary culture) in the early decades of the nineteenth century in Britain, and this was reflected in the periodical literature, monographs, lives and letters, and in a wide range of other writings.

2. There was a relatively homogeneous and satisfactory natural theology, best reflected in William Paley's classic *Natural Theology* (1802) and innumerable works reflecting the same point of view. These works were reviewed enthusiastically and at length in the periodical literature in the first four decades of the century.[5] The *Bridgewater Treatises* (1833–36) were an attempt to codify this tradition in the light of detailed findings in the several sciences, but they reflect the fact that an attempt to spell out the Paleyian point of view in detail led to difficulties which natural theology could not overcome without considerably modifying its view of the relationships among God, man and nature.[6]

3. The impact of scientific findings progressively altered this coherent natural theology until it was virtually devoid of content as a discipline in its own right.

4. The common intellectual context came to pieces in the 1870s and 1880s, and this fragmentation was reflected in the development of specialist societies and periodicals, increasing professionalisation, and the growth of general periodicals of a markedly lower intellectual standard.

It is at this point that my troubles begin. It would be folly to argue that the impact of science on natural theology provides an adequate *cause* for the alleged breakup of the (putative) common intellectual culture of the intelligentsia. Stated badly, the hypothesis only replaces internalist history of science with internalist intellectual history and offers a monocausal explanation. Surely this is simplistic in the extreme. For example, if one uses circulation figures and the growth of new periodicals with lower intellectual standards as a partial index, what weight should one attach to such factors as the emergence of serialised novels, the growth of literacy, the appeal of middle-brow periodicals to former readers of the heavier *Edinburgh* and *Quarterly Reviews*?[7] A more candid way of putting the problem is that I have come up against the limits of the legitimate explanatory power of the history of ideas, no matter how widely one casts the net into bodies of literature and their reception. The questions I have raised cry out for consideration from the point of view of social, political and economic historical research. These are perspectives which historians of science have yet to adopt in a serious and sustained way on most topics and certainly with respect to Victorian science. In order to carry my own research further I now realize that I must embark on a process of self-education for which there has been no preparation in eleven years as a student (and five as a teacher) of philosophy, science, medicine, and history and philosophy of science – all of which were taught without reference to the historical forces at work in the socio-economic order.

There is, of course, a complementary omission. Social, political and economic historians have been no less remiss, since they have failed to take seriously the structure and texture of the intellectual debates which fundamentally altered the perception of man's place in nature, and the role of science and technology in the productive, social and intellectual spheres. So I am not tempted to renounce intellectual history and plunge, amnesiacally, into other specialist studies in history. I hope eventually to be able to bring these approaches together as aspects of a totality, perhaps to reintegrate the fragments whose origins I am considering here.

Short of that long-range ambition, I should like to gather together the work I have done so far and put forward some evidence in support of the working hypothesis that the impact of science on natural theology was *a* factor, one which I believe was significant. That is, I wish to draw attention to a concomitant variation for the present, without in any way wanting to beg the question of what factors would add up to a satisfying causal account.

The remainder of the paper is in five parts: (1) Historiographic issues in connection with my approach to the problems mentioned above, with some remarks on the discipline of 'Victorian Studies'; (2) and (3) Two case studies in the changing interpretation of natural theology; (4) A case study, which I find symptomatic, of the attempt to retain a common intellectual context; (5) Some manifestations of the fragmentation of that common context, resulting in disciplinary and topic boundaries recognisable to our own demarcations.

I

First, some methodological problems involved in the definition of my domain and the problem of relating its constituent parts. My research on the debate on man's place in nature is related to at least eight sources of evidence in historical scholarship (which can be indicated by books which I have found more or less useful):

1. The use of literature to illuminate the history of ideas and assumptions in the period[8];
2. The history of scientific ideas, particularly in geology and biology[9];
3. The history of religion and theology[10];
4. The history of philosophy[11];
5. History of popular opinion and related issues[12];
6. Works attempting to build bridges between the history of scientific and philosophic ideas on the one hand and history of political and social ideas (these approaches are closest to my own current understanding)[13];
7. General political and historical context[14];
8. History of ideas in psychology, study of the nervous system, psychiatry and medicine, developments which were occurring in the period, no matter how small a place they found in the Victorian periodicals.[15]

Of course, this list could be extended indefinitely to embrace specialist studies in the histories of behavioural and social studies on the one hand and the history of the periodical press on the other,[16] but the purpose of making a list will have been served if it is obvious that the problem is that these are relatively isolated bodies of research, with occasional borrowings of the loosest generalisations from related disciplines. These are usually added on as rather desperate gestures in the direction of related disciplines whose literature one simply cannot master. An obvious book to use for this purpose is Gillispie's seminal work on *Genesis and Geology: The Impact of Scientific Discoveries upon Religious Beliefs in the Decades before Darwin*. In doing so, a non-specialist would be very likely to remain unaware that Gillispie has a very rudimentary grasp of the fine texture of the geological debate on which so much hung in the period 1820–50.[17] This is not a trivial point, since it turns out that the detailed proceedings of the Geological Society appeared in the quarterlies throughout this period, and every important geological work was reviewed in the major intellectual periodicals throughout the century. Geology was central to the debate on the relationship between God and Nature, the status of the Bible and man's place in nature. Thus, although Gillispie may not have exposed himself to these documents, the Victorian intelligentsia did.[18]

Is there any way out of this mess short of reading and sifting everything or collecting short essays from a wide variety of specialists (*Ideas and Beliefs of the Victorians*)?[19] In my opinion, Houghton in his *Wellesley Index to Victorian Periodicals* has shown us that the attempt to do everything need not be disastrous, while his bibliographical work since completing *The Victorian Frame of Mind* make it a great deal more orderly, if no less daunting, activity.

The approach which I have in mind, is to focus on a rather loosely-defined issue and to rely heavily on the Victorian periodicals, especially the quarterlies. Of course, as Ellegard has shown (without intending to) in his very ambitious study of the reception of Darwin's theory of evolution in the British periodical press, 1859–1872, this approach brings its own risks.[20] Foremost among these is the likelihood that one will have so many documents and such a vast network of related issues that the result is an impressively annotated bibliography, whereas one had set out to illuminate the ideas and assumptions of the period by a different method. Ellegard encountered these difficulties in dealing with two books – *Origin of Species* and *Descent of Man* – and Millhauser did an even less satisfactory job on the reception of Chambers' *Vestiges of Creation*.[21] With his example before us it really does seem absurdly ambitious to attempt to provide an interpretation of the debate within which these works and the controversies surrounding them played a very significant but (if one is considering the bulk of the literature) still relatively small part.

The way I have attempted to get round this difficulty is to be at once more and less ambitious. That is, one resigns onself to diminishing the number of essays and reviews which are to be considered with respect to a given work and at the same time to increasing the number of works considered. I would argue that this does not lead to an inevitable reduction of sense of nuance and depth. Rather, it points to the mindlessness of being excessively inclusive. Thus, this method requires a casualty at an early stage, i.e. the social history of ideas conceived as the study of low-brow popular opinion. There is some evidence that this sacrifice is not a crippling one; scholars who have studied certain aspects of the popular reactions are very reassuring. For example, Susan Budd read through the biographies of 150 members of the Secularist movement and found that ideas from geology, evolution and scientific historiography were, on the available evidence, influential in only three cases of loss of religious faith. Tom Paine and the comparison of the Bible with the institutions of church and society were much more important.[22] Similarly, Jenifer Hart and Edward Norman, respectively, have looked at the sermons of country parsons and the debates among the ecclesiastical Lords and assure me that in so far as scientific ideas were mentioned, this occurred at a level of crude, cliched unsophistication which would not repay detailed study. Finally, when Millhauser studied the writings of the Scriptural Geologists in the early decades of the century, he concluded that they did not illuminate the issues in the geological debate on miracles *versus* the uniformity of nature which were significant in the wider controversy on the impact of scientific ideas on the conception of nature and man's place in it.[23]

I am not suggesting that there is nothing to be learned from the history of popular opinions. On the contrary, I would do much to encourage detailed studies of popular phrenology, the ideology of the Mechanics' Institutes, and the activities of provincial literary and philosophical societies. I only want to argue that it is legitimate to demarcate this sort of activity from the study of the views of the intelligentsia. Thus, I have chosen to confine my attention to certain major works and the debates surrounding them in the more sophisticated periodicals and intellectual circles. The hard core of works which I consider central to the issue of man's place in nature and which I am in the process of investigating are the writings of Malthus, Paley, William Buckland, George Combe, Lyell, the authors of the *Bridgewater Treatises* (some more than others), Adam Sedgwick, William Whewell, Baden Powell, Robert Chambers, Darwin, Spencer, Wallace, Alexander Bain, Huxley, Tylor, Lubbock, and Tyndall.

Once one has made this decision and eliminated the certainty of being inundated by every conceivable opinion, it becomes impossible to retain certain other very convenient demarcations – those of subject and of specialized interests.

The existence of a growing number of scholars who call their discipline 'Victorian Studies' implies all that I am about to say, but I feel that we are all very slow to appreciate the intellectual consequences of this: we are dealing with a common context of issues which diminishes the importance of divisions between science and arts, economic and political and biological theory, poetry and geology, psychology and theology, education and the novel – indeed, any combination of twentieth-century academic divisions which we care to name. Of course, this situation drops us back into the morass which we attempted to avoid by setting out to study the views of the intelligentsia on scientific topics. However, certain aspects of Victorian Studies are relatively advanced; indeed, the field is dominated by students of literature, with social and political historians running second. This situation may make it possible to attempt to assess the role of scientific ideas in a body of literature which is already relatively well studied from other points of view.

It is acknowledged on all sides that scientific developments were at the centre of Victorian intellectual life, but it remains the case that, e.g. Knoepflmacher, Hillis-Miller, Ward, Haight and Houghton give little evidence of having read, e.g. Lyell, Chambers and Darwin, even though their works are utterly dependent on the reactions to these and related works which the Victorians about whom they write *did* read.[24] To put this point even more contentiously, I doubt that Kitchel, Everett or Haight[25] have read the works of Gall, Spurzheim, Combe, Bray and Spencer on phrenology and psychology, but the subjects of their studies – George Eliot, and George Lewes – took these works seriously enough to read them aloud to each other by the fire, and they included some of the authors among their most intimate friends. It is said that George Eliot even had her head shaved in order to obtain a more precise phrenological delineation.[26] The activities of George Eliot and George

Lewes perfectly exemplify the common context to which I refer. All of the following interests – and more – were intimately intermingled in their lives and works: drama, theatre, literature, positivism, zoology, the novel, the study of the nervous system, poetry, political journalism and biblical criticism.[27] Similarily, the study of the impact of science requires a much fuller appreciation of the congeries of ideas which is evoked by the terms 'empiricism', 'associationism', 'Utilitarianism', 'Philosophical Radicalism', and 'Positivism'. When one speaks of the impact of science on the intellectual culture of the period and sets out to examine the journals which provided a platform for scientific naturalism (e.g. the *Westminster Review*, the *Leader*, the *Fortnightly Review*, the *Nineteenth Century*) it is important to look much further into the relevant scientific and philosophical works.[28] These issues fail to figure prominently in the standard histories of the period. When they are discussed, they appear much more in the form of political programmes than as manifestations of a philosophy of nature which gave rise to the political, social and aesthetic movements which form the subjects of so many works in 'Victorian Studies'. Instead of burrowing deeper into the minutiæ of critical receptions of Victorian novels, it might be worthwhile to pay closer attention to the works and the movements which *evoked* so much Victorian writing.

In attempting to do this, it seems to me that we could do worse than to centre our studies in the Victorian periodicals and the members of the intellectual *milieux* who edited and wrote for them. It turns out that this network of journals, authors and friends includes everyone whose opinion seems relevant. Thanks to the activities of Professor Houghton, we can now begin to identify the authors of the essays and reviews in the major periodicals. As a result of an immense labour, we have a simple catalogue to tables of contents and authors which allows us to illuminate the issues from a wide variety of perspectives. This activity was formerly almost impossible because of the practice of anonymous reviewing. The *Wellesley Index* identifies the authors of 97 per cent of the articles in eight of the major reviews, and the subsequent volumes attempt to do the same for a further thirty-two periodicals. The contents of the main reviews and the activities of their (now identified) contributors confirm one's belief in the existence of a common intellectual context which poured out controversy at an astonishing rate and a most surprising level of sophistication. Neither the periodicals nor their regular contributors made significant distinctions among science, literature, philosophy, theology, political economy, etc. Thus, I heartily agree with Brown, who argues that 'the history of the English mind and English public opinion cannot be written without careful attention to the influence and history of periodical literature'.[29]

There was no dearth of organs for discussing these issues. Between 1800 and 1900 more than 1,000 new magazines of various kinds were started in London, and in the year in which *On the Origin of Species* appeared, 115 new periodicals were begun in Britain.[30] What were they reviewing? In the period in which I believe that natural theology was playing an important integrative function in the intellectual culture, theological works were pouring out at a great rate. Of the roughly 45,000 books published in England between 1816 and 1851, well over 10,000 were religious, far outdistancing the next largest category – history and geography – with 4,500 and fiction with 3,500. There was also an immense circulation of religious periodicals and tracts.[31]

II

Turning now to the relationship between the periodicals, the common context, and natural theology, I can only indicate a conclusion from detailed reading of many of the periodicals and cite some examples which indicate the general development as I see it. In the period from about 1800 to 1880, it seems to me that the role of theology seems to change from that of providing the *context* for the debate to that of acting as the point of view in a *conflict*. In the early debate the effort is to retain harmony between science and theology; after about 1850 increasing efforts are made to *separate* them or to make the claims of theology so abstract that they cannot come into conflict with the discoveries of science.

I would now like to move from these general hypotheses to a particular case. Much of the debate on man's place in nature can be interpreted in terms of the principle of the Continuity or Uniformity of Nature. As this conception was applied to successively larger domains, the role of non-material causes was increasingly diminished – miracles, geological catastrophes, special creations of species, special vital laws in biology, as well as a separate realm of mind, were challenged by scientific developments in the middle two quarters of the century. Of course, there was vehement opposition to each of these moves, and they were hotly debated in the quarterlies, in monographs, in newspapers, in the pulpit, the universities, literary and philosophical societies, mechanics' institutes, letters and drawing rooms. To concentrate on the views of the intelligentsia, the opponents of uniformitarian ideas in geology and biology who were most respected were (in increasing order of sophistication) William Buckland, Adam Sedgwick, and William Whewell. Their opponents were (with variations on particular issues) Lyell, Chambers, Spencer, Huxley, and Darwin, among others, but the most sophisticated philosopher of the Principle of Uniformity was Baden Powell.

John Burrow has observed that 'Much in nineteenth-century thought can be interpreted on the assumption that the Uniformity of Nature had acquired for many intellectuals a logical status and a numinous aura which made it a substitute for the idea of God.'[32] In his recollections on the intellectual atmosphere at Oxford in the period 1856–59, John Morley wrote, 'The forces of miracles and myth and intervening Will in the interpretation of the world began to give way before the reign of law.'[33] Morley was later the editor of the *Fortnightly Review*. Its first editor, G. H. Lewes, wrote in 1866 that science also offered 'a cure for souls. . . . Formerly the best indication of a nation's progress was in its religious conceptions. Now the surest indication is in its scientific conceptions.'[34] In its first ten years the *Fortnightly* contained an article on the relations between science and theology in almost every issue.[35] In the first number, George Eliot set the tone by pointing to 'the gradual reduction of all phenomena within the sphere of established law, which carries as a consequence the rejection of the miraculous. . . .'.[36] These views, along with those of Spencer, Huxley, and Tyndall, represent the advance of the positivist, phenomenalist, nominalist group. That is, they represent the impact of science. My object in the rest of this paper will be to lend plausibility to the hypotheses (1) that as theology moved towards the identification of God with the Uniformity of Nature (under pressure from science), there came to be little to choose between the devout and the agnostic, and (2) that theology ceased to serve the unifying function which it had in the early decades of the century.

The case studies which I have chosen are not concerned with the views of those who come under the heading of 'the impact of science'. Rather, I should like to draw attention to the changing views of theologies who were attempting to interpret the challenge from science. Since the writings of William Whewell extend over the crucial period 1830–60+, it will help to illustrate my view of the development of natural theology if we consider his writings in some detail and briefly contrast his views with those of Baden Powell, who argued for a very different view of the relations among God, man and nature.

William Whewell was the most articulate and sophisticated inter-preter of the providential view which was associated with the belief that God intervened in the course of nature with catastrophic alterations of its geology and its complement of species. He was also very influential in the community of scholars which embraced science, philosophy and theology as an integrated discipline and which was beginning to show signs of strain. Indeed, his life and work provide an excellent, if extreme, example of the common intellectual context which was characteristic of the period. He wrote 64 scientific papers, in addition to books and shorter works which are listed in 5 columns in the *Dictionary of National Biography*. He wrote on mechanics, dynamics, electricity, theory of the tides (his most original scientific work), mathematics, geology, architecture, theology, education, poetry, logic, and political economy. He also wrote and translated poetry (both classical and German) and Plato. When he wrote on the question of life on other worlds an unkind epigrammatist claimed that he wrote it to prove that 'through all infinity there was nothing so great as the Master of Trinity'. In his spare time from his writing, Whewell held posts as Tutor, then Master, of Trinity, was twice the Vice-Chancellor, and successively held chairs in Mineralogy and Moral Philosophy at Cambridge. He was also largely responsible for the introduction of both the Moral Sciences and the

Natural Science Triposes in Cambridge (1848). He was founder-member of the Cambridge Philosophical Society, Fellow and President (1837) of the Geological Society and successively Secretary (1833), Vice-President (1835) and President (1841) of the British Association. He twice essayed on the whole field of science – first in his *Bridgewater Treatise* (1833, the most popular of the series: it went through ten editions by 1864) and in his five-volume (continually revised) *History and Philosophy of the Inductive Sciences* (1837–40, etc.) Lyell once said that he regretted that Whewell had not concentrated his energies as a specialist in one science but later decided that he was most effective as an eminent 'universalist'. A less kind observer commented that 'Science is his forte, omniscience his foible.'

It is worth recalling that Whewell's position on the question of miracles in geology and biology was consistent with his general philosophical position, according to which the operations of nature required the constant sustenance and occasional direct intervention of the Deity: all force is will force and leads ultimately to God's efforts. This position on the philosophy of nature was intimately related to his views on the philosophy of scientific discovery, which led to a direct and protracted conflict between him and J. S. Mill, in which Mill argued that science investigated regularities in nature and could not provide ultimate explanations. Mill's empiricism is much closer to the positions of the Young Turks of scientific positivism – Huxley and Tyndall – than Whewell's views, which can be characterised as deeply anti-naturalist and even idealist. As a philosopher, Whewell set out to introduce an anti-Lockean philosophy to Cambridge. His views were influenced by Kant and can be described as transcendental and intuitionist.[37]

Whewell played a prominent part in the debates of geology and evolution and their relationship with theology, and his writings in the periodicals, his own works and his letters shed light on the reception of Lyell, Chambers, and Darwin. He reviewed the first volume of Lyell's *Principles of Geology* in the *British Critic* and the second volume in the *Quarterly*. In the former he pointed out that Lyell's principle of uniformity as applied to the history of the earth had implications for the question of the origin of species. If Lyell was going to 'give even a theoretical consistency to his system, it will be requisite, that Mr Lyell should supply us with some explanation by which we pass from a world filled with one group of species to one which contained different species'. Since Lyell could not supply such an explanation, Whewell confidently concluded that it was 'undeniable that we see in the transition from an earth peopled by one set of animals, to the same earth swarming with entirely new forms of organic life, a distinct manifestation of creative power, transcending the operation of known laws of nature: and, it appears to us, that geology has thus lighted a new lamp along the path to natural theology'.[38]

Later, in his *History and Philosophy of the Inductive Sciences*, Whewell set out to adjudicate the debate between uniformitarians and catastrophists in geology and the bearings of these positions on the origin of species. (Indeed, it was Whewell who dubbed the respective positions 'uniformitarian' and 'catastrophist', thereby polarising the debate in a way which was not so neatly divided in the minds of most practising geologists.)[39] In what follows, it is noteworthy that Whewell is arguing his case in the light of the existing scientific evidence. This was already becoming a very risky strategy by the time that he wrote: the evidence should go either way. Thus, the roles of the advocates of the principle of the Continuity of Nature and naturalist explanation on the one hand and those who advocated Divine intervention in the course of nature on the other were curiously reversed by Whewell, when he argued that 'Nothing has been pointed out in the existing order of things which has any analogy or resemblance, of any valid kind, to that creative energy which must be exerted in the production of a new species. And to assume the introduction of new species as a part of the order of nature, without pointing out any natural fact with which such an event can be classed, would be to reject creation by an arbitrary act.' There being no such natural means by which new species come to be, Whewell concludes that a creative power is exerted at the beginning of each cycle of new species. 'Thus we are led by our reasonings to this view, that the present order of things was commenced by an act of creative power entirely different to any agency which has been exerted since. None of the influences which have modified the present races of animals and plants since they were placed in their habitations on the earth's surface can have had any efficacy in producing them at first. We are necessarily driven to assume, as the beginning of the present cycle of organic nature, an event not included in the course of nature.'[40] Material science cannot offer any explanation of the origin of species. An abyss interposes itself between us and any intelligible beginning of things.[41]

Having convinced himself that science can say nothing of the origins of states of the earth or of the introduction of new species, Whewell turns to the topic of the relationship between the sciences of geology and biology on the one hand and the traditional Biblical account on the other. Man keeps records, but the sacred narrative is inevitably difficult to interpret. Science constantly changes the language for describing facts, while the language of Scripture is always the same and was written for the common man. 'Hence the phrases used by Scripture are precisely those which science soon teaches man to consider as inaccurate.' In this, and the next four sections of his analysis, Whewell places the relations between the Book of God's Word and the Book of God's Works on a slippery slope which could only lead to science claiming more and more as its legitimate domain, while interpretation of Scripture became increasingly abstract.[42]

He chooses as his example the conflict between Galileo and the Church (not, significantly, an issue in the contemporary debate). This episode taught that it was in the highest degree unjustifiable to extract astronomical or any other doctrines from the Scriptures and that any attempt to do so could lead 'to no result but a weakening of the authority of Scripture in proportion as its credit was identified with that of these modes of applying it. And this judgement has since been generally assented to by those who most reverence and value the study of the designs of Providence as well as that of the works of nature.' He then protects himself by flatly asserting that science can teach us nothing about creation: 'The thread of induction respecting the natural course of the world snaps in our fingers, when we try to ascertain where its beginning is. Since, then, science can teach us nothing positive respecting the beginning of things, she can neither contradict nor confirm what is taught by Scriptures on that subject; and thus, as it is unworthy timidity to fear contradiction, so it is ungrounded presumption to look for confirmation in such cases. The providential history of the world has its beginning, and its own evidence; and we can only render the system insecure, by making it lean on our material sciences.'[43]

This passage is a straightforward abandonment of one of the traditional aims of natural theology – to illuminate revelation by science. The trouble was, of course, that zealous scientist-theologians had claimed to find evidence which *confirmed* the Biblical account. When this evidence was found to be erroneous, the consequences for the Biblical story were alarming. Just this had happened in the case of William Buckland, Professor of Geology at Oxford, whose dramatic evidence in support of the Biblical Deluge had been proudly proclaimed in the 1820s, only to be quietly withdrawn in a footnote in his *Bridgewater Treatise* in 1836.[44] Whewell saw that supporting Scripture by scientific evidence could not be separated from *loss* of support from new evidence.

From this position he might well have gone on to argue that science and theology should never be mingled, as Bacon had urged. Instead, he placed the Scriptures in danger of constant reinterpretation in the light of new scientific findings. He makes this point in a complacent, reassuring spirit, little suspecting that in his own lifetime he would find this faith failing him, and by ten years after his death this doctrine had been hardened into the aggressive thesis that as science advances, religion retreats.

8. *Scientific views, when familiar, do not disturb the authority of Scripture.* – There is another reflection which may serve to console and encourage us in the painful struggles which thus take place, between those who maintain interpretations of Scripture already prevalent and those who contend for such new ones as the new discoveries of science require. It is this; – that though the new opinion is resisted by one party as something destructive of the credit of Scripture and the reverence which is due, yet, in fact, when the new interpretation has been gradually established and incorporated with men's current thoughts, it ceases to disturb their views of the authority of the Scripture or the truth of its teaching. When the language of Scripture, invested with its new meaning, has become familiar to men, it is found that the ideas which it calls up are quite as reconcilable as the former ones were with the most entire acceptance

of the providential dispensation. And when this has been found to be the case, all cultivated persons look back with surprise at the mistake of those who thought that the essence of the revelation was involved in their own arbitrary version of some collateral circumstance in the revealed narrative. At the present day, we can hardly conceive how reasonable men could ever have imagined that religious reflections on the stability of the earth, and the beauty and use of the luminaries which revolved around it, would be interfered with by an acknowledgement that this rest and motion are apparent only. And thus the authority of revelation is not shaken by any changes introduced by the progress of science in the mode of interpreting expressions which describe physical objects and occurrences; providing the new interpretation is admitted at a proper season, and in a proper spirit: so as to soften, as much as possible, both the public controversies and the private scruples which almost inevitably accompany such an alteration.[45]

As he says in another place, 'the results of true geology and astronomy cannot be irreconcilable with the statements of true theology'.[46] Whewell only warns that theology should not over-hastily accept a new interpretation of scripture, once again, lest it have to reverse its new interpretation when the scientific findings are shown to be unsound. Nevertheless, he goes on to say that '. . . when a scientific theory, irreconcilable with its ancient interpretation, is clearly proved, we must give up the interpretation, and seek some new mode of understanding the passage in question, by means of which it may be consistent with what we know; for if it be not, our conception of the things so described is no longer consistent with itself'.[47] 'It is impossible', he concludes, 'to overlook the lesson which here offers itself, that it is in the highest degree unwise in the friends of religion, whether individuals or communities, unnecessarily to embark their credit in expositions of Scripture on matters which appertain to natural science. By delivering physical doctrines as the teaching of revelation, religion may lose much, but cannot gain anything.'[48] But, of course, the problem was to determine just what pertained to natural science. Whewell felt confident that certain major changes in the history of the earth, the origin of species, the appearance of man, and his moral nature were not within the domain of science. Within four years, Robert Chambers' anonymous work on *The Vestiges of the Natural History of Creation* appeared and argued that astronomy, geology, the origin of species, man's moral nature and his social behaviour were all subject to natural law -- that there was no distinction between the physical and the moral. Chambers claimed that mind (and therefore morality) thereby passed into the category of natural law.[49] (Seven years later this interpretation was reinforced by Spencer's advocacy of evolution, massively backed up three years after that by a work interpreting psychology in evolutionary terms.[50] There rapidly followed a rash of works interpreting the Scriptures themselves in scientific terms,[51] followed by the works of Darwin and Wallace on evolution and the works of others on the applications of evolutionary ideas to history, sociology, anthropology, and, finally, all mundane things.)

Whewell was invited to review the *Vestiges of Creation* in the *Edinburgh Review*. He declined (the review was written by Adam Sedgwick, whose career and views might have served as a case study to complement Whewell's).[52] Instead, Whewell published a reply to *Vestiges* which took a curious form (no less curious; however, than Sedgwick's writings on the subject): he simply republished extracts from the *History and Philosophy of the Inductive Sciences* in a slim volume with a preface which alludes to the doctrine of *Vestiges* without mentioning the book![53] In a letter he acknowledges that the selection, under the title of *Indications of the Creator*, '. . . was published with some reference to the *Vestiges*, . . . and as an answer to it so far as truth must in some measure be an answer to falsehood on the same subject. . . . I have attempted to show that, dim as the light is which science throws upon creation, it gives us reason to believe that the placing of man upon the earth (including his creation) was a supernatural event, an exception to the laws of nature. The *Vestiges* has, for one of its main doctrines, that this event was a natural event, the result of a law by which man grew out of monkey.' The appeal of the book, he felt, lay in its 'bold, unscrupulous, and false scientific generalisations' which are attractive to the ignorant, while they enrage scientists.[54]

In the Preface to the second edition, Whewell does refer to a defence of *Vestiges*, *Explanations*, and to the offending book itself. Once again, his position depends on the state of contemporary science: '. . . according to the best scientific views hitherto obtained, the Origin of Man, and the Origin of Life upon the earth, were events of a different order from the common course of nature. The Origin of Man is the Origin of Language, of Law, of Social Relations, of Intellectual and Social and Moral Progress; and though in all these characteristics of humanity we can trace a constant series of changes and movements, we can discern in them no evidence of a beginning homogeneous with the present order of changes. If Science does not positively teach that man was placed upon the earth by a special act of his Creator, she at least shews no difficulty in the way of such a belief? She leaves us free to hold that the placing of man upon the earth was not an ordinary step in the natural course of the world; but an extraordinary step, the beginning of a providential and moral course of the world. She negatives the doctrine that men grew out of apes, that language is the necessary development of the jabbering of such creatures, and reason the product of their conflicting appetites.' Where Lyell and the author of *Vestiges*, respectively, had referred the history of the earth and of life to the uniform action of the same natural causes which were then operating, Whewell says that this is a gratuitous assumption, while his views are the '*result* of scientific investigations'. If we try to account for these changes by causes now in action, we fail.[55]

Whewell notices Chambers' claim that his system denies human agency and human virtue but need not shake one's faith in God, but argues that this is simply inconsistent with the doctrine of *Vestiges*.

Fortunately, these doctrines are fantastical, have been repudiated by all those competent to judge them, and can appeal only to the most credulous.[56] If the doctrine of *Vestiges* be false, what then is true? Whewell replies, 'To this question, men of real science do not venture to return an answer.' 'Since, as I have endeavoured to show in the following pages, the chain of existing causes does not, in any case, conduct us back to its origin; – not in the history of the mass of the earth, nor of its strata, nor of animal life, nor of man; – I must necessarily confess that we cannot obtain from science a complete view of the history of the universe. That human reason should thus be unable to fathom and comprehend the acts of the Creator and Governor of the world, is no surprise nor humiliation to me; and I think that those do well who learn, from the study of the sciences, this lesson of humility'.[57]

Looking back from the vantage point of the addresses by Huxley and Tyndall at the Belfast meeting of the British Association in 1874, these passages read like a dare. Find the mechanism of evolution, and you can be as arrogant as you please. And they were. Tyndall claimed that science had unrestricted right of search in all of nature, and Huxley argued that both men and animals are conscious automata.[58] A year later, William Draper's provocative book on *The History of the Conflict Between Religion and Science* appeared in the popular International Scientific Series.[59] This book accurately represents the state of the debate as it appeared in 1875. Whatever else one might wish to say about the results of Whewell's interpretation of the relationship between religion and science, it cannot be said that following it had the effect of making theology secure and scientists humble. Even earlier – in 1872 – statistical enquiries were being undertaken on the efficacy of prayer.[60]

Whewell's response to Darwin's theory has an unvigorous air about it, in marked contrast to his spirited treatments of Lyell and Chambers. He wrote a polite note to Darwin, saying, '. . . I cannot, yet at least, become a convert. But there is so much of thought and of fact in what you have written that it is not to be contradicted without careful selection of the ground and manner of the dissent.' Darwin was so pleased that he sent the letter to Lyell, noting that Whewell 'is not horrified with us'.[61] In the next four years many wrote on the question of the origin of species, the antiquity of man, and man's place in nature. Whewell fell back on the claim that the arguments in his *Indications of the Creator* still held. 'It still appears to me that in tracing the history of the world backwards, so far as the palaetiological sciences enable us to do so, all the lines of connection stop short of a beginning explicable by natural causes; and the absence of any conceivable natural beginning leaves room for, and requires, a supernatural origin. Nor do Mr Darwin's speculations alter this result. For when he has accumulated a vast array of hypotheses, still there is an inexplicable gap at the beginning of his series.' Darwin, of course, never denied this. His doctrine, like those of Hutton and Lyell before him, was not concerned with the origin of the earth or of life. Whewell had been making much stronger claims in the 1830s and 1840s for repeated miraculous, wholesale creations of species. Now he is reduced to arguing for a miraculous beginning to the series, a point which Darwin concludes in the last paragraph of the *Origin of Species*. Whewell continues rather feebly: 'To which is to be added, that most of his hypotheses are quite unproved by fact. We can no more adduce an example of a new species, generated in the way in which his hypotheses suppose, than Cuvier could.' (Again, a point conceded by Darwin, Huxley, and Tyndall.) Finally, Whewell, now badly out of touch, claims that his earlier conception of a uniformitarian–catastrophist debate still reflects the state of geological controversy: 'And though the advocates of uniformitarian doctrines in geology go on repeating their assertions, and trying to explain all difficulties by the assumption of additional myriads of ages, I find that the best and most temperate geologists still hold the belief that the great catastrophes must have taken place; and I do not think that the state of the controversy on that subject is really affected permanently. I still think that what I have written is a just representation of the question between the two doctrines.' He went on to say that he didn't believe that any cause would be served by his writing further on the subject; what he had written over twenty years earlier still seemed to him to apply to most of the questions under consideration in 1863. Finally, he says, he simply does not have the detailed knowledge to adjudicate the contemporary controversy.[62]

Three months later Whewell turned his attention to the question of the antiquity of man, which was then at the centre of the debate as a result of the appearance of Lyell's work on that subject and Huxley's essays on *Man's Place in Nature*.[63] There are signs in Whewell's remarks of a man who is simply out of touch and feeling nostalgic for a time when science and theology were more simply and reassuringly related. He says, 'I have myself taken no share in the discussions on the antiquity of man; but I will not conceal from you that the course of speculation on this point has somewhat troubled me. I cannot see without some regrets the clear definite line, which used to mark the commencement of the human period of the earth's history, made obscure and doubtful. There was something in the aspect of the subject, as Cuvier left it (Cuvier died in 1832), which was very satisfactory to those who wished to reconcile the providential with the scientific history of the world, and this aspect is now no longer so universally acknowledged. It is true that a reconciliation of the scientific with the religious view is still possible, but it is not so clear and striking as it was. But it is weakness to regret this; and no doubt another generation will find some way of looking at the matter which will satisfy religious men.' He adds that he should be glad to see his way to such a view and is hoping to do so soon.[64] He lived for two more years, and it is a rather pathetic footnote to his career that his last gesture in the great debate on the relations among God, Man, and nature was to refuse to allow a copy of Darwin's *Origin of Species* to be placed in the Library of Trinity College.[65]

III

I should like to contrast the views of Whewell with those of Baden Powell. Where Whewell was involved in a rather tired and increasingly disorderly retreat, Powell was running headlong into the arms of the enemy. He was a sort of diminutive Whewell. He held only one Chair, that of Geometry at Oxford. As a scientist he wrote on light and radiant heat and elementary books on curves and on differential calculus. He wrote chiefly on optical questions and managed only one slim volume on the entire history of science. He was active in doctrinal questions from a latitudinarian point of view and took an active part in university reform. (On the other hand, he did sire thirteen children, one of whom founded the Boy Scouts.) In spite of his obscurity in recent secondary literature, Powell was the most clear-sighted of the interpreters of the uniformitarian point of view and wrote a series of monographs, essays and reviews which nicely parallel Whewell's works from the interventionist or catastrophist point of view. Lest it be thought that I am rescuing an obscure figure for my own polemical purposes, I should add that his essays were commonly reviewed with Darwin's *Origin of Species*,[66] and he was one of the contributors to *Essays and Reviews* – a work in the naturalist tradition which was causing more controversy in 1860 than Darwin's.[67] (*Essays and Reviews* went through six editions by 1865.) Darwin also acknowledges Powell's work in the Preface to later editions of the *Origin*.[68] Powell suits us, because he interpreted Lyell in the 1830s in a monograph *On the Connection of Natural and Divine Truth*, defended *Vestiges* in an essay on 'The Philosophy of Creation', and was an early advocate of Darwin in his 'Study of the Evidences of Christianity', his contribution to *Essays and Reviews* (which evoked a flood of replies). Powell promptly died, thereby missing out on the prosecution of some of his fellow contributors and giving rise to snide remarks about his punishment for blasphemy from one quarter and about his being blissfully free from a lot of Bibliolaters from the others.[69]

Powell's position throughout his writings was consistent: the only rational and safe solution to the problem of reconciling science and theology is to keep them *completely separate* and to give up entirely the concept of miracle as having anything to do with science. This was a radical, uncompromising solution to the problems posed for natural theology by science: the physical and moral departments of nature should be divorced once and for all. The Bible was an inspirational document, having nothing to do with science. Any attempt to stretch scripture to accommodate science would harm both faith and reason.[70] In 1855 he wrote that any solution short of this has always resulted 'either in a lamentable antagonism and hostility, or in futile attempts to combine them in incongruous union, upon fallacious principles.'[71] The foundations of scientific and religious truth are, he insisted, entirely separate.[72]

One might feel that the complete separation of moral and physical phenomena hardly *solves* one of the problems of natural theology: it either abrogates the discipline or merely restates its main problem. However, this was the solution which he, along with Combe and Chambers, proposed, and this was the basis for Adam Sedgwick's fulminations against phrenology and the doctrine of *Vestiges*.[73] Powell's essay made him much angrier than Darwin's hypothesis, since it explicitly advocated the very separation of moral and physical with which Sedgwick said was implied by Darwin's book.[74] It could be argued that Sedgwick and Wilberforce had clearer heads – that the only consistent theological position which did not entail slow, deliberate theological suicide was to stand and fight.[75] However, the constraints which operate on a given thinker in a given historical context do not always – or even usually – lead to his holding an internally consistent position. Thus, Combe argued that the brain is the organ of the mind but that this did not imply materialism and fatalism. Chambers claimed that phrenology and social statistics did imply fatalism and annulled the distinction between the physical and the moral and then argues blithely that physical and moral laws are independent. Powell supports uniform natural laws in all of geology and biology but considers man's mental and spiritual nature independent. Surely this was what was at issue in the evolutionary debate! In any case, Powell claims that there can be no real hiatus in the physical operations of nature and that all of science depends on this assumption.[76]

The complete separation of science and theology was cold comfort to many, and it avoided the crucial claims of science to account for mental and moral phenomena and the status of the Bible as an historical document. Nevertheless, this was the line that was taken as an alternative to Whewell's idea of an orderly withdrawal. It is in this context that one can make sense of Huxley's claim that the theory of evolution 'is neither Anti-theistic nor Theistic. It simply has no more to do with Theism than the first book of Euclid has.' 'The doctrine of Evolution, therefore, does not even come into contact with Theism, considered as a political doctrine. That with which it does collide, and with which it is absolutely inconsistent, is the conception of creation, which theological speculators have based upon the history narrated in the opening of the book of Genesis.'[77] Well, if evolution had nothing to do with theism but merely contradicts the creation of the world by God, Huxley's *laissez-faire* view does not seem to have carried him as far as the end of the next sentence. Huxley, of course, goes on to claim the domain which Powell had reserved from the Principle of Uniformity. He did not deny that men and (*pace* Descartes) animals are conscious *automata*.[78] Similarly, the term 'creation' came increasingly to be used in an utterly emasculated sense to imply ignorance of the natural mode of production of a phenemon.[79] It is not surprising, on this account, that Lyell, Chambers, and Darwin could argue that their theories implied a *grander* view of the Creator – One who operated by general laws. Granting this cost them nothing as scientists. We know that Darwin had imperceptibly lost his religious faith somewhere along the way, but he could still be quite conventionally pious in *On the Origin of Species*.[80] For all practical purposes this view was indistinguishable from positivism and agnosticism. It had been anticipated in Charles Babbage's uninvited *Ninth Bridgewater Treatise*, in which the inventor of the calculating engine (the lineal ancestor to the modern computer) argued that a miracle was equivalent to a sequence of numbers which the Programmer had instructed the calculating engine to produce and which only appeared to be an exception to the orderly sequence.[81] This well-meaning offer of help to those troubled by the relationship between natural laws and miracles had the effect of making the claims of theism vacuous by making them so abstract that they had no empirical reference.[82] By the time, fifty years later, that Frederick Temple argued in his Bampton Lectures that theology had nothing to fear from the evolutionary theory, the author was half-way between being the writer of one of the shocking *Essays and Reviews* and being the Archbishop of Canterbury.[83] Perhaps theologians had become convinced that they had nothing to fear at the expense of making no claims for theism.

This situation is reminiscent of seventeenth-century developments in which the exponents of a mechanical view of nature claimed that this involved no threat to theism. Commenting on these 'Mechanick Theists', John Ray wrote in *The Wisdom of God* (1690), 'Wherefore these Atomick Theists utterly evacuate that grand argument for a God, taken from the phenomena of the artificial frame of things, which hath been so much insisted upon in all ages, and which commonly makes the strongest impression of any other upon the minds of men &c. the atheists in the mean time laughing in their sleeves, and not a little triumphing to see the cause of Theism thus betray'd by its profess'd friends and assertors, and the grand argument for the same totally slurred by them, and so their work done, as it were, to their hands.'[84]

Now, to review. The tension between theodicy and the requirements of the scientific method had been harmonized in Paley by means of simple examples and the device of begging all questions to do with origins.[85] As detailed findings of science in areas which bore directly on the history of the earth, of life and of man were brought into contact with natural theology, the theologians and theologian-scientists made three sorts of moves:

1. Separate the moral from the physical (Combe, some passages in Chambers and Powell).
2. Reinterpret scripture in the light of new scientific findings (Buckland, Sedgwick, Whewell – even some of the most ardent literalists, e.g. William Kirby and Thomas Chalmers, did this to some degree).[86]
3. Make the concept of God more abstract until He is identified with the Uniformity of Nature.

Whatever move was made (and there are innumerable intermediate positions), one result was common: these adaptations involved the abandonment of the traditional claims of natural theology by emasculating its theodicy. It was no longer a strong enough doctrine to serve a unifying function in the intellectual culture of the period. If the justification of the ways of God to man is that God works by general laws and/or that His intention was that moral and physical laws should be separate, then this would seem to guarantee God's indifference to a careless man of piety who, while praying fervently, stepped over a cliff and perished while obeying the laws of gravity. Of course, this view of God's relationship to nature does not preclude a very personal God serving essentially psychological functions of inspiration and consolation, but it was necessary to confine claims for His efficacy to the afterlife, since science, scientific historiography and statistics eliminated the claims of Divine creation and the efficacy of prayer.

IV

At the outset I was very tentative about the relationship between three issues: (1) the common context of ideas, the periodicals in which they were discussed and the individuals who wrote about them; (2) the impact of science on natural theology; (3) the breakdown of the common context in the 1870s and later. One way of providing evidence for a causal relationship between these would be to show that the principal figures who wrote on various aspects of the debate were so concerned that they came together to discuss the influence of science on the philosophy of nature and the implications for theology, literature, ethics, legislation, etc. If these developments are related, these men should find increasing difficulty in communicating with one another and, should eventually lose interest and go their separate ways. Of course, it would help if this fantasy discussion group included the editors of, and the most distinguished contributors to, the major periodicals. They should be very interested in the aims of the group but should draw the lesson that they could no longer meet on common ground and go away to found new, specialized (or less sophisticated) societies and periodicals. We could then argue with some confidence that natural theology had become too abstract and/or too separated from scientific and moral issues to act as a cohesive force in the intellectual life of the period.

It would be very obliging of, say Huxley, Cardinal Manning, Gladstone, William Carpenter, Walter Bagehot, Tyndall, Froude, Frederic Harrison, R. H. Hutton, A. P. Stanley, Leslie Stephen, J. S. Mill, Darwin, Wallace, and Spencer to find time to conduct this *experimentum crucis* for us, but unfortunately my list is a tiny bit over-ambitious. That is, Mill and Spencer declined to join the Metaphysical Society, and Darwin and Wallace were not members, but perhaps we can be reconciled to their absence by the presence of Tennyson, Ruskin, James Martineau, F. D. Maurice, James Knowles, James Fitzjames Stephen, the Duke of Argyll, Sir John Lubbock, A. J. Balfour, W. K. Clifford, Mark Pattison, J. R. Seeley, Henry Sidgwick, George Croom Robertson, James Sully and St George Mivart, among others, making up sixty-two of the most eminent thinkers, churchmen, editors, men of affairs, writers and scientists of the period.

Just about everything concerning this society realizes an historian's daydreams, including the fact that someone has done a very informative but not very incisive study of it.[87] There are some minutes, almost all their papers were printed and are preserved in some form (many in Victorian periodicals), and the activities of the society can be illuminated by various lives and letters.[88] It is true that one cannot attend the meetings, but they thoughtfully recorded their views in *The Nineteenth Century* in a series entitled 'The Modern Symposium.'[89] Finally, several of the papers are reflections on the society itself, and R. H. Hutton has written a very evocative reminiscence on the origins of the society, including a typical meeting.[90] The Metaphysical Society was founded in 1869, a year after one of the great holdouts on evolutionism had reluctantly accepted Darwin's view. In reviewing Lyell's final acceptance of evolution, A. R. Wallace, the co-discoverer of the theory, told the readers of the *Quarterly Review* that he was excepting some aspects of man from the process of evolution by natural selection.[91] Things were getting very messy indeed, and the enunciation of the doctrine of Papal Infallibility a year later did little to improve the atmosphere.

Ninety meetings were held, and the Metaphysical Society voluntarily disbanded – for lack of interest – in 1880.[92] It was first conceived as a Theological Society, but A. P. Stanley argued for *rapprochement* with the scientists, and the conception was broadened and the name was changed to 'Metaphysical and Psychological Society' (the 'Psychological' was soon dropped). The form was adopted from an amalgam of practices drawn from scientific societies and the Cambridge Apostles. The initiative came from the theological side, but there was a successful campaign to recruit scientists. The terms of reference of the society are just what one would hope. They set out to investigate 'mental and moral phenomena, the faculties of lower animals, the grounds of belief, the logic of the physical and social sciences, the immortality and identity of the soul, the existence and personality of God, conscience, and materialism'.[93] I wish that there was space to go through the papers read before the society in detail, but I can really only mention one or two and refer you to A. W. Brown's account, to which I am heavily in debt for this portion of my paper. At the first meeting, Tennyson's poem, 'The Higher Pantheism' was read by Knowles. It is characteristic in that many of the lines referring to the Deity end in queries, and it concludes,

God is law, say the wise; Oh Soul, and let us rejoice,
For if He thunder by law the thunder is yet His voice.

Law is God, say some; no God at all, says the fool,
For all we have power to see is a straight staff bent in a pool;

And the ear of man cannot hear, and eye of man cannot see;
But if we could see and hear, this Vision – were it not He?[94]

This uncertainty about the faith which can be based on such a pale God nicely reflects my general points. Sixteen years later, in his reminiscence about the society, R. H. Hutton makes the same point, with the same ambiguity: 'The uniformity of nature is the veil behind which, in these latter days, God is hidden from us.' Indeed, he chose as the paper read at his imaginary meeting of the society, 'Can Experience Prove the Uniformity of Nature?'[95] Between these two quotations lie 90 papers, beginning with Hutton 'On Mr Spencer's Theory of the Gradual Transformation of Utilitarian into Intuitive Morality by Hereditary Descent'. I am tempted simply to reproduce the titles, since they convey the troubled and irreconcilable views of the members. They might have clashed and hammered out something. Indeed, Ward expressed Christian horror at Huxley's views in an early meeting, and in reply Huxley mentioned the intellectual degradation which would result from general acceptance of Ward's views, but from that time on, no words of that kind were heard.[96]

Brown diagnoses the disintegration in a number of ways, but they add up to a surfeit of tolerance coupled with such irreconcilable views that the various persuasions slipped past one another and never grappled. The only common ground available to them, as I have tried to show, was too abstract to be of any use. Where they used the same terms, they lacked common meanings. Let me cut the analysis short and say that they simply lacked a common context of ideas, and they were the last generation to attempt to maintain one. In a way it is remarkable that they spent eleven years arriving at the conclusion.

I should like to leave a detailed examination of the process of this discovery to a later occasion and draw your attention once again to the periodicals and other manifestations of the intellectual and cultural consequences of the Metaphysical Society. First, let us note the ages of the members at the time of its dissolution. Of the total membership since 1869, eight were dead, six were over seventy, eighteen were over sixty, fourteen were over fifty, ten were over forty, and six were over thirty (five of these elected in the last year),[97] i.e. a waning generation.

It seems to me that the Metaphysical Society was something of a turning point. It attracted some of the most influential members of the cultural and political élite, men who were attempting to embrace all of the aspects of culture and society. At the same time it was the cradle of a number of specialized societies and periodicals which, on my reading of the evidence, were symptoms of the fragmentation of the common intellectual context of the early decades of the century. Its most active members included an astonishing group of editors and contributors: Hutton – *Spectator*; Alford – *Contemporary*; Knowles – *Contemporary and Nineteenth Century*; Ward – *Dublin*; Bagehot – *National* and *Economist*; Froude – *Fraser's*; Leslie Stephen – *Cornhill* and *D.N.B.*; Robertson – *Mind*. A. C. Fraser had edited the *North British* (1850–7), and two relatively inactive members edited major periodicals: Grove (*Macmillan's*) and Morley (*Fortnightly* – 1867–82 – and later, for brief periods, *Pall Mall Gazette*, and *Macmillan's*). Among the prominent and prolific contributors were Stanley, Huxley, Mivart, Argyll, Harrison, Tyndall, Carpenter, Pattison, Gladstone, J. Martineau, Tennyson, Ruskin.[98]

Before turning to some symptoms of the breakdown of the common context, I should like to cite some statistics which strike me as interesting and probably significant, but I don't know how to interpret them. Between 1860 and 1870 there were 170 new periodicals launched in London (an average of about one every three weeks; most were short-lived); between 1870 and 1880 there were 140; 1880–90 – 70; 1890–1900 – 30 (less than 1791–1800). There was also a great increase in the availability of cheap books in the last decades of the century.[99] Meanwhile, the great quarterlies were in decline. The circulation of the *Edinburgh Review*, was 13,500 in 1818 and about 7,000 between 1860 and 1870; the analogous figures for the *Quarterly* were 14,000 and 8,000. The *Westminster Review* (the third of the great quarterlies) was founded in 1824, never had a large circulation, and averaged 4,000 in the period 1860–70. Scientific and religious monthlies and quarterlies sold 1,000–2,000 copies, while the general monthlies sold 3,000–20,000, although the *Cornhill* sold 80,000 in 1860 (2,500 in 1871 and later 12,000) and *Cassell's Magazine* sold 200,000 in 1870. *Macmillan's* and the *Cornhill* sold 100,000 in their best years. The weeklies had circulations of up to 300,000, but 40,000–60,000 was more usual. The *Fortnightly* (the main platform for radical opinion) sold 2,500 in 1873.[100] Radical periodicals never had large circulations.

In order to make sense of these figures beyond saying that a lot of periodicals were being sold (and presumably read), one would have to make detailed studies of which popular novelists wrote for which periodical in which year. However, one can draw the unsurprising conclusion that the lower the intellectual standard, the higher the circulation. At the same time, growing literacy did not swell the circulation figures of the *Edinburgh* and *Quarterly*. The new formula for success lay with the *Nineteenth Century*, and this does seem significant, since its editor, Knowles, persuaded very eminent contributors to write in a more popular style, with great success. (It would be interesting to examine the circulation figures of the *Contemporary Review*, since it was founded as an organ making the same kind of appeal to liberal opinion in the Church that the *Fortnightly* made to a more secular audience and then underwent a transformation in Knowles' hands before he founded the *Nineteenth Century*. However I have been unable to find any figures.)[101]

V

In this concluding section I should like to indicate briefly a number of things which seem to me to be symptoms of the fragmentation of the common intellectual context which was thriving in the early decades of the century and which the Metaphysical Society was attempting to maintain. The examples which I have chosen are the decline of the British Association, a confused book which was much praised by Darwin, the founding of *Nature* as a general scientific periodical outside of the context of the quarterlies, and the founding of a number of specialist journals and societies which emerged from the matrix of the Metaphysical Society.

The British Association was founded partly as a protest against corruptions in the Royal Society and partly as an attempt to give British science the sort of status which it had on the Continent, but its main function was to interpret science to the intelligent layman. The level of exposition of its general and specialist papers, and the interest which its meetings generated on the part of the public (as shown by their attending meetings and by the detailed reports of meetings in the newspapers and periodicals), shows that the Association was playing an important part in the common intellectual culture. In the same period that the 'heavy' quarterlies ceased to reflect a common culture, the British Association was ceasing to serve the function of the expositor of the best science to a glittering audience. After an uncertain start the first fifty years (since 1831) had been the best. A jubilee meeting was held at York in 1881; three years later the first overseas meeting was held in Montreal, and somehow by this time the Association had become the 'Ass'. Thus a verse from 'Red Lions'' songs entitled 'The Travelled Ass':

> At York they thought she was sure to die
>   For she didn't seem to enjoy age;
> But at last the doctors bade her try
>   The effects of an ocean voyage.[102]

In the same year that the Metaphysical Society was founded, and in the same period that the British Association was ceasing to serve as the main vehicle for communication between specialists in the various sciences and for the dissemination of their findings to the general public, a periodical was founded, the existence of which, it seems to me, was very significant. *Nature*, a general scientific weekly, began publication soon after the Metaphysical Society began meeting. The appearance of *Nature* diluted the role of science in the general periodical literature and diverted the attention of scientists from the general periodical culture. Coupled with the decline of the British Association and the development of much more popular expositions of science in the *Nineteenth Century* a few years later, the existence and popularity of *Nature* among increasingly professional scientists represents a withdrawal from the common matrix in which important scientific findings found their main reviews and discussions in the quarterlies.

In the 1870s and 1880s there were still long essays in the quarterlies on scientific monographs, but the range of monographs which were being reviewed was narrower, and the authors of the reviews were no longer the most eminent scientists. Similarly, if one looks at the lives and letters of the leading scientists, they were no longer tremulously awaiting the reviews in the quarterlies and discussing them in detail. Perhaps it is not too great an exaggeration to say that 'Nature' had withdrawn from the common intellectual culture and had joined medicine as a domain for specialists.

Who was left to interpret science to the layman and to discuss the large issues raised by science? There were, as I have said above, still reviews in the quarterlies, but their authors were less competent than, e.g. Lyell, Whewell, Sedgwick and Herschel, had been in the earlier decades. Those who still wrote for a non-specialist audience, e.g. Huxley, Wallace, and Tyndall, were self-consciously involved in popularization. The field was left to pretentious hacks and to more or less competent amateurs. A detailed study of this new sort of interpreter needs to be made, and I have begun reading the works of Leslie Stephen, Lecky, Draper, Le Conte, and others, but it will be apparent that in the period after 1875 it is impossible to follow the methodology outlined in the first sections of this paper. That is, one is perforce involved in the study of popularizations in this period.[103] Confining one's attention to the intelligentsia no longer works, and this discovery has played a significant part in developing the line of argument used in this paper. What is required is a study analogous to Millhauser's essay on popular geology in the earlier decades, but the difference is that these writings are not complemented by extended essays for the intelligentsia written by the relevant specialists. Thus, one should begin looking at the writers who played the role, in the period 1870–1900, which Wells and Eddington played in the early decades of the twentieth century. (A study of the International Scientific Series would seem a promising place to begin, and Becker's *Scientific London* – 1874 – can serve as a guide to the scientific scene.)[104]

One example of the sort of book which I have in mind is *The Creed of Science, Religious, Moral and Social*, by William Graham. The author was Professor of Jurisprudence at Belfast and set out to review the implications of physical, evolutionary and psychological science and to consider 'the old eternal questions in their present aspect'. The book was aimed especially at the general reader who was troubled by the threat of 'intellectual and moral nihilism.'[105] The point about the book is that it was muddled but reassuring. As one reviewer put it, '. . . his efforts to save the remnants of older speculations seem almost wilful perversity. . . . [He] seems to have a fondness for raising spectres in order apparently to have the satisfaction of laying them. . . . His Transcendentalism and his Experimentalism are not harmonised; they are simply juxtaposed, and he slips from the critical to the dogmatic vein almost without knowing it.' These remarks, by W. P. Coupland, appeared in the new professional philosophical journal, *Mind* (of which more below) and represent the way in which the book strikes the modern reader.[106] However, the popular audience of the period welcomed the book (a second edition soon appeared), and the author wrote popular books on other aspects of the philosophical, social and political aspects of the new wave of science, including a volume for the exceedingly popular

International Scientific Series.[107] My point is that the public craved such works and that the sort of overview which Graham attempted was achieved by means of casuistry.

Another aspect of the apparent fragmentation of the common context is the change of status of various specialist societies. It is very difficult to be sure about such things, but one has the impression that a number of societies which had been in existence for many decades, began to become more self-consciously professional. This had the effect of making amateurs less at home and led the professionals to devote more of their energies to a narrower conception of their interests. Once again, detailed study of individual societies is required, and a number of other variables are surely involved. I will cite one example which will help me to say something which I want to say at the end. In 1869 various disparate societies and publications in anthropology came together as a self-conscious discipline merging the *Journal of the Ethnological Society*, the *Memoirs Read before the Anthropological Society*, and the *Anthropological Review*. More recently, the relevant *Journal of the Anthropological Institute* has been renamed *Man*.[108]

When the Metaphysical Society disbanded, it left its residual funds to a new journal, *Mind*, whose foundation in 1876 seems to me to reflect the fact that psychology and philosophy of mind had never really found a home in the common context. It is true that works on psychology had been reviewed – beginning with a fulminating attack on phrenology by Thomas Brown in the second volume of the *Edinburgh* (1803) which was followed by others in 1815 and 1826 by John Gordon and Francis Jeffrey. The first fair hearing for the phrenologists came in 1828 in the *Foreign Quarterly Review*. For the most part, however, there was no debate on psychology in the quarterlies.[109] I believe that that this was so because the allowance of such a debate would involve conceding the point at issue – that there *could be* a science of psychology. Instead one finds characteristic fulminating paragraphs in reviews of the major evolutionary works.[110] This belief is linked with another undemonstrable theory of mine – that concentrating on geological and biological issues constituted a holding action against the real issue: Is mind a part of the natural world, subject to scientific laws? In any case, *Mind* was conceived in the year of Tyndall and Huxley's Belfast Addresses and first appeared in the year (1876) in which David Ferrier summarized the experimental findings on the physiological basis of mind, a study which had been developing dramatically since 1870. Its founder was Alexander Bain, who had done more than any other scholar to establish the identity of psychology as an academic discipline, relatively free from the epistemological, logical, educational and social issues which had preoccupied earlier writers, (e.g. Locke, Brown, J. Mill, J. S. Mill).[111]

Mind was the first professional journal in the field in any country, and the Prefatory words in the first number included the following remarks. 'Even now the notion of a journal being founded to be taken up wholly with metaphysical subjects, as they are called, will little commend itself to those who are in the habit of declaring with great confidence that there can be no science in such matters, or to those who would only play with them now and again [a reference to the Metaphysical Society?] . . . MIND intends to procure a decision of this question as to the scientific standing of psychology.'[112] Thus, *Mind* represents the growing specialization of psychology, a discipline related to philosophy and physiology but not dependent on either. More important, it reflects an attitude which was antithetical to that of Chalmers' *Bridgewater Treatise* of 1833, in which the Moral and Intellectual Constitution of Man was wholly absorbed by natural theology.[113] *Mind* went on to become – and remain – the most distinguished philosophical periodical.

The first editor of *Mind* was George C. Robertson, a pupil of Bain and an active member of the Metaphysical Society.[114]

It is worth noting in passing that a separate journal *Brain* for the study of neurology and neurophysiology was founded in 1878, thus reflecting the relative independence of these studies from general medicine on the one hand and from psychology on the other. Two years earlier, a separate Physiological Society was founded, with Huxley, Ferrier, and G. H. Lewes on its council.[115]

In 1879, the Aristotelian Society for the Systematic Study of Philosophy was founded. Shadworth Hodgson was the prime mover, with generous advice from the positivist Frederic Harrison. Hodgson had also been one of Croom Robertson's most active supporters in the establishment of *Mind*, and the two ventures remained closely connected. (Since 1918 they have met jointly once a year.) The members of the Aristotelian Society included Bain, Romanes (the heir to Darwin in the comparative study of psychology), William James (the prime mover in American psychology), J. M. Cattell, L. T. Hobhouse – and later Bertrand Russell, Alfred North Whitehead, G. E. Moore – bringing us to recent work. In the first twelve years (i.e. until 1890–1, when its Proceedings were first published), psychological and methodological problems dominated the fortnightly meetings of the Society, and philosophy proper occupied a comparatively small place, even though its study was the main purpose of the Society. The Executive Committee complained in this vein in 1890–1.[116]

The decline of the 'heavy' quarterlies and the growth of specialized journals and societies were complemented by the success of a self-consciously 'middle-brow' periodical; *The Nineteenth Century* was founded in 1877 and catered for a more popular audience. Its editor, Knowles, was able to persuade a number of eminent men to change their loyalties and write for it. Indeed, the members of the Metaphysical Society contributed a quarter of the articles in the first four volumes and a fifth of the next four, including twenty-three by Gladstone, thirteen by Huxley, nine by Froude, and eight poems by Tennyson. Huxley also devoted a good deal of energy to the popularization of science in its pages. *The Nineteenth Century* was the most widely respected of the monthly reviews and remained so until the end of the century.[117] Its success lay in its presenting a different *level* of analysis and exposition, not unlike that exhibited in our own time in *Encounter* and *The New York Review of Books*, with perhaps an element of something between *New Scientist* and *Scientific American*.

After the disbanding of the Metaphysical Society, the only society which was very like it was the Synthetic Society, founded in 1896. However, its aims were really those of the initial conception of the Metaphysical Society before the scientists were invited to join: it was specialized in its own way and was frankly theological. So little did its members disagree that they came to be known as the 'Sympathetic Society'.[118] The attempt to maintain a common context of ideas had failed, and the theologically-minded had joined their other specialized colleagues in recognizing a separation. Theology was no longer the context; it was but one element in a fragmented culture.

Thus, by the 1880s Mind, Brain, Nature, Man (anthropology), and Philosophy had separated themselves off. These are examples pointing to the break-up of a common intellectual context and its associated literature, leading to the development of specialist careers, societies and periodicals. Gladstone epitomized the position in a conversation with Tennyson, Froude and Tyndall in 1881: 'Let the scientific men stick to their science, and leave philosophy and religion to poets, philosophers, and theologians.'[119] What remained common was popularization and uncertain generalization.

Source: Paper presented to the King's College Research Centre Seminar on Science and History, 1969; to appear in *Darwin's Metaphor and Other Studies of Nature's Place in Victorian Society.* Cambridge University Press, 1981.

Notes
1. R. M. Young, 'Scholarship and the history of behavioural sciences', *History of Science*, **5** (1966): 1–51.
2. R. M. Young, 'The Functions of the brain: Gall to Ferrier (1808–86)', *Isis*, **59** (1968): 251–68; *Mind, Brain and Adaptation in the Nineteenth Century.* Oxford, 1971.
3. R. M. Young, 'Malthus and the evolutionists: the common context of biological and social theory', *Past and Present* (May 1969).
4. R. M. Young, 'The impact of Darwin on conventional thought', in A. Symondson (ed.), *The Victorian Crisis of Faith.* London, 1970.
5. See, for example, Francis Jeffrey, 'Dr Paley's *Natural Theology*', *Edinburgh Rev.*, **1** (1803): 287–305; Anon., 'Paley's *Sermons* and *Memoirs*', *Quarterly Rev.,* **2** (1811): 75-88; Anon., 'Meadley's *Memoirs of Dr Paley*', ibid., **9**; (1813): 388–400; T. D. Whitaker, 'Gisbourne's natural theology', ibid., **21** (1819): 41–66; W. H. Fitton, 'Geology of the Deluge', *Edin. Rev.*, **39** (1823): 196–234; E. Copleston, 'Buckland – *Reliquiae Diluvianae*', *Quart. Rev.*, **29** (1823): 138–65; H. Brougham, 'Natural theology – Society of Useful Knowledge', *Edin. Rev.*, **46** (1827): 515–26; J. J. Blunt, 'Works and character of Paley', *Quart. Rev.*, **38** (1828): 305–35; Anon., 'Crombie's

Natural Theology', ibid., **51** (1834): 213–28; A. Crombie, '*Paley's Natural Theology Illustrated* – Lord Brougham's *Preliminary Discourse*', ibid., **55** (1836): 387–416; W. F. Cannon, 'The problem of miracles in the 1830s', *Vict. Stud.*, **4** (1960): 5–32, and 'Scientists and broad churchmen: an early Victorian intellectual network', *J. Brit. Stud.*, **4** (1964): 65–86.

6. Anon., '*The Bridgewater Treatises* – the Universe and its Author', *Quart. Rev.*, **50** (1833): 1–34; D. Brewster, 'The Bridgewater bequest – Whewell's *Astronomy and General Physics*', *Edin. Rev.*, **58** (1834): 422–57; G. P. Scrope, 'Dr Buckland's *Bridgewater Treatise*', *Quart. Rev.*, **56** (1836): 31–64; D. Brewster, 'Dr Buckland's *Bridgewater Treatise – Geology and Mineralogy*', *Edin. Rev.*, **65** (1837): 1–39; W. H. Brock, 'The selection of the authors of the *Bridgewater Treatises*', *Notes & Rec. Roy. Soc.*, **21** (1966): 162–79.

7. A. Ellegard, 'The readership of the periodical press in mid-Victorian Britain', *Goteborgs Universitets Arsskrift*, **63** (1957): 1–41; R. G. Cox, 'The reviews and magazines', in B. Ford (ed.), *The Pelican Guide to English Literature*, rev. edn, Harmondsworth, 1960, **6**: 188–204; R. K. Webb, 'The Victorian reading public', in ibid.: 205–26.

8. W. E. Houghton, *The Victorian Frame of Mind, 1830–1870*. New Haven, 1957; L. J. Henkin, *Darwinism in the English Novel, 1860–1910*. 1940, reprinted New York, 1963; B. Willey, *Nineteenth-Century Studies*. London, 1949 and *More Nineteenth-Century Studies*. London, 1956; U. C. Knoepflmacher, *Religious Humanism and the Victorian Novel*. Princeton, 1965; R. Williams, *The Long Revolution*. London 1961; J. Hillis-Miller, *The Disappearance of God*. Cambridge, Mass., 1963; J. H. Buckley, *The Victorian Temper*. Cambridge, Mass., 1951 and *The Triumph of Time*. Cambridge, Mass., 1967. Gordon S. Haight, *George Eliot: A Biography*. Oxford, 1968.

9. L. Eiseley, *Darwin's Century*. London, 1959; Sir G. de Beer, *Charles Darwin*. London, 1963; C. C. Gillispie, *Genesis and Geology*. Cambridge, Mass., 1951; F. C. Haber, *The Age of the World: Moses to Darwin*. Baltimore, 1959; J. T. Merz, *A History of European Thought in the Nineteenth Century*, 4 vols. London, 1904–12.

10. O. Chadwick, *The Victorian Church, Part I*. London, 1966; A. W. Benn, *The History of English Rationalism in the Nineteenth Century*, 2 vols. 1906, reprinted New York, 1962; J. M. Robertson, *A History of Free Thought in the Nineteenth Century*. London, 1929; J. Hunt, *Religious Thought in the Nineteenth Century*. London 1896; L. E. Elliott-Binns, *Religion in the Victoria Era*, 2nd edn. London, 1946 and *English Thought 1860–1900. The Theological Aspect*. London, 1956; P. d'A. Jones, *The Christian Socialist Revival, 1877–1914*. Princeton, 1968.

11. J. Passmore, *A Hundred Years of Philosophy*. London, 1957; R. Metz, *A Hundred Years of British Philosophy*, trans. J. W. Harvey et. al., J. H. Muirhead (ed.). London, 1938; M. Richter, *The Politics of Conscience. T. H. Green and his Age*. London, 1964.

12. R. D. Altick, *The English Common Reader*. Chicago, 1957; R. K. Webb, *The British Working Class . eader, 1790–1848*. London, 1955; A. Cruse, *The Victorians and Their Books*. London, 1935.

13. E. Halévy, *The Growth of Philosophic Radicalism*, rev. edn, trans. M. Morris. London, 1952; J. W. Burrow, *Evolution and Society*. Cambridge, 1966.

14. C. Brinton, *English Political Thought in the Nineteenth Century*. London,

1933; Sir E. Barker, *Political Thought in England, 1848 to 1914*. 1915, 2nd edn, revised. London, 1963; W. L. Burn, *The Age of Equipoise*. London, 1964; A. Briggs, *The Age of Improvement*. London, 1959; G. Kitson Clark, *The Making of Victorian England*. London, 1962; F. J. C. Hearnshaw (ed.), *The Social and Political Ideas of Some Representative Thinkers of the Victorian Age*. London, 1933; E. Halévy, *A History of the English People in the Nineteenth Century*, 6 vols trans. E. I. Watkin and D. A. Barker. 1913, 2nd edn, 1949; J. Hamburger, *James Mill and the Art of Revolution*. New Haven, 1963 and *Intellectuals in Politics: John Stuart Mill and the Philosophic Radicals*. New Haven, 1965; E. P. Thompson, *The Making of the English Working Class*. London, 1963; R. H. Murray, *Studies in the English Social and Political Thinkers of the Nineteenth Century*, 2 vols. Cambridge, 1929; J. M. Robson, *The Improvement of Mankind: The Social and Political Thought of John Stuart Mill*. Toronto, 1968; R. Hofstadter, *Social Darwinism in American Thought*, rev. edn. Boston, 1955; G. M. Young, *Victorian England*, 2nd edn. Oxford, 1953.

15. T. Ribot, *English Psychology*, trans. J. Fitzgerald. London, 1873; H. C. Warren, *A History of the Association Psychology*. London, 1921; E. G. Boring, *A History of Experimental Psychology*, 2nd edn. New York, 1950; G. Murphy, *Historical Introduction to Modern Psychology*, rev. edn. New York, 1949; G. S. Brett, *Brett's History of Psychology* (3 vols, 1912–21) ed. and abridged by R. S. Peters. London, 1953; O. Temkin, 'Gall and the phrenological movement', *Bull. Hist. Med.*, **21** (1947): 275–321; L. S. Hearnshaw, *A Short History of British Psychology, 1840–1940*. London, 1964; Sir G. Jefferson, *Selected Papers*. London, 1960.

16. For example, see G. W. Stocking, *Race, Culture, and Evolution*. New York, 1968; J. Clive, *Scotch Reviewers. The 'Edinburgh Review' 1802–1815*. London, 1957; H. and H. C. Shine, *The Quarterly Review under Clifford*. Chapel Hill, 1949; E. M. Everett, *The Party of Humanity. The Fortnightly Review and its Contributors 1865–1874*. Chapel Hill, 1939.

17. Chadwick, *The Victorian Church*, Ch. 8; Gillispie, *Genesis and Geology*.

18. I am not intending to be rude about Prof. Chadwick or Prof. Gillispie but to make a point about the complexity of the subject. Gillispie's is still the best book on the subject and was the inspiration of much of my own research. However, his focus on the *religious* aspect involves a corresponding diminution of a sense of how much the readers of the periodicals knew about *geology*. See, for example, F. Jeffrey, 'Playfair's *Illustrations of the Huttonian Theory*', *Edin. Rev.*, **1** (1802): 201–16; J. Playfair, 'Cuvier on the *Theory of the Earth*', ibid., **22** (1814): 454–75; W. H. Fitton, '*Transactions of the Geological Society*, vol. 2 and vol. 3', ibid., **28** (1817): 174–92, and **29** (1817): 70–94; C. Lyell, 'Scope's *Geology of Central France*', *Quart. Rev.*, **36** (1827): 437–83; J. Fleming, 'On the geological Deluge', *Edin. Philos. J.*, **14** (1825): 203–39; C. Lyell, '*Transactions of the Geological Society*, [N.S.] vol. I'. *Quart. Rev.*, **34** (1826): 507–40; T. Thomson (?), '*Transactions of the Geological Society*, [N.S.] vols 1–3 – progress of geological science', *Edin. Rev.*, **52** (1830): 43–72; G. P. Scrope, 'Lyell's *Principles of Geology*', *Quart. Rev.*, **43** (1830): 411–69 and **53** (1835): 406–48; W. H. Fitton, 'Lyell's *Elements of Geology*', *Edin. Rev.*, **69** (1839): 406–66; G. P. Scrope, 'Murchison's *Silurian System*', *Quart. Rev.*, **64** (1839): 102–20; Anon., 'The earth and man', *Westminster Rev.*, **8** (1855): 151–70; Anon., '*Testimony of the Rocks*', ibid., **12** (1857): 176–85; F. R. Condor, 'Scepticism in geology',

Edin. Rev., **147** (1878): 354–86; E. H. Bunbury, *'Life, Letters and Journals of Sir Charles Lyell'*, *Quart. Rev.*, **153** (1882) 96–131. Others reviews of geological works are referred to in other parts of this paper. I have cited these reviews to emphasize my point. It could be driven home *ad nauseam* with a *very* long bibliography. See *The Wellesley Index to Victorian Periodicals*. Toronto, I, 1966.

19. London, 1949. Paperback edn; New York, 1966.
20. A. Ellegard, *Darwin and the General Reader*. Goteborg, 1958.
21. M. Millhauser, *Just Before Darwin. Robert Chambers and 'Vestiges'*. Middletown Conn., 1959.
22. S. Budd, 'Reasons for unbelief among members of the Secular Movement in England, 1850–1950', *Past and Present* No. 36 (1967): 106–25, 110n.
23. M. Millhauser, 'The Scriptural geologists, an episode in the history of opinion', *Osiris* **11** (1954): 65–86.
24. See above, note 8 and A. Ward, *Walter Pater: the Idea in Nature*. London, 1966.
25. A. T. Kitchel, *George Lewes and George Eliot. A Review of Records*. New York, 1933; Everett, *The Party of Humanity*; Haight, *George Eliot: A Biography*.
26. J. W. Cross, *George Eliot's Life as Related in Her Letters and Journals*, new edn. London, 1885: 135–7, 259. In 1856, George Eliot wrote to a friend, 'We are reading Gall's 'Anatomie et Physiologie du Cerveau', and Carpenter's 'Comparative Physiology', aloud in the evenings; and I am trying to fix some knowledge about plexuses and ganglia in my soft brain . . .' (p. 196); In 1844 Charles Bray persuaded her to have a phrenological cast made of her head (Jefferson, *Selected Papers*, p. 41, cf. p. 40). See also G. H. Lewes, 'Phrenology in France', *Blackwoods Edin. Mag.*, **82** (1857): 665–74, and the chapters on Gall in the second and third editions of his *Biographical History of Philosophy* (London, 1857, 1867–71); D. Duncan (ed.) *The Life and Letters of Herbert Spencer* (London, 1908): 63–4, 542; F. E. Mineka (ed.) *The Earlier Letters of John Stuart Mill 1812–1848* (2 vols. London, 1963) for information on Mill's relationship with Lewes and Eliot; R. E. Ockenden, 'George Henry Lewes (1817–1878)', *Isis*, **32** (1940): 70–86.
27. See Everett, *The Party of Humanity*, Ch. 2; Kitchel, *George Lewes and George Eliot, passim*.
28. See Everett, *The Party of Humanity*, Chs 3 and 4.
29. A. W. Brown, *The Metaphysical Society: Victorian Minds in Crisis, 1869–1880*. New York, 1947: 169.
30. Ibid.: 167, 168.
31. Webb, In: *Pelican Guide*, 6: 206.
32. Burrow, *Evolution and Society*: 169.
33. Quoted in Everett, *The Party of Humanity*: 80.
34. Ibid.: 115–16.
35. Ibid.: 106.
36. Ibid.: 115.
37. Mrs Stair Douglas, *The Life and Selections from the Correspondence of William Whewell, D.D.*, 2nd edn. London, 1882, p. 248; I. Todhunter, *William Whewell, D.D. An Account of His Writings with Selections from his Literary and Scientific Correspondence*, 2 vols. London, 1876; R. M. Blake *et al.*, 'William Whewell's philosophy of scientific discovery', in *Theories of Scientific Method*. Seattle, 1960, Ch. 9; E. W. Strong, 'William Whewell and

John Stuart Mill: their controversy about scientific knowledge', *J. Hist. Ideas*, **16** (1955): 209–31; A. Ellegard, 'Darwin's theory and nineteenth-century philosophies of science', ibid. (June, 1957): J. F. W. Herschel, 'Whewell on *Inductive Sciences*', *Quart. Rev.*, **68** (1841): 177–238.

38. W. Whewell, 'Lyell – *Principles of Geology*', *Brit. Critic*, **9** (1831): 180–206, 194 and 'Lyell's *Geology*, vol 2 – changes in the organic world now in progress', *Quart. Rev.*, **47** (1832): 103–32.

39. W. F. Cannon, 'The Uniformitarian–Catastrophist debate'. *Isis*, **51** (1960): 38–55.

40. W. Whewell, *The Philosophy of the Inductive Sciences*, 2 vols. London, 1840, 2: 134.

41. Ibid.: 135, 137.

42. Ibid.: 137–52.

43. Ibid.: 144–5; cf. W. Whewell, *History of the Inductive Sciences*, 3 vols. London, 1837, 3: 601–2.

44. W. Buckland, *Reliquiae Diluvianae; or, Observations on the Organic Remains contained in Caves, Fissures, and Diluvial Gravel, and on Other Geological Phenomena, attesting to the Action of a Universal Deluge.* London, 1823; *Geology and Mineralogy Considered with Reference to Natural Theology.* London, 1836: 94n–95n; G. P. Scrope, *Quart. Rev.*, **56** (1836): 34n.

45. Whewell, *Philosophy of the Inductive Sciences*, 2: 146–7.

46. Whewell, *History of the Inductive Sciences*, 3: 586.

47. Whewell, *Philosophy of the Inductive Sciences*, 2: 148.

48. Ibid., 2: 150, cf. 92–3.

49. London, 1844; 11th edn. London, 1860. Chambers' authorship was acknowledged in a new edition which appeared in 1884.

50. H. Spencer, 'The development hypothesis', *The Leader* (20 March 1852), reprinted in *Essays: Scientific, Political, and Speculative*, 3 vols. London, 1901, 1: 1–7; *The Principles of Psychology.* London, 1855.

51. Anon., 'The present state of theology in Germany', *Westmin. Rev.*, **11** (1857): 327–63; Anon., '*The Life of Jesus* by Strauss' [trans. George Eliot!], ibid., **26** (1864): 291–322; Cruse, *The Victorians and their Books*, Ch. 5.

52. Sedgwick successively opposes Lyell, Chambers and Darwin. A. Sedgwick, 'Natural history of Creation', *Edin. Rev.*, **82** (1845): 1–85; *A Discourse on the Studies of the University of Cambridge*, 5th edn with Additions and a Preliminary Discourse. Cambridge, 1850 (Sedgwick attacks *Vestiges*, phrenology, German historiography, Tract XC, Utilitarian morals, and Lockean sensationalism); J. W. Clark and T. M. Hughes, *The Life and Letters of the Reverend Adam Sedgwick*, 2 vols. Cambridge, 1890.

53. W. Whewell, *Indications of the Creator*. London, 1845.

54. Douglas, *Life of Whewell*: 316–19. Chambers' book *did* enrage scientists. Chambers' facts were usually wrong, he confused the concept of law with that of cause, enraged Herschel and Huxley, and convinced Darwin not to base his own argument on the geological evidence. Nevertheless, Chambers grasped the general idea that *all* of nature (including man and society) is under the domain of law, more clearly than his more cautious and scientifically respectable contemporaries.

55. Whewell, *Indications of the Creator*, 2nd edn. London, 1846: 6–9; cf. 43–4.

56. Ibid.: 20–1.

57. Ibid.: 21–4.

58. J. Tyndall, 'The Belfast address', in *Fragments of Science*, 7th edn, 2 vols. London, 1889, 2: 135–201, esp. 200; cf. 202–50 (the separately printed version of *The Belfast Address* – London, 1874 – was modified after the protest evoked by the original version, but Tyndall restored the omissions in his *Fragments* after being criticized for modifying the printed version): T. H. Huxley, 'On the hypothesis that animals are automata and its history' (1874). Reprinted in *Collected Essays*, 9 vols. 1894, 1: 199–250, esp. 236–44; Brown, *The Metaphysical Society*: 231–8.

59. J. W. Draper, *History of the Conflict between Religion and Science*. London, 1875; cf. *A History of the Intellectual Development of Europe*. New York, 1861, which Tyndall used as a major source for the argument of his Belfast address. Draper claimed to have originated the historiographic thesis of scientific determinism – progressing from darkness to light. Although this claim would have amused the Positivists, among others, it represents a new awareness of the relationship between religion and science as one of conflict, with science progressively advancing as religion retreats. An earlier paper by Draper at the Oxford meeting of the British Association was the occasion of the famous Huxley–Wilberforce confrontation over the respective merits of simian grand-parents and debased intelligence. See D. Fleming, *John William Draper and the Religion of Science*. Philadelphia, 1950, Ch. 7. See also Anon., 'The intellectual development of Europe', *Westmin. Rev.* **83** (1865): 94–142; Anon., 'The intellectual development of Europe', *Saturday Rev.*, **17** (1864): 726–7.

60. F. Galton, 'Statistical inquiries into the efficacy of prayer', *Fortnightly Rev.*, N.S. **12** (Aug. 1872); Tyndall initiated this bit of coat-trailing in the *Contemporary Review* in July 1872, where he suggested controlled experiments in hospitals – praying for some patients and not others. See Brown, *The Metaphysical Society*: 177–80. For an earlier, theological, point of view, see W. G. Ward, 'Science, prayer, free will, and miracles'. *Dublin Rev.*, N.S. **8** (1867): 255–98.

61. F. Darwin (ed.), *The Life and Letters of Charles Darwin*, 3 vols. London, 1887, 2: 261 and 261n. During his undergraduate days at Cambridge, Darwin had met Whewell. See N. Barlow (ed.), *The Autobiography of Charles Darwin, 1809–1882*, with original omissions restored. London, 1958: 66, 104.

62. Todhunter, *Life of Whewell*, 2: 433–5.

63. C. Lyell, *The Geological Evidences of the Antiquity of Man with Remarks on Theories of the Origin of Species by Variation*. London, 1863; Anon., 'The Antiquity of Man', *Westmin. Rev.*, **79** (1863): 517–51; T. H. Huxley, *Man's Place in Nature*. London, 1863. (Although these essays are reprinted in Huxley's *Collected Essays*, there are omissions, and the best available edition is an Ann Arbor Paperback, Michigan, 1959.)

64. Todhunter, *Life of Whewell*, 2: 435–6.

65. F. Darwin (ed.), *Life of Darwin*, 2, 261n.

66. R. Owen, 'Darwin on *The Origin of Species*', *Edin. Rev.*, **111** (1860): 487–532, esp. 487 and 532; William Carpenter, 'Darwin on the *Origin of Species*', *National Rev.*, **10** (1860): 188–214; and 'The theory of development in nature', *Brit. & For. Medico-Chir. Rev.*, **25** (1860): 367–404.

67. F. Temple *et al.*, *Essays and Reviews*. London, 1860; F. Harrison, 'Neo-Christianity', *Westmin. Rev.*, **18** (1860): 293–332; S. Wilberforce, '*Essays and Reviews*', *Quart. Rev.*, **109** (1861): 248–305; A. P. Stanley,

'Essays and Reviews', *Edin. Rev.*, **113** (1861): 461–500; Anon., *'Essays and Reviews* – Dr Lushington's judgement', *Westmin. Rev.*, **22** (1862): 301–15; B. Willey, *More Nineteenth-Century Studies*, Ch. 4; collection of replies and tracts on *Essays and Reviews* in Cambridge University Library, class marks **8.33.31, 8.33.34–36, 8.33.45.** M. A. Worden of Somerville College Oxford has recently completed a doctoral dissertation on conflicts over religious inquiry among the Anglican clergy in the 1860s. Cf. P. Appleman *et al.* (eds), *1859: Entering an Age of Crisis*. Bloomington, 1959, Part I. Michael Wolff's essay on Victorian reviewers and cultural responsibility is the best treatment of the issues raised in this paper which I have seen (ibid.: 267–89).

68. M. Peckham (ed.), *The Origin of Species by Charles Darwin. A Variorum Text*. Philadelphia, 1959: 69; Sir G. de Beer (ed.), 'Some unpublished letters of Charles Darwin', *Notes and Rec. Roy. Soc.*, **14** (1959): 12–66, 51–4. Darwin wrote, 'Permit me to add that I read your Philosophy of Creation with great interest: it struck me as excellently and vigorously argued and written with a clearness, which I remember excited my warmest admiration' (p. 53). Darwin had also acknowledged Powell in the preface to his never-published great work, *Natural Selection* (ibid.: 54).

69. B. Powell, *Revelation and Science*. Oxford, 1833; *The Connection of Natural and Divine Truth*. London, 1838; *Tradition Unveiled*. London, 1839; *Essays on the Spirit of the Inductive Philosophy, the Unity of Worlds, and the Philosophy of Creation*. London, 1855; 2nd edn entitled *The Unity of Worlds and of Nature*. London, 1856; *Christianity without Judaism, a Second Series of Essays*. London, 1857; *The Order of Nature considered in reference to the Claims of Revelation. A Third Series of Essays*. London, 1859; R. J. Mann, 'The plurality of worlds', *Edin. Rev.*, **102** (1855): 435–70; Anon., 'The order of nature', *North Brit. Rev.*, **31** (1859): 353–83; Anon., 'Recent latitudinarian theology', *Christian Remembrancer*, **38** (1860): 389–427; Anon., 'Recent rationalism in the Church of England', *North Brit. Rev.*, **33** (1860): 217–55. For a biographical account of Powell, see W. Tuckwell, *Pre-Tractarian Oxford*. London, 1909, Ch. 7. There are fifteen tracts and essays aimed exclusively or largely at Powell in the Cambridge U.L. collection. See note 67 above; cf. Ellegard, *Darwin and the General Reader*: 112, 158.

70. B. Powell, *Natural and Divine Truth*: 237–56, 297–301.

71. B. Powell in *Essays and Reviews*: 440.

72. B. Powell, *Natural and Divine Truth*: 229, 231, 268–9.

73. G. Combe, *The Constitution of Man*. Edinburgh, 1827, Introduction; 8th and later (Henderson Trust) editions, Edinburgh, 1841, etc.: 20 and Ch. 5; R. Chambers, *Vestiges of Creation*, 12th edn: 405; see also letters to Combe and from Powell: xxx–xxxi; Clark and Hughes, *Life of Sedgwick*, 2: 83–4; cf. above note 52.

74. N. Barlow (ed.), *Darwin and Henslow, the Growth of an Idea. Letters 1831–1860*. London, 1967: 206; F. Darwin (ed.), *Life of Darwin*, 2: 247–50.

75. S. Wilberforce, 'Darwin's *Origin of Species*', *Quart. Rev.*, **108** (1860): 225–64, esp.: 258–60. I am indebted to Professor Sydney Eisen, who has helped me to see the merits of Wilberforce's position by contrasting it with Charles Kingsley's attempts to ingratiate himself with the evolutionists.

76. B. Powell in *Essays and Reviews*: 354–5.

77. T. H. Huxley, 'On the reception of the "Origin of Species"', in F. Darwin (ed.), *Life of Darwin*, 2, Ch. 5: 202–3.

78. Huxley, *Collected Essays*, 1: 236–8, 244.

79. Lyell in F. Darwin (ed.), *Life of Darwin*, 2: 193n; cf. Huxley's comments: 192–3; Powell, *Philosophy of Creation*: 118–19, 439 and in *Essays and Reviews*: 139; R. Owen in *British Association Reports* (1858). London, 1859: xc.

80. M. Mandelbaum, 'Darwin's religious views', *J. Hist. Ideas*, **19** (1958): 363–78.

81. C. Babbage, *The Ninth Bridgewater Treatise. A Fragment*. London, 1837, 2nd edn, 1837: 92–9, 191 and Chs 10–13.

82. Chambers used Babbage's argument in an attempt to settle religious doubts about his evolutionary theory (*Vestiges of Creation*, 12th edn: 203–11).

83. F. Temple, *The Relations between Religion and Science*, Bampton Lectures. London, 1884 (reprinted 1885 and 1903): 188; Anon., 'Dr Temple on religion and science', *Westmin. Rev.*, **123** (1885): 366–85; St G. J. Mivart, 'Bishop Temple's lectures on religion and science', *Edin. Rev.*, **162** (1885): 204–33; F. Temple, 'The education of the world', in *Essays and Reviews*; cf. Elliott-Binns, *English Thought 1860–1900*: 25. It is worth recalling that both Lyell (d. 1875) and Darwin (d. 1882) were buried in Westminster Abbey.

84. J. Ray, *The Wisdom of God manifested in the Works of Creation* (1690), 5th edn. London, 1709: 47.

85. W. Paley, *Natural Theology: or, Evidences of the Existence and Attributes of the Deity collected from the appearances of Nature* (1802), new edn. London, 1816. Chs 1–5 and: 239–40, 265, 268, 308–10, 361, 373–5; cf. Blunt in *Quart. Rev.*, **38** (1828): 312–13, where Paley is praised for his simple examples and his failure to use technical language.

86. See the *Bridgewater Treatises* of Kirby and Chalmers where liberties are taken with the Hebrew language and the days of the Creation.

87. Brown, *The Metaphysical Society*, Appendix A (membership list).

88. Ibid., Appendix C: The papers of the Metaphysical Society. There are more or less complete sets of the papers in the Bodleian and Manchester College Libraries, Oxford.

89. A modern symposium. [1] 'The influence upon morality of a declining religious belief', *Nineteenth Century*, **1** (1877): 331–58, 531–46; [2] 'The soul and future life', ibid., **2** (1877): 329–54, 497–536; [3] 'Is the popular judgement in politics more just than that of higher orders?' ibid., **3** (1878): 797–822, **4** (1878): 174–92.

90. H. E. Manning, 'A diagnosis and a prescription' (No. 36, 10 June 1873); Mark Pattison, 'Double truth (No. 74, 12 Feb. 1878); M. Boulton, 'Has the Metaphysical Society any *raison d'être*? (No. 75, 9 April 1878); R. H. Hutton, 'The Metaphysical Society: a reminiscence', *Nineteenth Century*, **18** (1885): 177–96.

91. A. R. Wallace, 'Sir Charles Lyell on geological climates and the origin of species', *Quart. Rev.*, **126** (1869): 359–94, 1869, esp.: 379–96.

92. Brown, *The Metaphysical Society*, Appendix D.

93. Ibid., Chs 1–3, esp.: 21–2, 26.

94. Ibid.: 42–3.

95. Hutton, in *Nineteenth Century*, **18** (1885): 180; W. G. Ward did give a paper under this title (No. 30, 10 Dec. 1872), and the topic – as my hypothesis requires – was the most common theme in the Society's deliberations. Towards the end of the series, Leslie Stephen also gave a paper entitled 'The uniformity of nature' (No. 82, 11 Mar. 1879).

96. Brown, *The Metaphysical Society*: 29.

97. Ibid.: 250n–251n.
98. Ibid.: 169–70.
99. Ibid.: 168.
100. Ibid., *passim*; Ellegard in *Goteborgs Universitets Arsskrift*, **63** (1957): 27–8, 32–4.
101. Brown, *The Metaphysical Society*, Chs 9 and 10.
102. O. J. R. Howarth, *The British Association for the Advancement of Science: A Retrospect, 1831–1931*, 2nd edn. London, 1931: 252.
103. It is important to make a distinction between two sorts of popularizer. The first group consists of professional scientists and philosophers writing for a popular audience, e.g. Huxley's *Collected Essays* (especially the *Lay Sermons*. London, 1870 and 13 reprintings and new editions by 1903); Tyndall's *Fragments of Science*; St G. J. Mivart's *Essays and Criticisms*, 2 vols, 1892 and his 'Examination of Mr Herbert Spencer's philosophy', *Dublin Rev.*, (1874–80), (9 parts); *Spencer's Essays*; A. R. Wallace's *Studies Scientific and Social*, 2 vols. London, 1900; and *The Wonderful Century*. London, 1898, 4th edn, 1901. Cf. the writings of W. B. Carpenter, G. H. Lewes, and G. J. Romanes. For an example of the new sort of popularization of science, see Anon., 'Recent science', *Nineteenth Century*, **4** (1878): 765–84. 'Professor Huxley has kindly read, and aided the Compilers, and the Editor with his advice upon, the following article.' (ibid.: 755).

The writers who come into increasing prominence in the late 1860s are not always professionally qualified in science but are attempting to interpret its implications and consider its applications in ethics, politics, history, and social theory. The International Scientific Series is interesting because it embraces both groups. Some of the works which seem to me to represent the growing significance of this sort of writing are: W. K. Clifford, *Lectures and Essays*, 2 vols, F. Pollock and L. Stephen (eds). London, 1879 and *Common Sense of the Exact Sciences*, edited and partly written by K. Pearson. London, 1885; G. D. Campbell (8th Duke of Argyll), *The Reign of Law*. London, 1866, 6th edn, 1871, and 'The unity of nature', *Comtemp. Rev.*, **38** and **39** (1880–1): (10 parts); W. Bagehot, *Physics and Politics or Thoughts on the Application of the Principles of 'Natural Selection' and 'Inheritance' to Political Society*. London, 1869, 2nd edn, 1873; R. H. Hutton, *Aspects of Religious and Scientific Thought*. London, 1899, and *Criticisms on Contemporary Thought and Thinkers*, 2 vols. London, 1894; Samuel Butler, *Life and Habit*. London, 1878; *Evolution, Old and New*. London, 1879; *Unconscious Memory*. London, 1880; *Luck or Cunning as the Main Means or Organic Modification?* London, 1879; H. T. Buckle, *History of Civilization in England*, 2 vols. London, 1857–61; W. E. H. Lecky, *History of the Rise and Influence of the Spirit of Rationalism in Europe*, 2 vols. London, 1865 (19 editions and reprints by 1910), and *The Map of Life*. London, 1899 (Buckle and Lecky are symptomatic of the development of a naturalist, determinist historiography). Sir L. Stephen, *Essays in Freethinking and Plainspeaking*. London, 1873; *The Science of Ethics*. London, 1882; *An Agnostic's Apology*. London, 1893; Joseph Le Conte, *Evolution. Its Nature, Its Evidences, and Its Relation to Religious Thought* (1888), 2nd edn. London, 1893; Benjamin K: dd, *Social Evolution*. London, 1894 (11 editions and reprints by 1898); J. Watson, *Comte, Mill and Spencer*. Glasgow, 1895; R. Mackintosh, *From Comte to Benjamin Kidd*.

The Appeal to Biology or Evolution for Human Guidance. Many of these works are collections of essays from Victorian periodicals. The lists of works in this (putative) *genre* could be extended indefinitely; I have cited those which I happen to have consulted. See also the following essays: J. Hannah, 'The attitude of the clergy towards science', *Contemp. Rev.*, **9** (1868): 395–404, and 'One word more on the clergy and science', ibid., **10** (1869): 74–80; F. W. Farrar, 'The attitude of the clergy towards science', ibid., **9** (1868): 600–20; J. B. Mozley, 'The argument of design', *Quart. Rev.*, **127** (1869): 134–76; A. Grant, 'Philosophy and Mr Darwin', *Contemp. Rev.*, **17** (1871): 274–81; J. Martineau, 'The place of mind in nature and intuition in man', ibid., **19** (1872): 606–23, and 'Modern materialism: its attitude towards theology', ibid., **27** (1876): 323–46; C. Elam, 'The gospel of evolution', ibid., **37** (1880): 713–40.

It is difficult in this period to draw lines between different philosophical positions by means of attitudes towards the Uniformity of Nature. As I have suggested the principle had become so vague as to embrace positions extending all the way from belief in miracles to orthodox Comtism and beyond. An important *desideratum* in late-Victorian studies is to establish some sort of classification of theories. A series of papers by S. Eisen helps to chart one network of relations with respect to Comtism: 'Huxley and the positivists', *Vict. Stud.* 7 (1964): 337–58; 'Frederic Harrison and the religion of humanity', *South Atlantic Quart.*, **66** (1967): 574–90; 'Herbert Spencer and the spectre of Comte', *J. Brit. Stud.*, (1967): 48–67; 'Frederic Harrison and Herbert Spencer: embattled unbelievers', *Vict. Stud.*, **12** (1968): 33–56. See also W. M. Simon, *European Positivism in the Nineteenth Century*. Ithaca, 1963, Chs 7–8.

104. There is a short discussion of the International Scientific Series in Fleming, *John William Draper*; B. H. Becker, *Scientific London*. London, 1874, is a rather chatty treatment of the institutional aspect of science in London.

105. W. Graham, *The Creed of Science. Religious, Moral and Social* (1881), 2nd edn, 1884. Preface: cf. J. S. Crone, *A Concise Dictionary of Irish Biography*. Dublin, 1937: 80.

106. W. C. Coupland, *The Creed of Science. Mind*, **6** (1881): 563–74, 568, 572.

107. Anon., 'The province of scepticism and the limits of free thought', *Church Quart. Rev.*, **14** (1882): 412–34. Graham's other works: *Idealism, An Essay Metaphysical and Critical*. London, 1872; *The Social Problem in its Economical, Moral, and Political Aspects*. London, 1886; *Socialism, Old and New*. London, 1890 (International Scientific Series); *English Political Philosophy from Hobbes to Maine*. London, 1899; *Free Trade and the Empire, A Study in Economics and Politics*. London, 1904.

108. T. K. Penniman, *A Hundred Years of Anthropology*, 3rd edn. London, 1965: 91. The relevant historical study is: Sir A. Keith, 'How can the Institute best serve the needs of anthropology?' *J. Roy. Anthropol. Inst.*, **48** (1917): 12–30.

109. T. Brown, '*Villers sur une nouvelle theorie du cerveau*', *Edin. Rev.*, **2** (1803): 147–60; J. Gordon, 'The doctrines of Gall and Spurzheim', ibid., **25** (1815): 227–68. (This article brought Spurzheim to Edinburgh and led to George Combe's conversion and to the establishment of the phrenological movement in Britain. The numerous societies, the *Phrenological Journal* and successors until 1966, and the role of phrenology in the development of naturalism, education and various reform movements, as well as its

influence on Chambers, Spencer, Lewes, Wallace, and numerous other figures – all of these deserve further study.) F. Jeffrey, 'Phrenology', *Edin. Rev.*, **44** (1826): 253–318, 515 and **45** (1826): 248–53; R. Chevenix, 'Gall and Spurzheim – phrenology', *For. Quart. Rev.*, **2** (1828): 1–59; Sir G. Jefferson, 'The contemporary reaction to phrenology', in *Selected Papers*: 94–112; see the articles on phrenology in *The Encyclopedia Britannica*, 8th–11th edns. Combe's *Constitution of Man* was not reviewed in the major periodicals, although it sold 85,000 copies in Britain by 1850 (*Leader*, **1** (1850)): 24; cf. A. R. Wallace, 'The neglect of phrenology', in *The Wonderful Century*: 159–93. Some of Combe's later works and his biography were noticed in the periodicals.

The first edition of Spencer's *Principles of Psychology* (London, 1855) was not widely reviewed. Lewes and J. D. Morell did help out a friend and noticed it, while R. H. Hutton discussed it under the title 'Modern atheism', but Spencer's works did not receive much critical attention until the *Synthetic Philosophy* was well under way. (H. Spencer, *An Autobiography*, 2 vols. London, 1904, 1: 468–72.) See, e.g. St G. J. Mivart, 'Herbert Spencer', *Quart. Rev.*, **135** (1873): 509–39, and above, note 103. When Spencer's *Autobiography* and his *Life and Letters* D. Duncan (ed.). London, 1908), appeared, he received the full treatment in the *Edinburgh, Contemporary* and *Fortnightly*.

A third leading contributor to psychology in the period was Alexander Bain. The first volume of his major work (*The Senses and the Intellect*. London, 1855) was not reviewed and lost money. He had trouble getting the second volume published. J. S. Mill and George Grote guaranteed £100 against loss by the publisher. When *The Emotions and the Will* (London, 1859) appeared, Mill helped it along with a very laudatory review: 'Bain's psychology', *Edin. Rev.*, **110** (1859): 287–321. The two volumes went through four editions by the turn of the century, and some of Bain's other works were very popular and influential. However, they were hardly noticed in the general periodicals.

It might be worthwhile to make a detailed comparison of the sales of psychological works with their treatment in the periodical press in the period. The works themselves reveal a network of ideas including phrenology, mesmerism, positivism, associationism, utilitarianism, physiology (and later) evolutionism, anthropology, and agnosticism. A good example of this in the earlier period is H. G. Atkinson, and H. Martineau, *Letters on the Laws of Man's Nature and Development*. London, 1851, cf. Kitchel, *George Lewes and George Eliot*: 79–80. In the post-1859 period, representative works are H. Spencer, *Principles of Psychology*, 2 vols, 2nd edn. London, 1870–2; G. H. Lewes, *The Physical Basis of Mind*. London, 1877.

110. Sedgwick in *Edin. Rev.*, **82** (1845): 11–12, 63–4; Wilberforce, in *Quart. Rev.*, **108** (1860) 258; St G. J. Mivart, 'Darwin's *Descent of Man*', ibid., **131** (1871): 47–90, 89–90; W. B. Dawkins, 'Darwin on *The Descent of Man*', *Edin. Rev.*, **134** (1871): 195–235, 195–6, 200, 207–8; T. S. Baynes, Tylor on *Primitive Culture*', *Edin. Rev.*, **135** (1872): 88–121, 113–15, 117–18.

111. These developments are discussed in detail in Young, *Mind, Brain and Adaptation*: 94–100 and Chs 3–9.

112. J. C. Robertson, 'Prefatory words', *Mind*, **1** (1876).

113. T. Chalmers, *On the Power, Wisdom and Goodness of God as Manifested in*

the Adaptation of External Nature to the Moral and Intellectual Constitution of Man, 2 vols. London, 1833; cf. Anon., in *Quart. Rev.*, **1** (1833): 4–5.

114. Brown, *The Metaphysical Society*: 197–206.

115. Kitchel, *George Lewes and George Eliot*: 276–7 [cf.p. 21]. Where can one learn about the 'Philosophical Club', the '*x* Club' and 'The Club' (L. Huxley, *Life and Letters of Thomas Henry Huxley*, 2 vols. London, 1900, 2: 255–61).

116. Brown, *The Metaphysical Society*: 247–52.

117. Ibid., Chs 9–10.

118. Ibid.: 252–60.

119. Quoted in ibid.: 106. The growing indifferences (coupled with arrogance) of scientists to theology is nicely reflected in the spread of the designation 'agnostic'. Huxley's account of its origin conveys the change in atmosphere very well: 'When I reached intellectual maturity and began to ask myself whether I was an athiest, a theist, or a pantheist; a materialist or an idealist; a Christian or a freethinker; I found that the more I learned and reflected, the less ready was the answer; at last, I came to the conclusion that I had neither art nor part with any of these denominations, except the last. . . . This was my situation when I had the good fortune to find a place among the members of that remarkable confraternity of antagonists, long since deceased, but of green and pious memory, the Metaphysical Society. Every variety of philosophical and theological opinion was represented there, and expressed itself with entire openness; most of my colleagues were -*ists* of one sort or another; and, however kind and friendly they might be, I, the man without a rag of a label to cover himself with, could not fail to have some of the uneasy feelings which must have beset the historical fox when, after leaving the trap in which his tail remained, he presented himself to his normally elongated companions. So I took thought, and invented what I conceived to be the appropriate title of "agnostic". It came into my head as suggestively antithetic to the "gnostic" of Church history, who professed to know so much about the very things of which I was ignorant; and I took the earliest opportunity of parading it at our Society, to show that I, too had a tail, like other foxes. To my great satisfaction, the term took; and when the *Spectator* had stood godfather to it, any suspicion in the minds of respectable people, that a knowledge of its parentage might have awakened was, of course, completely lulled.' (T. H. Huxley, 'Agnosticism' (1889), in *Collected Essays* 5, Ch. 7: 237–9).

2.3 John Farley and Gerald L. Geison, Science, Politics and Spontaneous Generation in Nineteenth-century France: The Pasteur–Pouchet Debate

Introduction

Between 1859 and 1864, Louis Pasteur engaged in a celebrated debate over spontaneous generation with Félix Pouchet, a respected naturalist from Rouen and a corresponding member of the French Academy of Sciences. Traditional accounts of his debate have focused almost exclusively on the experimental issues dividing the combatants.[1] In our view, this approach ignores the very real significance of the extra-scientific, political aspects of the debate. It illegitimately suggests that the debate was resolvable solely at the level of experimental fact and that Pasteur's ultimate triumph can be ascribed entirely to his experimental skill and devotion to the true principles of the 'experimental method'. It is these assumptions which have filtered down into the modern biological community through those notorious 'historical' sections which one finds in many elementary and not-so-elementary textbooks. Without much exaggeration, one can say that every one of these textbooks which finds room for some historical 'humanism' gives an account of Pasteur's experimental victory over the myth of spontaneous generation.

And while Pasteur has been uncritically celebrated for his methodology, Pouchet has been castigated for his *a priori*, 'metaphysical' approach, as proof of which repeated use has been made of the following passage from the preface to his *Hétérogenie*: 'When by meditation it was evident to me that spontaneous generation was one of the means employed by nature for the reproduction of living things, I applied myself to discover the methods by which this takes place.'[2] By quoting this passage, while making a passing allusion to the religio-philosophical implications of the debate, standard accounts have conveyed the impression that Pouchet's belief in spontaneous generation derived from his (presumably unorthodox) philosophical and political position. By reading beyond his preface and by discussing what the objects of his 'meditation' in fact were, we shall show how utterly misleading that impression is. More generally, we believe that our reexamination of the Pasteur–Pouchet debate reveals the direct influence of extrinsic factors on the conceptual content of serious science[3]. At the very least, we can add a new dimension to the standard accounts of that debate as we seek to refute the simplistic view that Pouchet 'meditated' and was wrong, while Pasteur 'experimented' and was right.

The scientific and political background to the Pasteur–Pouchet debate
Although advanced in a variety of more or less sophisticated forms, the doctrine of spontaneous generation rests ultimately on the notion that living organisms can arise independently of any parent, whether from inorganic matter (abiogenesis) or organic debris (heterogenesis). Following an erratic historical career in which it long enjoyed the support both of natural philosophers and of Christian theology, only to be declared heretical by both in later eras, this doctrine reached its zenith of popularity during the first three decades of the nineteenth century, particularly in Germany when the early parasitologists and the *Naturphilosophen* argued forcefully in its favor.[4] In France, too, spontaneous generation received support through the writings of the materialist Cabanis, the transformist Lamarck, and the putative *Naturphilosophen* Geoffroy St Hilaire and his student Antoine Dugès. But the popularity of spontaneous generation was short-lived in France. There, by its presumed association with the doctrines of materialism and

transformism, it became not only scientifically discredited, but also politically, socially and theologically suspect.

This tendency to associate spontaneous generation with transformism derived in large measure from the eventual commitment of Lamarck and Geoffroy St Hillaire to both notions. In the full-fledged version of his theory of transformism, Lamarck insisted that a continuous spontaneous generation was necessary in order to replenish the lowest forms which had evolved into more complex organisms. Without such continuous spontaneous generation, he argued, the earth would be devoid of primitive life. Especially after Geoffroy revealed his allegiance to similar ideas, the French tended to associate spontaneous generation with any evolutionary theory.

Beginning about 1802, Georges Cuvier launched a vigorous campaign against the doctrines of Lamarck and Geoffroy, culminating in his celebrated debate with Geoffroy in the 1820s and early 1830s. Most witnesses to this debate awarded the palm of victory to Cuvier [. . .]

Cuvier did not hesitate to buttress his scientific arguments against Geoffroy with religio-philosophical and political supports forged for him by his influential post in the *Académie des Sciences* and by events in the national arena. With the rise to power of Napoleon Bonaparte, followed by the Restoration under Charles X, Cuvier hastened to associate his opponents and their doctrines with the speculative and supposedly pantheistic *Naturphilosophie* of the German enemy and with the materialism of the late eighteenth-century *philsophes* and *idéologues*, on whom the public placed much of the blame for the chaos and terror of the French Revolution [. . .] Nor does it seem to have made much difference that Geoffroy repeatedly and explicitly tried to dissociate himself from *Naturphilosophie*, materialism and impiety.[5] Whether consciously or not, Cuvier and much of the public displayed a convenient disregard for the complexity of the relationships among spontaneous generation, transformism, pantheism, *Naturphilosophie* and materialism. What mattered was the public perception and *belief* that spontaneous generation somehow belonged with these politically and religiously dangerous doctrines, and ought therefore to receive its share of guilt for the turmoil of the recent past.

A generation later, when Louis Pasteur waged his famous battle against spontaneous generation, the scientific and political situation bore a remarkable resemblance to that which had obtained during the Geoffroy–Cuvier debate. In the scientific arena, the similarities reflect in part the continuing influence on French scientists of Georges Cuvier, dead since 1832 [. . .]

In the political arena, France had again entered a period of conservatism following the republican experiment of the 1830s and 1840s.[6] As Cuvier had launched his campaign against transformism and spontaneous generation during the First Empire, so did Pasteur launch his – more strictly against spontaneous generation – during the Second.

Napoleon Bonaparte's nephew, Louis Napoleon, had been elected president of the Republic in 1848, thanks in part to the support of the Catholic Church which effectively controlled the votes of the newly enfranchised French peasants. In 1850 the new president had signed the notorious Falloux Law which allowed religious teaching in the state schools and gave the Church the right to its own secondary schools. In 1852, his power having been fortified by the *coup d'état* of 1851 and by the associated plebiscite, in which he won an overwhelming popular mandate, Louis Napoleon declared himself Emperor, once again with the general support of the Catholic Church. Thus, from the outset of the Second Empire, religious issues were simultaneously political issues. The forces of Church and State united in the face of the common enemy – republicanism and atheism. For opposition to the Church and State came not only from republican or liberal ranks but also from positivists, materialists and atheists, all of whom associated themselves with the scientific movement of the nineteenth century. Indeed, for many the new scientific movement became a sort of religion in its own right and Taine looked 'forward to the time when it will reign supreme over the whole of thought and over all man's actions'.[7]

In response to this liberal undercurrent, the Church became increasingly authoritarian and reactionary, culminating in the Papal Encyclical of Pope Pius IX in 1864, which emphasized the dangers of religious tolerance and of accommodation with the forces of liberalism and republicanism [. . .]

This climate was further exacerbated by the appearance of Clemence Royer's translation of Darwin's *Origin of Species* in 1862 and of Ernst Renan's *Vie de Jesus* in 1864. The latter attempted to rewrite the life of Christ on the basis of historical criticism and scientifically verifiable events. The former was even more significant since Royer adhered simultaneously to every doctrine the conservative forces loathed: atheism, materialism and republicanism. Her preface to the *Origin* was an extended diatribe against the Catholic Church, which she described as a 'religion spread by an ignorant, domineering and corrupt priesthood' and which she identified as the major cause of all social ills. It is scarcely surprising, then, that Darwinian evolution was regarded in France as a politico-theological doctrine allied with the forces which threatened Church and State. Nor is it surprising that so many French critics of Darwinian evolution focused on the issue of spontaneous generation. For besides its historical association in France with evolutionary theories, spontaneous generation was perceived as a threat to the belief in a providential Creator.[8]

Against this background, the outcome of the Pasteur–Pouchet debate over spontaneous generation carried implications of enormous importance to the political fabric of the Second Empire [. . .]

Félix Pouchet's views prior to the debate

When the debate between them began, Pasteur was 37 years old, while

Pouchet was nearly 60.[9] Pasteur had only recently entered the study of biological problems, before which his training, interest and expertise lay in the fields of crystallography and chemistry. Pouchet, on the other hand, entered the debate after a long career in traditional biology, his major interest having been animal generation. At what point precisely he became an advocate of spontaneous generation is less than clear [. . .]

In any case, Pouchet spent a significant portion of his book insisting that his views on spontaneous generation had nothing in common with the atheistic and dangerous versions of the doctrine so familiar from.the past. Indeed, *Hétérogenie* opened with a 137-page historical and metaphysical justification for a belief in spontaneous generation, and Pouchet insisted throughout that his version of the doctrine was in complete accord with orthodox biological, geological and religious beliefs.

Heterogenesis, he argued, was not the chance doctrine of the ancient atomists; instead it supposed that 'under the influence of forces still inexplicable, and which ... will remain inexplicable, a plastic manifestation is produced, whether in the animals themselves, or elsewhere, which tends to group molecules and to impose on them a special mode of vitality from which finally a new being results'. This 'plastic force', which takes part in the organization of animals and plants, 'can also manifest itself in plant and animal debris'. But, argued Pouchet – and this is the most distinctive feature of his version of spontaneous generation – it is *not* adult organisms which are thereby spontaneously generated, but rather their *eggs*:

Spontaneous generation does not produce an adult being. It proceeds in the same manner as sexual generation, which, as we will show, is initially a completely spontaneous act by which the plastic force brings together in a special organ [the egg] the primitive elements of the organism.[10]

In the second chapter of the book Pouchet comes to grips with the religious arguments put forth against the doctrine of spontaneous generation. He agreed that the first appearance of life was 'a true spontaneous generation operating under divine inspiration', but to deny any further spontaneous generation was an 'illegitimate fear, for if the phenomenon exists, it is because God has wished to use it in his design'. 'Where is the verse in the sacred text', he asked, 'which tells us that he imposed on himself never to resume his work? Or where is it said that after this rest, he has broken his molds and annihilated his creative faculty?' He argued that God, having fashioned the germ of things, had also imposed laws of matter and of life which determined when the organizational forces gave rise to new beings. 'The laws of heterogenesis', he insisted, 'far from weakening the attributes of the Creator, can only augment the Divine Majesty'[. . .][11]

Pouchet sought to demonstrate that his belief in heterogenesis was compatible with current geological thinking. This aspect of his belief was to grow more significant in the 1860s, for he maintained that heterogenesis was compatible with the theory of successive creations but not with evolutionary transformation. These arguments carry a greater cogency than one might expect. He agreed with Pictet that 'the theory of successive creations is the only one which agrees with the law that the species are entirely different from one strata to another'.[12] As suggested above, the problem of the origin of these successive species was by now considered outside the scope of geology. But Pouchet's theory of heterogenesis provided a mechanism for the successive creation of new species, since the 'plastic force' could be retained in organic debris and give rise to the first primordia of new creations. Moreover, Pouchet now had an answer to the objection that such complex and complete new forms could hardly be produced by heterogenesis – for it was the undifferentiated *eggs* of such forms and not the adult organisms themselves which were produced in this way. Then, linking his geological defense of heterogenesis with his earlier religious defense, Pouchet portrayed as irrational the belief that 'this great work, so frequently repeated . . . ought to stop'. Instead, he argued that just as the intensity and universality of geological catastrophes had diminished over time – the last universal catastrophe being that which raised the Andes and caused the Mosaic Flood – so the power of generating new beings 'no longer attains the same proportions as in ancient times'. Just as present 'catastrophes' were limited to local minor upheavals, so the plastic force was now limited to the production of the eggs of infinitely small organisms.

Such, in brief, was the theory of spontaneous generation which Pouchet offered in developed form in his book of 1859. What makes it particularly interesting is its essentially vitalistic and providential character [. . .] In fact, Pouchet emphatically opposed materialistic doctrines and explicitly denied the possibility of abiogenesis. He went to great pains to insist on the compatibility of his views with Christian teaching and with the 'laws of successive creations' endorsed by most French geologists and biologists. Against this background, it seems manifestly absurd that Pouchet should have been associated with the forces of materialism, transformism and atheism, but in the political climate of the Second Empire that is exactly what happened. Like Geoffroy before him, Pouchet found his name attached to heresies which he explicitly repudiated.

Pasteur and spontaneous generation before 1859
By the time Pasteur turned his attention to the problems of fermentation and spontaneous generation, his prior work in crystallography had already convinced him that life was intimately associated with molecular asymmetry (observable as optical activity) and could not be produced artificially by ordinary chemical procedures. The precise basis and origin of these beliefs is the subject of some controversy, but Pasteur's commitment to them seems undeniable from 1852, and he may even have held them implicitly from the outset of his scientific career. For even then it was clear (especially from the work of Biot, Pasteur's

own mentor) that optical activity was generally present in organic substances and uniformly absent from inorganic compounds [. . .]

Given his tendency to associate optical activity with life, it seems natural that Pasteur should then have adopted the position that fermentation (and its optically active products) must depend upon the activity of living microorganisms. In his initial paper of 1857 on the lactic-acid fermentation, he described the appearance of a grey deposit which increased in amount as the fermentation progressed and which consisted, like brewer's yeast, of minute globules. He more nearly assumed than proved that this 'lactic yeast' was a living organism.[13]

In a series of subsequent papers, culminating in a long memoir of 1860,[14] Pasteur extended this view of fermentation as a result of vital activity to ordinary alcoholic fermentation, which had always been viewed as the archetypical fermentative process and which had been the chief battleground for the debates between chemical and biological theories of fermentation. With these papers Pasteur largely overwhelmed the previously dominant chemical theory and laid the basis for the doctrine of specific ferments by showing that the same medium could give rise to different fermentations depending on the nature of the microorganisms sown into it – thus brewer's yeast induced only alcoholic fermentation, while 'lactic yeast' induced only lactic acid fermentation.

Implicit in the notion of the specificity of fermentation microorganisms is a commitment to an ordinary sort of generation for them. Only if they arise by an ordinary sort of reproduction, it would seem, could they retain the specific hereditary properties which must account for the specificity of their actions during fermentation.[15] On this ground alone, Pasteur may have been predisposed from the outset against the spontaneous generation of fermentative microorganisms, but the argument never emerges clearly and explicitly in his works [. . .]

From this point of view, Pasteur's subsequent campaign against the heterogenesis of fermentative microorganisms seems to flow naturally from the internal logic of his research. It seems hardly less natural that he should soon extend this campaign by implication to an indictment of 'spontaneous generation' in general. What is remarkable, however, and what does deserve immediate examination, is the fact that Pasteur could so vigorously prosecute the case against Pouchet and heterogenesis (and against spontaneous generation in general) while harboring in silence his own belief in the possibility of abiogenesis and his own earlier attempts to produce life artificially.

Indeed, that Pasteur ever made such attempts may itself seem remarkable in view of his conviction that asymmetry, optical activity and life were intimately associated, and that none of them could be produced artificially by ordinary chemical procedures. From this perspective, Pasteur looks for all the world like an ordinary 'vitalist', eager to maintain a barrier between the living and non-living. And yet Pasteur himself tried more than once to leap this barrier. The way out of this apparent paradox is to recognize the crucial import in Pasteur's mind of the phrase 'by ordinary chemical procedures'. For his own attempts to leap the barrier between the living and non-living bore no resemblance to ordinary chemical procedures. They derived instead from his notion of 'asymmetric forces', the intervention of which he considered essential to the production of asymmetric molecules (and hence of life). Like his conviction of the intimate association between **asymmetry and life, Pasteur's belief in asymmetric forces has an** uncertain origin, and the two notions may well be coeval, both dating from 1852 if not before [. . .][16]

In two famous lectures of 1860, 'On the Asymmetry of Natural Organic Products', Pasteur asked:

Why this asymmetry? Why any particular asymmetry rather than its inverse? . . . Why not only non-asymmetric [substances], like those in dead nature.

There are evidently causes of this curious behavior of the molecular forces. To indicate them precisely would certainly be very difficult. But I do not believe I am wrong in saying that we know one of their essential characteristics. Is it not necessary and sufficient to admit that at the moment of the elaboration of the immediate principles in the vegetable organism, an asymmetric force is present? [. . .]

Can these asymmetric actions be connected with cosmic influences? Do they reside in light, electricity, magnetism, heat? Could they be related to the movement of the earth, the electrical currents by which physicists explain the terrestrial poles?[17]

In other words, Pasteur boldly compared his 'asymmetric forces' to physical forces at work in the universe at large. Moreover, at one point in these lectures, Pasteur even suggested that 'it seems logical to me to suppose that [artificial or mineral substances] can be made to present an asymmetric arrangement in their atoms, as natural products do'.[18] From these two passages, considered in isolation, it might be supposed that Pasteur was here pressing his belief that asymmetric molecules (and thus life) could be produced artificially under the influence of physical asymmetric forces – that abiogenesis could occur under purely 'mechanistic' conditions.

Nonetheless, the dominant thrust of the 1860 lectures was to emphasize the distinction between (asymmetric) 'living nature' and (symmetric) 'dead nature', and to insist that asymmetric molecules could not be produced from symmetric starting materials by ordinary chemical procedures [. . .]

Nor in fact did he express much conviction that the barrier between living and non-living would soon (if ever) fall. If it seemed 'logical' to suppose that symmetric molecules could become asymmetric, it remained to be discovered how this could be accomplished.[19] If it seemed plausible to ask whether asymmetric forces might be related to physical forces in the universe, it was 'not even possible at present to offer the slightest suggestions' as to the answer.[20] If it was 'essential' to conclude that 'asymmetric forces exist at the moment of the elaboration of natural organic products', it was equally clear that these forces 'would be absent or without effect in our laboratory reactions, whether because of the violent action of these phenomena or because of some other unknown circumstances'.[21] In the end, the molecular asymmetery of natural organic products remained 'perhaps the only well marked line of demarcation that we can at present draw between the chemistry of dead nature and the chemistry of living nature'.[22]

Because Pasteur used his concept of asymmetric force in such a context, and because he stated it so tentatively and elusively, others regarded it as little more than another 'vitalistic' attempt to erect a barrier between the living (or asymmetric) and the non-living (or symmetric). No doubt his speculations would have attracted vastly more interest had he emphasized that while he believed asymmetric forces to be 'absent or without effect in our laboratory reactions', he did not consider them beyond the scope of experimental inquiry. Almost certainly, his lectures would have created a sensation had he then described his own earlier attempts to modify and even to create life.

In fact, however, it was not until 1883 that Pasteur briefly described for the first time in public his attempts to 'imitate nature' and to 'introduce asymmetry into chemical phenomena'.[23] As early as 1852, while still in Strasbourg, he had tried to bring asymmetric influences to bear upon crystallization by means of powerful magnets built according to his own specifications. On the basis of manuscript evidence, it seems clear that these attempts were made at least in part upon *inorganic* substances, more specifically upon the elements sulphur, potassium, copper, hydrogen, oxygen, chlorine and carbon (in the form of diamond) [. . .][24]

His friend and patron, Biot, tried early on to dissuade Pasteur from these sorts of experiments,[25] and Pasteur himself admitted to his father late in 1853 that 'one has to be a little mad to undertake what I am now trying to do'.[26] That he nonetheless gave such work high priority is clear from his wife's remark in a letter, also to his father, 'that the [experiments Louis] is undertaking this year will give us, should they succeed, a Newton or a Galileo'.[27]

In fact, of course, Pasteur did not succeed in creating asymmetry or life and temporarily abandoned these experiments. But he continued to believe that abiogenesis should be possible under some such experimental conditions. He thus came into the debate over spontaneous generation faced with a curious scientific dilemma. On the one hand, his work on fermentation led him to discount the possibility of heterogenesis, while on the other his theoretical views on asymmetry and life led him not only to believe in the possibility of abiogenesis but actually to attempt such a feat experimentally. If it seems illogical simultaneously to believe that life can be produced artificially from inorganic elements but not from a rich organic soup, it is essential to recall that Pasteur reached this paradoxical position as the result of two quite separate research problems and to emphasize the distinction in his mind between symmetric chemical influences and asymmetric physical forces. Nevertheless, and this is the central point, Pasteur could deny the possibility of spontaneous generation only by suppressing part of his own scientific beliefs.

The debate

The Pasteur–Pouchet debate began in private and quite politely, in contrast to the stormy reception given Pouchet's initial 1858 memoir on the subject by members of the French *Académie*, especially Milne Edwards. In this memoir Pouchet described the appearance of microorganisms in boiled hay infusions under mercury after exposure to artificially produced air or oxygen.[28] A year later Pouchet's controversial book *Hétérogenie* appeared, which elaborated in great detail a coherent view of nature in which the doctrine of spontaneous generation formed a central theme. These publications appeared just as Pasteur was reaching the conviction that fermentation depended on living organisms which could not arise heterogenetically. In February of 1859, in a note on the lactic acid fermentation, Pasteur asserted that the lactic 'yeast' in his experiments always came 'uniquely by way of the atmospheric air'. On this point, he wrote, 'the question of spontaneous generation has made an advance'.[29] This note prompted a letter from Pouchet, which has apparently not survived, but Pasteur's reply has: 'The experiments I have made on this subject', he began, 'are too few and, I am obliged to say, too inconsistent in results . . . for me to have an opinion worth communicating to you.' Nevertheless, he repeated the conclusion he had just announced in his published note and advised Pouchet that if he repeated his experiments with the proper precautions, he would see 'that in your recent experiments you have unwittingly introduced [contaminated] common air, so that the conclusions to which you have come are not founded on facts of irreproachable exactitude'. Thus, wrote Pasteur, 'I think . . . you are mistaken – not for believing in spontaneous generation, for it is difficult in such a question not to have a preconceived idea – but rather for affirming its existence.' He concluded by apologizing for 'taking the liberty of telling you what I think on so delicate a subject which has taken only an incidental and very small part in the direction of my studies'.[30]

Within a year, however, the question of spontaneous generation had taken a central – indeed dominant – place in Pasteur's research. Beginning in February 1860, Pasteur presented a series of five notes to the *Académie* on the subject, the results of which were brought together in his prize-winning essay, 'Mémoire sur les corpuscules organisés qui existent dans l'atmosphère', published in the *Annales des sciences naturelles* in 1861. Recognizing that the existence of atmospheric germs was as yet undemonstrated, Pasteur set out to show that air did contain living organisms and to deny that 'there exists in the air a more or less mysterious principle, gas, fluid, ozone, etc., having the property of arousing life in infusions'.[31]

Pasteur showed, by a series of brilliant and well-known experiments whose details need not concern us here,[32] that the microorganisms which appeared in a variety of decomposing media arose not from the presence of air alone but from particles in it, which he supposed to be living germs. To this, the heterogenesists objected that the smallest quantity of air invariably sufficed to induce all of these varied decompositions. So if, as Pasteur claimed, each decomposition resulted from a specific germ carried in the air, then the atmosphere must be so thick with a variety of germs as to appear foggy or even solid. Pasteur answered this objection by showing that flasks of boiled yeast-water exposed briefly to air in different localities and at different altitudes did not always putrefy and indeed at high altitudes very rarely did. Thus of 20 flasks he exposed to air at 2,000 meters on a glacier in the French Alps, only one underwent subsequent alteration.[33] In 1863 Pouchet travelled with two collaborators (Joly and Musset) to the Pyrenees to repeat Pasteur's experiments precisely except for the substitution of hay infusions for Pasteur's yeast-water. All eight of their flasks underwent subsequent alteration, as would be expected if the organic infusion required only oxygen to engender life.[34] In the face of Pasteur's contemptuous attitude toward these Pyrenees experiments, Pouchet and his collaborators insisted on the accuracy of their results and issued a challenge which ended in the appointment by the *Académie* of a second commission on spontaneous generation, the first commission having completed its work just two years earlier.

The Académie des Sciences and the Pasteur–Pouchet debate
In the highly centralized structure of post-revolutionary French science, the outcome of any scientific controversy was determined chiefly by the reaction of the Paris-based *Académie des Sciences*. This body typically responded to controversies by appointing commissions to adjudicate between the conflicting parties in order to arrive at a presumably objective decision, which thereby became the quasi-official viewpoint of the French scientific community. To a very large extent, the victory of Pasteur over Pouchet was determined by the response of the two commissions set up in the 1860s to examine the question of spontaneous generation.

The controversy aroused by the appearance of Pouchet's *Hétérogenie* in 1859 no doubt stimulated the *Académie* to propose a prize of 2500 francs to be awarded in 1862, 'to him who, by well-conducted experiments, throws new light on the question of so-called spontaneous generation'. The commission appointed to award the prize consisted initially of Geoffroy St-Hilaire, Serres, Milne Edwards, Brogniart and Flourens, but before a judgment could be rendered Geoffroy died and Serres was dropped from the panel. Their places were taken by Claude Bernard and Coste, thereby producing a panel unanimously unsympathetic to spontaneous generation from the outset.[35] Milne Edwards and Bernard had already responded critically to Pouchet's initial experimental paper of 1858; Brogniart and Flourens were disciples of Cuvier; and Coste opposed Pouchet's embryological views on the origin of infusorians in hay infusions. In addition, all of them, with the possible exception of Coste, were Catholics. Nonetheless, according to Georges Pennetier, Pouchet and his collaborators entered the competition, only to withdraw when some members of the commission announced their decision before even examining the entries, thereby leaving Pasteur to receive the prize uncontested, on the strength of his 1861 memoir.[36]

Pouchet's subsequent experiments in the Pyrenees and the resulting challenge aroused indignation in the *Académie*, most of whose members felt that the issue was closed. Flourens expressed these views publicly in 1863, by insisting that 'the experiments of M. Pasteur are decisive'.[37] Nevertheless, the *Académie* named a new commission in 1864, whose members were again united in their agreement with Pasteur: Milne Edwards, Brogniart, Dumas, Balard and, incredibly enough, Flourens himself. Balard, one of two members of the new panel not on the earlier commission, had been Pasteur's mentor in chemistry and had long taken an active interest in his career. Moreover, according to Duclaux, Balard played an important direct role in Pasteur's work against spontaneous generation by suggesting to him his famous 'swan-necked' flask experiments.[38] The second new member of the 1864 commission, Jean-Baptiste Dumas, was perhaps even more likely to be predisposed toward Pasteur's position. Like Balard, Dumas had long and actively promoted Pasteur's interests and may even have felt a special additional bond with his protégé since he, like Pasteur, had been subjected to abuse at the hand of Liebig. Dumas, moreover, was an important political figure in the Second Empire, having been appointed minister of agriculture and senator by Louis Napoleon, and later occupying the presidency of the Paris Municipal Council.

In the face of this patently biased commission, Pouchet and his collaborators displayed a rather precipitous loss of nerve and engaged the commission in a long and complicated dispute over the timing and nature of the experiments they would be allowed to present before it. In general, Pouchet and his collaborators sought to expand the scope of the inquiry and of the experimental program, while Pasteur and the commission insisted that the issue be very strictly confined to the question, does the least quantity of air invariably suffice to induce decomposition in fermentable media.[39] In the end, Pouchet and his collaborators once again withdrew in the belief that they would be denied a fair hearing.[40]

But the biased composition of these commissions and the uncritical acclaim which they accorded Pasteur's experiments formed only one aspect of the 'official' position of the French scientific community on spontaneous generation. Concurrently the French scientific elite devoted considerable effort to a refutation of Darwinism precisely on the basis of Pasteur's overthrow of spontaneous generation. In fact, Flourens – who had succeeded Cuvier as Perpetual Secretary of the

Académie at the latter's own request – published his *Examen du livre de M. Darwin sur l'origine des espèces* in the very year the second commission was constituted. The major tenet of the book was that Darwinism depended on the occurrence of spontaneous generation and could no longer be maintained since 'spontaneous generation is no more. M. Pasteur has not only illuminated the question, he has resolved it.'[41] Other leading scientists rallied to the cause and did so in terms which left no doubt as to the political and religious danger of evolutionary ideas.[42] In this way the issue of spontaneous generation became intimately linked with political and theological issues in the public mind. In 1860 Ernst Faivre remarked with passion that the problem of spontaneous generation 'excites at the moment the best minds, for it touches science, philosophy and religious beliefs'. The destruction of spontaneous generation, he concluded, 'is capable of lifting us from the consideration of physical laws to that of general truths, which enlighten our reason and confirm our religious beliefs'.[43]

Remarks such as this suggest that members of the French scientific community may have chosen Pasteur over Pouchet at least in part for socio-political reasons, especially since many who joined the two-pronged attack against Darwinism and spontaneous generation were dubiously qualified to do so. Their complete disregard for Pouchet's insistence on the orthodoxy of his version of spontaneous generation leads one to wonder if they had even seriously considered his work before denouncing it. Most importantly, the powerful *Académie des Sciences* reacted to the issue of spontaneous generation in ways that seem at least partly political in character – not only by appointing obviously biased commissions to consider the issue, but also by failing to perceive the oddly superficial way in which these commissions carried out their charge (see p. 126 below). In the end, one suspects that the *Académie* and the French scientific community allowed 'external' factors to shape their judgment at least as much as the internal scientific issues themselves.

The influence of external factors on Pouchet

But did such factors influence the actual scientific research of the contending parties themselves? Did either Pouchet or Pasteur permit external factors to shape their own scientific judgment? Insofar as Pouchet is concerned, the question in one sense reduces to whether his insistence on the orthodoxy of his version of spontaneous generation was sincere, or merely a façade behind which he hid unorthodox views on evolution, materialism, etc. In opposition to the latter view stands his background as a member of a well-known Protestant family and his detailed familiarity with scripture. His treatise on Albertus Magnus hardly seems the work of a covert atheist. But above all, his views on spontaneous generation, incorporated into a vitalistic, catastrophic view of nature are far too coherent and sustained to be dismissed as mere façade.

Throughout the 1860s, Pouchet repeatedly emphasized that his views were not those of an evolutionary atheist. In 1862, he wrote the geologist Desnoyers of his beliefs that 'geologists are necessarily heterogenesists', and that 'to prove that a given century is not and will not be able to produce new organisms, is tacitly to advance that in the past all is derived from a single and unique creation'.[44] This letter merely restated in forceful fashion the views in his book of 1859, although his geological arguments perhaps gained in relative significance as his experimental clash with Pasteur proceeded. He now claimed, for example, that the solution to the problem of spontaneous generation could come only from an examination of creation in its entirety and not from chemists examining the contents of their phials. In 1864, following the appearance of the French edition of the *Origin of Species*, he affirmed that 'the fixity of species, as put forward by Flourens, is the most important and best demonstrated fact in natural history, notwithstanding the assertions of Darwin'.[45] In 1865, he published his popular *L'univers, les infinement grands et les infiniment petits*, also based on catastrophic geology and the concept of the fixity of species. In 1868, in a preface to Pennetier's *L'origine de la vie*, he again insisted that heterogenesis followed logically from the doctrine of successive creations.

Only in 1870, two years before his death, did Pouchet seem to modify his views slightly, as revealed in a letter to Darius Rossi, one of the few French supporters of Darwin. In his book, *Le Darwinism et les générations spontanées*, Rossi sought to refute Flourens' notion that Pasteur had disproved spontaneous generation and thereby dealt a mortal blow to the Darwinian theory. Instead, Rossi argued that the denial of spontaneous generation was the work of scientists 'jealous of conserving certain immutable religious traditions that are believed to be endangered by taking the side of science'.[46] He concluded that heterogenesists must accept the transformation of species and Lamarck's explanation for the continual occurrence of primitive forms.

In his letter to Rossi (which the latter included in his book), Pouchet now agreed that spontaneous generation could be used in support of Darwinian theory, thus modifying his earlier insistence that heterogenesis was incompatible with transformism. But he did so without at all embracing the increasingly popular Darwinian theory – indeed, he declared his own continued belief in successive creations. At the same time, he reiterated his belief in spontaneous generation. By this point, if not long before, a politically motivated Pouchet almost certainly would either have openly endorsed evolutionism (if a radical) or have abandoned the presumably dangerous doctrine of spontaneous generation (if a conservative).

This is not to insist that Pouchet's version of spontaneous generation was entirely free of socio-political influences – indeed, his very eagerness to emphasize the orthodoxy of his views suggests otherwise. Nonetheless, he resisted to the end the temptation to modify the essential content of his scientific ideas themselves. Ultimately, one is most impressed by the tenacity with which Pouchet adhered to his belief in spontaneous generation in spite of its alleged threat to orthodox religious and political views which he seems fully to have shared. From this perspective, the picture of Pouchet that emerges is one more traditionally associated with Pasteur. T. S. Hall writes of Pasteur that he:

did not permit his religious convictions to influence his scientific conclusions. His canon, as a scientist, he firmly stated, was to stand aloof from religion, philosophy, atheism, materialism and spiritualism. . . . It was Pasteur's point that the material conditions of life had momentous implications for religion, but that these implications must not interfere with the objective interpretation of experimental results.[47]

Here we might equally well substitute the name of Pouchet. The question now is whether Pasteur's own response to the issue of spontaneous generation can also be described in these terms.

Pasteur's religious and political beliefs
If Pasteur entered the debate with Pouchet curiously divided in his own private scientific position on the issue of spontaneous generation, he also entered it secure in his political and religious convictions. In the latter areas, Pasteur's views conformed perfectly to the reigning orthodoxies of the Second Empire. From his father, who had served with distinction in the Napoleonic armies, Pasteur learned to hope for a day when France would regain the glory, internal stability and external prestige it had lost since the fall of the First Empire. As a symbol of that hope, Louis Napoleon enjoyed Pasteur's support from the beginning and held it to the end. Immediately after the coup d'état of 2 December 1851, Pasteur declared himself a 'partisan' of the new leader. During the 1860s, his support gained new force from personal acquaintance with the emperor, who encouraged his research – especially the more practical aspects of his studies on wine and on silkworm diseases – and who helped him to secure the facilities and financial resources needed to pursue it. In 1865, Pasteur enjoyed a lavish week as a personal guest of the emperor at one of his most elegant residences, the Palais de Compiègne. Three years later the emperor promoted Pasteur to Commander of the Imperial Order of the Legion of Honour – having already named him Chevalier in 1853 – and intervened to secure 60,000 francs toward the construction of a large new laboratory for him at the Ecole Normale Supérieure. The emperor's abdication in 1870 profoundly saddened Pasteur and nullified an imperial decree of 27 July 1870, by which he was to have been made a senator and awarded a national pension.[48]

In return for these and other expressions of Imperial favor, Pasteur dedicated his 1866 book on wines to the emperor and his 1870 book on silkworm diseases to Empress Eugénie. In 1867 he denounced and sought to expel from the Ecole Normale two students who had referred ironically and with seeming approval to an assassination attempt against the emperor while leading a student agitation in support of free speech. Although such organized political activity clearly violated the by-laws of the Ecole Normale, Pasteur's rigid, authoritarian handling of this and other disciplinary matters precipitated further student protest, culminating in the closing of the Ecole and in Pasteur's removal from the administrative posts he had held there.[49] As a matter of fact, this episode can serve to illustrate Pasteur's general preference for order and stability over free speech, civil liberty or even democracy, whose potential for anarchy and mediocrity he feared.

Despite his public insistence that he was apolitical, Pasteur readily acceded to the wishes of friends from Arbois that he run for the Senate in the election of 1876. He ran as a conservative, presenting himself as the champion of science and patriotism and making his central political pledge 'never to enter into any combination whose goal is to upset the established order of things'.[50] But lest that be taken as a commitment to the new republican form of government, he reminded the electorate that the Republic was by law a temporary experiment whose continued existence should depend upon its success at restoring domestic order and international prestige. Unable effectively to deny his close ties to the Second Empire and suspicions of his continued Bonapartist inclinations, Pasteur was overwhelmed in the election by two republican candidates.[51] He then withdrew entirely from the political arena, taking refuge in his scientific research so long as his strength for it remained. In the 1880s he described politics as sterile and ephemeral compared to science and at least twice during that decade declined appeals to try again for the Senate.[52] On a return visit to his native Arbois in 1888, Pasteur was jeered and in fact physically assaulted because of his unpopular political conservatism.[53] In short, Pasteur found the political climate of the Third Republic notably less congenial than that of the Second Empire.

Not surprisingly, Pasteur's religious and philosophical positions were of a piece with his political instincts. Although not quite so 'devout' a Catholic as is sometimes supposed, he had been reared in the teachings of the Church and died with the benefits of its last rites. He never doubted the existence of God or the immortality of the soul. Insisting on a sharp distinction between matters of knowledge and matters of faith, he described his own 'philosophy' as one 'entirely of the heart'.[54] Openly contemptuous of materialism, atheism and 'free thought', he also criticized positivism for its failure to take account of that 'most important of positive motions, that of the Infinite', one form of which is the idea of God.[55] In fact, of the traditional philosophical schools, only spiritualism (in the sense of anti-materialism) seems ever to have held any significant appeal for him. In his inaugural discourse of 1882 at the *Académie Française*, he somewhat cryptically suggested that his scientific work had perhaps contributed to the spiritualist cause, 'so neglected elsewhere, but certain at least to find a glorious refuge in your ranks'.[56] The conformity of Pasteur's political and religious views to those of the Second Empire scarcely requires further elaboration. More importantly, there exists explicit evidence that he consciously perceived spontaneous generation as a threat to those views.

Pasteur's Sorbonne lecture of 1864

On the evening of 7 April 1864 in a striking departure from his usual *modus operandi*, Pasteur gave a wide-ranging public lecture at the Sorbonne on spontaneous generation and its religio-philosophical implications.[57] His lecture was strategically timed, coming two months after the second commission on spontaneous generation had been named but two months before it actually met. The large audience consisted of the Parisian social, scientific and political elite, and there can be little doubt what they hoped and expected to hear. Pasteur did not disappoint them. He opened the lecture with a list of the great problems then agitating and dominating all minds: 'the unity or multiplicity of human races; the creation of man several thousand years or several thousand centuries ago; the fixity of species or the slow and progressive transformation of one species into another; the reputed eternity of matter . . .; and the notion of a useless God'. In addition to these questions – indeed transcending them all, since it impinged directly or indirectly on each of the others and since it alone could be subjected to experimental inquiry – was the question of spontaneous generation: 'Can matter organize itself? In other words, can organisms come into the world without parents, without ancestors? There's the question to be resolved.'

After a brief historical sketch of the controversy – in which his aim was to show that the doctrine of spontaneous generation 'has followed the developmental pattern of all false ideas' – Pasteur struck to the heart of the matter:

Very animated controversies arose between scientists, then as now – controversies the more lively and passionate because they have their counterpart in public opinion, divided always, as you know, between two great intellectual currents, as old as the world, which in our day are called *materialism* and *spiritualism*. What a triumph, gentlemen, it would be for materialism if it could affirm that it rests on the established fact of matter organizing itself, taking on life of itself; matter which already has in it all known forces! . . . Ah! If we could add to it this other force which is called life . . . what would be more natural than to deify such matter? Of what good would it then be to have recourse to the idea of a primordial creation, before which mystery it is necessary to bow? To what good then would be the idea of a Creator – God? . . .

Thus, gentlemen, admit the doctrine of spontaneous generation, and the history of creation and the origin of the organic world is no more difficult than this. Take a drop of sea water . . . which contains some nitrogenous material, some sea mucus, some 'fertile jelly' as it is called, and in the midst of this inanimate matter, the first beings of creation take birth spontaneously, then little by little are transformed and climb from rank to rank – for example, to insects in 10,000 years and no doubt to monkeys and man at the end of 100,000 years.

Do you now understand the link which exists between the question of spontaneous generations and those great problems I listed at the outset?

But if Pasteur thus explicitly recognized the religio-philosophical implications of the spontaneous generation controversy, he hastened to deny that his scientific work had been motivated or influenced by such concerns:

But, gentlemen, in such a question, enough of poetry . . ., enough of fantasy and instinctive solutions. It is time that science, the true method, reclaims its rights and exercises them.

Neither religion, nor philosophy, nor materialism, nor spiritualism belongs here. I may even add: as a scientist, it does not much concern me. It is a question of fact. I have approached it without preconceived idea, equally ready to declare – if experiment had imposed the view on me – that spontaneous generations exist as I am now persuaded that those who affirm them have a blindfold over their eyes.

Pasteur then reviewed what he considered to be 'the most important experiments' on both sides of the issue. Crucial among these on the side of the heterogenesists were Pouchet's initial experiments on boiled hay infusions exposed only to artificial air or oxygen under mercury. Pasteur described these experiments as irreproachable in every regard except one. Pouchet had failed to recognize that ordinary laboratory mercury is itself filled with germs, as could be demonstrated by dropping a globule of this mercury into a fermentable medium in an atmosphere of calcined air. Within two days, the medium swarmed with a variety of microorganisms. But if the same experiments were conducted with precalcined mercury, not a single microbe appeared.

Having thus disposed of what he claimed to be the most persuasive evidence in favor of heterogenesis, Pasteur rehearsed his own famous experiments in which yeast-water was prevented from alteration by denying the access of any atmospheric dust. By depriving the medium of germs from the air, wrote Pasteur, 'I have removed from it the only thing that it has not been given to man to produce . . . I have removed life, for life is the germ and the germ is life.' Indeed, the simple unqualified conclusion to be drawn from his experiments in the French Alps was that 'spontaneous generation does not exist'. Thus far Pasteur had merely repeated the substance of his prize-winning memoir of 1861. Toward the end of his lecture, however, he informed his Sorbonne audience of yet another crucial blow he had more recently delivered against the doctrine of spontaneous generation. Until 1863 Pasteur had relied solely upon experiments in which organic infusions had been vigorously heated, leaving him vulnerable to the charge that such high temperatures might modify the media in such a way as to destroy any vital force it contained and thereby render it incapable of generating life.[58] But in March 1863, he had finally succeeded in preserving two highly alterable natural liquids – blood and urine – without heating them at all, but rather merely by collecting them directly and hermetically from the veins or bladder of healthy animals and then exposing them only to germ-free air.[59] Here, then, was another powerful proof that 'spontaneous generation is a chimera'.

Pasteur and the 'experimental method'

By this impressive experimental critique of heterogenesis, Pasteur sought to justify his claim that the question was one of fact, which he had approached 'without preconceived idea'. If so, one can only marvel at some oddly inconsistent aspects of his 'scientific' behavior and some surprising lapses in his supposedly pure commitment to the precepts of the 'experimental method'. At one point in his prize-winning memoir of 1861, Pasteur admitted that his own repeated attempts to prevent the appearance of microbial life in infusions under mercury succeeded only rarely, perhaps less than 10% of the time. But rather than draw the seemingly obvious conclusion that this microbial life had originated spontaneously, Pasteur refused to accept the experimental evidence at face value and pressed relentlessly toward an alternative explanation. 'I did not publish these experiments', Pasteur wrote, 'for the consequences it was necessary to draw from them were too grave for me not to suspect some hidden cause of error in spite of the care I had taken to make them irreproachable.'[60] Although Pasteur failed to specify what 'grave consequences' he feared, it seems obvious that the very existence of spontaneous generation must have been chief among them. As a matter of fact, throughout the spontaneous generation controversy, Pasteur virtually *defined* as 'unsuccessful' any experiments – including his own – in which life mysteriously appeared and as 'successful' any experiments which gave an opposite result. Happily for him, he managed to indict contaminated mercury as the source of microbial life which appeared in the many 'unsuccessful' experiments conducted with the mercury trough.

If this achievement seems to justify Pasteur's approach – if indeed it might even be used to emphasize his allegiance to the precept that one should suspend judgment until 'all facts are in' – no such interpretation can be applied to other aspects of his scientific behavior. Most remarkable among these is his apparent failure to repeat Pouchet's disputed experiments in the Pyrenees. The most striking feature of these experiments was the *absence* of mercury from Pouchet's flasks of boiled hay infusions. In his 1864 Sorbonne lecture, Pasteur entirely ignored this problem, choosing to discuss only Pouchet's earlier experiments with mercury. Only once did Pasteur attempt directly to challenge Pouchet's experiments in the Pyrenees. In a note of November 1863, he criticized Pouchet and his collaborators for limiting their flasks to so small a number as eight (thereby introducing the possibility that their results were due to mere 'chance') and on the really quite desperate ground that they had broken their sealed flasks with a heated file rather than a pair of pincers as Pasteur had.[61] Not even before the commission appointed to adjudicate their dispute did Pasteur repeat or directly refute Pouchet's experiments. Instead, he chose merely to repeat his own secure experiments with *yeast* infusions, in spite of which the commission praised his exactitude in a report which scarcely veiled its contempt for the opposite side.[62]

That Pasteur should thus have violated one of the presumably fundamental precepts of the 'experimental method' – namely, the duty to 'falsify' his opponents' experiments – seems no less remarkable than the failure of any member of the commission to perceive the violation. In the case of spontaneous generation, moreover, this violation was particularly serious since a single uncontroverted experiment in support of the doctrine automatically carried greater weight than any number of experiments against it. Advocates of spontaneous generation did not need to show that they could produce life artificially under a variety of circumstances, nor even that they could do so consistently. They needed only to show that the feat was possible – a situation Pasteur himself later recognized by emphasizing how difficult his task was compared to the heterogenesists and by noting that 'in the observational sciences, unlike mathematics, the absolutely rigorous demonstration of a negation [namely, that spontaneous generation does not exist] is impossible'.[63]

And in fact, as Duclaux emphasized long ago,[64] the Pasteur–Pouchet debate might have ended quite differently had Pasteur carefully repeated Pouchet's experiments, or had Pouchet and his collaborators maintained their nerve in the face of Pasteur's self-assurance and the commission's contempt. Thanks chiefly to continued experimental work outside of France, where scientists were relatively isolated from the presumed social dangers of spontaneous generation and from judiciary commissions of the Parisian *Académie des Sciences*, it became clear by the early 1870s that microbial life often appeared in boiled hay infusions even in experiments conducted with irreproachable exactitude and Pasteur-perfect technique. In 1876, the German botanist Ferdinand Cohn and the English physicist John Tyndall were able to offer an explanation for such cases of seeming spontaneous generation. They showed that the life cycle of the hay bacillus (Cohn's *Bacillus subtilis*) included a highly resistant endospore phase which could survive boiling and develop into the usual form of the bacillus upon the introduction of oxygen.[65] For this reason, Pouchet's flasks of boiled hay infusions might well have produced life even in Pasteur's hands upon exposure to the atmosphere, and might therefore have lent crucial support to spontaneous generation during the 1860s.

Curiously enough, Pasteur himself had argued as early as 1861 that the heat-resistance of certain microbes increased in alkaline media[66] (hay infusions are alkaline), and in his Sorbonne lecture of 1864 briefly raised the possibility that Pouchet's hay infusions might contain some unknown heat-resistant microorganism.[67] But he mentioned this possibility only in passing and seemed fully satisfied that Pouchet's precautions were sufficient to preclude it. For Pouchet's early experiments, this posed no serious problems for Pasteur, since he was able to indict contaminated mercury as the cause of the supposedly 'spontaneous' appearance of microbial life in Pouchet's flasks. But this explanation could not be applied to the mercury-free Pyrenees experiments of Pouchet and his collaborators. Perhaps because he remained satisfied with Pouchet's preliminary precautions, Pasteur did not now even mention the possibility that his opponents' media may have contained some unknown heat-resistant microorganism from the outset. Once again, but now on highly dubious grounds, he preferred instead to accuse them of having contaminated their flasks through sloppy technique.[68] If Pasteur ever did repeat Pouchet's experiments without mercury, he kept the results private.

The influence of external factors on Pasteur
In the absence of corroborating evidence, one might be reluctant to draw the conclusion that Pasteur's experimental work was influenced – consciously or unconsciously – by socio-political factors. But another body of evidence reinforces this conclusion. For throughout the 1860s – the politically most sensitive phase of the spontaneous generation debate – Pasteur gave no public indication of the scientific dilemma he faced in his own mind. While privately convinced that asymmetric forces might hold the key to the artificial production of life, he gave no public voice to this conviction, emphasizing instead the experimental errors of all who affirmed spontaneous generation. Never during this period did he refer in public to his own experimental attempts of the 1850s to 'imitate nature' by creating asymmetry or life. Perhaps something besides his initial lack of success and Biot's discouraging advice led Pasteur temporarily to abandon such experiments and to focus his energies elsewhere throughout the most sensitive phase of the debates in France over Darwinism and spontaneous generation.

In any case, Pasteur's public posture underwent a gradual and subtle change after the collapse of the Second Empire. Perhaps in some sense foreshadowing this public shift was his return, in the very year of Louis Napoleon's abdication, to those private experiments on asymmetry and life which he had abandoned nearly twenty years before. In manuscript notes written at Arbois in autumn 1870, while Paris was embroiled in the Commune, Pasteur speculated about the origin of life and projected a new series of experiments designed to create or modify life by means of magnets or other asymmetric influences.[69] If these experiments were ever carried out, their results never became public property. Soon after projecting these new attempts to create life, Pasteur began to modify his public position on spontaneous generation, but this shift took place so gradually and so subtly as to be virtually imperceptible. For the most part, in fact, he continued his uncompromising campaign against heterogenesis, and in 1872 even discovered a new political mode of discrediting his opponents. Now describing spontaneous generation as a 'German' theory, he impugned the patriotism of those Frenchmen who dared to defend it in the wake of the recent Franco–Prussian War.[70] By the mid-1870s, however, the tone of Pasteur's opposition seems rather less decisive. To his earlier bald assertions that spontaneous generation does not exist or is a chimera, he began to add such qualifying phrases as 'in the current state of science' and now sometimes confessed that it was impossible to disprove spontaneous generation experimentally.[71] In one discussion of February 1875, he even described as a 'capital discovery which perhaps contains the solution of spontaneous generation' Albert Bergeron's discovery of microorganisms in human abcesses which had never been exposed to the external air.[72] Even though that remark appears in a context which reveals Pasteur's doubt that Bergeron had in fact observed spontaneous generation, one may reasonably wonder whether he would even have uttered it during the 1860s.

Rather more positive evidence of Pasteur's public shift to spontaneous generation emerges from his discussions of asymmetry and life in the 1870s and 1880s. For he began at last to develop and articulate his previously tentative notion of asymmetric forces – forces about which he had been utterly silent since introducing them in his famous lecture of February 1860. He now made it clear, as he had not before, that these asymmetric forces should be considered within the bounds of experimental inquiry and as the means by which the barrier between asymmetric (living) and symmetric (dead) nature might actually be crossed. Such an achievement, he wrote in 1874, 'would give admittance to a new world of substances and reactions and probably also of organic transformations'. 'It is there . . .', he continued, 'that it is necessary to place the problem not only of the transformation of species but also of the creation of new ones.'[73] In a note of July 1875, he restated this position in essentially identical terms.[74]

By 1883, when Pasteur admitted for the first time in public that he had attempted thirty years before to 'imitate nature' and to produce 'the immediate, essential principles of life', he had apparently begun to look rather more favorably upon the Third Republic. In June 1880, writing to a friend of a governmental increase of 50,000 francs in his annual research budget, Pasteur declared, 'Finally, I have dealt with an amiable and confident Republic.'[75] In his lecture of 1883, delivered to the Chemical Society of Paris, Pasteur gave a brief description of his early **experiments with magnetism and light at Strasbourg and Lille and** suggested that to create life, 'it is necessary to manufacture some asymmetric forces, to resort to the actions of a solenoid, of magnetism, of the asymmetric movements of light . . .'.

The line of demarcation of which we speak is not a question of pure chemistry or of the obtaining of such or such products. It is a question of forces. Life is dominated by asymmetric forces which present themselves to us in their enveloping and cosmic existence. I would even urge that all living species are primordially, in their structure, in their external form, functions of cosmic asymmetry. Life is the germ and the germ life. Now who can say what the *destiny* of germs would be if one could replace the immediate principles of those germs –

albumin, cellulose, etc., etc. – by their inverse asymmetric principles? The solution would consist in part in the discovery of spontaneous generation, if such is within our power: on the other hand, in the formation of asymmetric products with the aid of the elements carbon, hydrogen, nitrogen, sulphur, phosphorus, if in their movements these simple bodies may be dominated at the moment of their combination by asymmetric forces. Were I to try some asymmetric combinations from simple bodies, I would make them react under the influence of magnets, solenoids, elliptically polarized light – finally, under the influence of everything which I could imagine to be asymmetric actions.[76]

How different, how almost 'materialistic', is Pasteur's tone here compared to his Sorbonne lecture of 1864! In fact, after this account of Pasteur's shifting and inconsistent views on spontaneous generation, does he any longer conform to that standard picture of the man who, in the words of T. S. Hall, 'did not permit his religious convictions to influence his scientific conclusions', and who did not allow external factors 'to interfere with the objective interpretation of experimental results'? Let it be clear from the outset that the answer to this question is subtle and complex. On the one hand, Pasteur's religious and political views obviously did not prevent him from conceiving of the possibility of spontaneous generation. As a matter of fact and in spite of its tortured quality, his divided position on spontaneous generation can be seen as having its genesis entirely in his scientific research – for his work in crystallography led him to believe in the possibility of abiogenesis, while his study of fermentation led him on equally scientific grounds to discount the possibility of heterogenesis. On the other hand, a highly persuasive – if not absolutely decisive – case can be made that Pasteur allowed 'external' factors to influence the timing and even the content of his public pronouncements on the spontaneous generation issue and, more remarkably, to affect his experimental work itself. It is hard otherwise to explain his total silence about asymmetric forces during the 1860s; it is hard otherwise to understand how during that same period he could have led a frontal assault against a doctrine in which privately he partly believed; it is hard otherwise to explain why he limited his published experimental work to those areas and to those media which he had good reason to believe would discredit that doctrine; and it is hard otherwise to imagine that so justly celebrated an experimentalist would have minimized the significance of Pouchet's experiments in the Pyrenees and failed in his duty to refute them.

Conclusion

Remarkably enough, we are led to a conclusion precisely the opposite of that usually attached to the Pasteur–Pouchet debate. For we are persuaded that external factors influenced Pasteur's research and scientific judgment more powerfully than they did the defeated Pouchet. Having formulated his version of spontaneous generation prior to the politically significant Darwinian controversy in France, Pouchet maintained his views with striking consistency in spite of their presumed threat to orthodox religious and political beliefs which he fully shared. By contrast, Pasteur's public posture on the issue seems to reveal a quite high degree of sensitivity to reigning socio-political orthodoxies. This is not necessarily to claim that Pasteur behaved this way deliberately or consciously. Nor is it to claim that unfavorable social circumstances robbed Pouchet of a victory that rightly belonged to him. For the fact remains that Pasteur was a more ingenious and more skillful experimentalist as well as a more persuasive advocate for his point of view [. . .]

Nonetheless, because Pasteur is so generally considered a leading exponent of the true 'experimental method', and because microbiology is so widely assumed to have attained maturity under his influence, the impact of socio-political factors on his scientific work deserves special emphasis. Indeed, by focusing so much on Pasteur we hope to have communicated a fairly clear sense of the specific manner in which external factors affected the actual experimental work and theoretical positions of one whose allegiance to 'the experimental method' has rarely if ever been challenged.

In writing this paper, we have quite naturally considered the possible influence of external factors on our own interpretation of the spontaneous generation debate, and it seems only proper to examine this issue in closing. Especially because we find distasteful many of Pasteur's religious, political and personal attitudes, our interpretation may well differ from that of historians and scientists of more conservative persuasions. In the end, the judgment one applies to an issue as subtle and complex as this depends on nuances and matters of emphasis in the interpretation of all the available evidence. In such matters there is no *experimentum crucis*. Nonetheless, we have subjected our personal beliefs to as critical a scrutiny as possible and we emerge convinced that the balance of evidence suggests that the denigrated Pouchet was at least as 'objective', if not more so, than the triumphant Pasteur. If we have championed Pouchet at the expense of Pasteur, it must be seen chiefly as a response to earlier distortions of the historical record so extreme as to approach caricature.

Source: Bulletin of the History of Medicine, **48** (1974), No. 2: 161–98.

Notes

1. See esp. William Bulloch, *The History of Bacteriology.* London, Oxford Univ. Press, 1938: 92–106; Emile Duclaux, *Pasteur: the History of a Mind,* trans. Erwin F. Smith and Florence Hedges. Philadelphia, W. B. Saunders, 1920: 85–111; and René Dubos, *Louis Pasteur: Freelance of Science.* Boston, Little, Brown, 1950: 165–77.
2. Félix A. Pouchet, *Hétérogenie, ou traité de la génération spontanée.* Paris, 1859: vi. This passage is quoted in Bulloch (n. 1): 92–3; Duclaux (n. 1): 104–5; and Dubos (n. 1): 165.
3. Although assertions of social influence on the conceptual content of mature science are neither new (having long been associated with Marxist or quasi-

Marxist historians) nor rare (indeed they are again quite fashionable), cogent demonstrations of such assertions have been conspicuous by their absence. A notable exception is Paul Forman's recent study of the adoption of acausality by physicists in Weimar, Germany. See P. Forman, 'Weimar culture, causality, and quantum theory, 1918–1927: Adaptation by German physicists and mathematicians to a hostile intellectual environment', *Hist. Studies in Phys. Sci.*, **3** (1971): 1–115.

4. For details of this period see J. Farley, 'The spontaneous generation controversy (1700–1860): The origin of parasitic worms', *J. Hist. Biol.*, **5** (1972): 95–125.

5. For examples of Geoffroy's attempts to defend himself from charges of allegiance to *Naturphilosophie*, materialism and impiety, see *Notions synthetiques*: 26, 33, 82, 110; *Comptes Rendus*, **5** (1837): 183–94; *ibid.*, **7** (1839): 489–91; and 'Hérésies panthéistiques,', *Dictionnaire de la conversation et de la lecture*, **31** (1836): 481ff.

6. Our brief summary of the politico-theological issues during the Second Empire derives chiefly from D. G. Charlton. *Secular Religions in France, 1815–1870.* Oxford, University Press, 1963; A Dansette, *Religious History of Modern France.* New York, Herder and Herder, 1961; and G. Wright, *France in Modern Times.* Chicago, Rand McNally, 1966.

7. Quoted in Dansette (n. 6): 311.

8. Clemence Royer, *De l'origine des espèces par sélection naturelle.* Paris, 1862. Details of the French Darwinian debate and its association with the issues of spontaneous generation are given in J. Farley, 'The initial reactions of French biologists to Darwin's *Origin of Species*', *J. Hist. Biology*, **7** (1974): in press.

9. Félix Pouchet (1800–72) was the son of a highly respected Rouen manufacturer. After obtaining his M.D. from Paris in 1827, he accepted the directorship of the Natural History Museum of Rouen and also filled the chair of zoology at the local medical preparatory college. In 1843 he received the Legion of Honour.

10. F. Pouchet, *Hétérogenie*: 7–9.

11. Ibid.: 97–8.

12. F-J. Pictet, *Traité de paléontologie ou histoire naturelle des animaux fossiles.* 2nd ed. Paris, 1853: 93.

13. Pasteur Vallery-Radot, (ed.), *Oeuvres de Pasteur* (hereafter cited as *OP*), 7 vols. Paris, 1922–39, II: 3–13.

14. Ibid., II: 51–126.

15. Cf. Duclaux (n. 1): 86–7.

16. See esp. Dorian Huber, 'Louis Pasteur and Molecular Dissymmetry: 1844–1857'. Unpublished MA thesis. Johns Hopkins University, 1969.

17. Ibid., I: 341.

18. Ibid., I: 337.

19. Ibid., I: 337.

20. Ibid., I: 341.

21. Ibid., I: 342.

22. Ibid., I: 343.

23. Ibid., I: 376.

24. See ibid., VII: 23. Although these manuscript notes date from 1870, the experiments they project are so similar in conception to those of 1852, that it seems fair to conclude – in the absence of contradictory evidence – that

they are essentially identical. It is particularly noteworthy that Pasteur refers to the Ruhmkorff (magnetic) apparatus in these 1870 notes, as he does when first describing his Strasbourg experiments of 1852. Cf. ibid., I: 376.

25. See Pasteur Vallery-Radot, (ed.), *Pages illustres de Pasteur*. Paris: Hachette, 1968: 10–11.

26. See Pasteur, *Correspondance*, 4 vols. Paris, B. Grasset, 1940–51, 1: 326.

27. Ibid., 1: 324.

28. F. Pouchet, 'Note sur des proto-organismes végétaux et animaux, nés spontanément dans l'air artificiel et dans le gaz oxygène', *Comptes Rendus*, **47** (1858): 979–84.

29. *OP*, II: 34–6.

30. Ibid., II: 628–30.

31. Ibid., II: 246.

32. For lucid accounts of these experiments, see Bulloch (n. 1): 96–102; and Duclaux (n. 1): 85–104.

33. *OP*, II: 202–5.

34. F. A. Pouchet, N. Joly, and Ch. Musset, 'Expériences sur l'hétérogenie exécutées dans l'intérieur des glaciers de la Maladetta', *Comptes Rendus*, **57** (1863): 558–61.

35. See Georges Pennetier, *Un débat scientifique, Pouchet et Pasteur (1858–68)*. Paris, 1907: 10.

36. Ibid.

37. M. J. P. Flourens. *Comptes Rendus*, **57** (1863): 845.

38. Duclaux (n. 1): 107.

39. On this episode, see Bulloch (n. 1): 103–5; Duclaux (n. 1): 104–9; Pennetier (n. 35): 10–12; and *OP*, II: 321–7, 637–47.

40. Pennetier (n. 35): 12. In 1877, H. Charlton Bastian also complained about the biased composition of the *Académie* Commission appointed in that year to adjudicate between him and Pasteur on spontaneous generation. See H. C. Bastian, 'The Commission of the French Academy and the Pasteur–Bastian Experiments', *Nature* (1877), **16**: 277–9.

41. M. J. P. Flourens, *Examen du livre de M. Darwin sur l'origine des espèces*. Paris, 1864: 170.

42. See Farley (n. 8).

43. E. Faivre, 'La question des générations spontanées', *Mémoires de l'Académie des Sciences, Belles Lettres et Arts, Lyon*, (1860), **10**: 172.

43. E. Faivre, 'La question des générations spontanées', *Mémoires de l'Académic des Sciences, Belles Lettres et Arts, Lyon*, (1860), **10**: 172.

44. F. Pouchet, *Les créations successives et les soulèvements du Globe. Lettres à M. Jules Desnoyers*. Paris, 1862, letter 1, 5 Jan. 1862.

45. F. Pouchet, *Nouvelles expériences sur la génération spontanée et la résistance vitale*. Paris, 1864: 199.

46. Darius C. Rossi, *Le Darwinisme et les générations spontanées*. Paris, 1870: vi.

47. T. S. Hall, *Ideas of Life and matter*. Chicago, Univ. of Chicago Press, 1969, 2: 294.

48. See, *inter alia*, Louis Pasteur, *Correspondance* (n. 26), I: 228, 230; II: 216–36, 345–51, 368, 375, 385, 484–5, 489, 567–8.

49. See ibid., II: 136–42, 332–39; and Pasteur Vallery-Radot, *Pasteur inconnu*. Paris, Flammarion, 1954: 36–58.

50. Pasteur, *Correspondance* (n. 26), II: 612.
51. On this campaign and election, see ibid., II: 611–30; and Pasteur Vallery-Radot, *Pasteur inconnu* (n. 49): 203–15.
52. Pasteur, *Correspondance* (n. 26), II: 424; IV: 34, 268–70.
53. See E. Ledoux, *Pasteur et la Franche-Comté; Dole, Arbois, Besançon*. Besançon, Chaffanjon, 1941: 55ff.
54. Pasteur, *Correspondance* (n. 26), II: 213–14.
55. See *OP*, VII: 338.
56. Ibid.: 326–39, quote on: 326.
57. For the text of this lecture, which the next several pages analyze in considerable detail, see ibid., II: 328–46.
58. Cf. Pasteur, *Correspondance* (n. 26), II: 134.
59. *OP*, II: 345–6,
60. Ibid., II: 236.
61. Ibid., II: 321–3.
62. See ibid., II: 637–47.
63. See ibid., II: 459; VI, 25fn. 1, 41, 54, quote on: 54.
64. See Duclaux (n. 1): 108–11.
65. See Glenn Vandervliet, *Microbiology and the Spontaneous Generation Debate during the 1870's*. Lawrence, Kans., Coronado Press, 1971: 43–54.
66. *OP*, II: 253–9.
67. See ibid., II: 337.
68. Ibid., II: 321–3.
69. See ibid., VII: 21–9.
70. Ibid., II: 379, 396–7.
71. See ibid., II: 459; VI: 25fn. 1, 41, 54, 57. But cf. ibid., VI: 59–60.
72. Ibid., VI: 20.
73. Ibid., I: 362.
74. Ibid., I: 364–5.
75. See Pasteur, *Correspondance* (n. 26), III: 140.
76. Ibid., I: 377–8.

Chapter 3

3.0 Introduction

Both articles in this chapter deal with an aspect of the history of the human sciences. The inclusion of these articles could be seen as a recommendation that 'mainstream' historians of science embrace studies of that kind (see the first section of Robert M. Young's article). Independently of that issue, it also happens that each article relates science and belief, even in the former's narrower sense of the physico-chemical and biological sciences.

Barbara S. Heyl's paper deals with a group of Harvard scholars united in the 1930s and early 1940s by a common interest in the sociological theories of the Italian Vilfredo Pareto (1848–1923). She concentrates on 'interaction networks' and 'patterns of influence' among four men, Crane Brinton, Lawrence J. Henderson, George C. Homans and Talcott Parsons, and establishes that Henderson, a physiologist, was dominant. For our purposes, two points stand out, (i) that several of the Harvard circle adopted Paretan sociology as a defence against Marxist notions prevailing on campus during the Great Depression, and (ii) that Pareto's sociology appealed especially to Henderson because it applied scientific methods of analysis to social phenomena. Henderson himself drew up a detailed analogy between Willard Gibbs' physico-chemical system and the idea of a 'social system', and compared the central notion of 'social equilibrium' to the body's self-healing powers, once known as *vis medicatrix naturae*.

In Robert M. Young's paper, we return to a theme already canvassed in Chapter 2, the Victorian debate on man's place in nature. This time we look at it from the point of view of the evolutionists, rather than the theologians and theologian-scientists. Young starts from the retro-spectively surprising fact that during the course of the debate, none of the supporters of evolutionary theory appealed to associationist psychology, phrenology or brain physiology to counter the objections of the anti-evolutionists. He is unable to explain this omission and compares it with the way in which for much of the debate geological evidence of man's antiquity was ignored; but he argues that at another level, psychology was at the heart of the debate. The point is made by tracing Darwin's debt to the associationist pleasure–pain theory of learning mediately through the influence of Malthus, and Malthus' own relationship with the Utilitarians. More direct links with either associationism or phrenology are put forward in the case of Wallace and of Spencer. Accordingly, Young concludes that evolutionary theory can in many respects be seen as 'applied psychology'.

3.1 Barbara S. Heyl, The Harvard 'Pareto Circle'

The concepts of social system and social equilibrium have been key ideas in sociological theory and have been debated from pre-Comtian times to the present. But these concepts enjoyed a special importance during the 1930s and early 1940s for a group of scholars at Harvard. These men were interested in Vilfredo Pareto, whose sociological writings were based on a mechanical model of society 'as a system of mutually interacting particles which move from one state of equilibrium to another'.[1] The group consisted of such men as George C. Homans, Charles P. Curtis, Jr., Lawrence J. Henderson, Joseph Schumpeter, Talcott Parsons, Bernard DeVoto, Crane Brinton, and Elton Mayo.[2] This paper concentrates on four of these men – Homans, Henderson, Parsons, and Brinton – in an effort to study in depth the sources of their conceptions of society as a system with a built-in homeostatic tendency.

The four men under study had all read Pareto's sociological writings, and his ideas became an integral part of their scholarly production during the 1930s. Two of them, Henderson and Homans, wrote books which dealt solely with Pareto's writings.[3] A third, Parsons, wrote a lengthy analysis of Pareto's ideas in *Structure of Social Action*.[4] And Brinton wrote two books in the 1930s in which he relied heavily on Paretan ideas.[5] In addition, three of the men – Parsons, Henderson and Brinton – wrote articles during the 1930s which were either explicitly on the works of Pareto or made reference to him.[6] A later part of this paper will trace the use of the specific social system, and equilibrium concepts from Pareto in the writings of the four scholars.

The impact of Pareto's ideas on these men was clearly important, not only as an impetus for the use of social system concepts, but also as a foundation for much of their more general theoretical thinking. But while attempting to uncover the intellectual influences on the four men, one should ask why Pareto won such a warm reception at Harvard during this time.

One aspect of the university climate during the 1930s was the widespread popularity of large-scale historical frameworks employed to describe socio-political phenomena. The Marxist framework was particularly in vogue. But Pareto's writings, too, posited such a grand historical theory, and, moreover, it seemed to provide an alternative to the Marxist approach.[7] One of the four men under study, Homans, has made explicit that one reason for his enthusiasm for Pareto was precisely the fact that Pareto gave him a response to the Marxists during this period. Homans writes:

As a Republican Bostonian who had not rejected his comparatively wealthy family, I felt during the thirties that I was under personal attack, above all from the Marxists. I was ready to believe Pareto because he provided me with a defense.[8]

Crane Brinton, too, indicated that the group of Harvard faculty members keenly interested in Pareto was somewhat harassed by leftist groups. He has recently recalled:

At Harvard in the thirties there was certainly, led by Henderson, what the then Communists or fellow-travelling or even just mild American style liberals in the University used to call 'the Pareto cult'. The favourite smear phrase for Pareto . . . was 'Karl Marx of the bourgeoisie'. The Pareto cult was never one that influenced a majority of the faculty, but it had fairly wide repercussions.[9]

Henderson, whose method in discussion was said to be only 'feebly imitated by the pile-driver',[10] was apparently often in debate with people he felt over-emphasized the rationality in men's behavior – people he called 'liberals'.

In his warfare with these liberals Henderson would use Paretan, indeed Machiavellian, terms which seemed to cast doubt on his own belief in any real goodness in men, in any validity for the great traditions of American democracy. The liberals, of course, replied with their then favorite word of abuse – 'fascist'.[11]

It appears, then, that the Marxists and the Paretans frequently confronted one another at Harvard in the 1930s. Moreover, it is likely that others, with Homans, embraced Pareto's ideas, if not in reaction to the prevailing Marxist notions on campus, then as a useful defense in debate against them.

Another source of the enthusiasm for Pareto felt by the four scholars discussed here – Homans, Henderson, Parsons and Brinton – was their contact with one another at Harvard in this period. If a group of men share an interest in the ideas of one man, and if they meet occasionally to discuss those ideas, their interest in their subject will tend to grow. Thus, the theses of this paper are that the reliance on social system and equilibrium concepts developed through the interaction among these men, and more specifically, that a careful reading of their works, combined with a study of their friendships during the 1930s, reveals that the individual who most influenced Homans, Brinton and Parsons with regard to social system concepts was Lawrence J. Henderson.

The paper has three parts. The first establishes that the four scholars met periodically during the 1930s and early 1940s to 'talk shop', and also that they had read one another's writings. The second part traces the use of the concepts of social system and equilibrium and the analogies employed by the authors in their major works during this period. Finally, in the third part, the characteristic approaches of these authors are compared to determine whether Henderson was indeed the one who shaped the thinking of the other scholars regarding these concepts.

I. Interaction within the 'Pareto Circle'

Membership in at least three social–intellectual associations brought these four men into regular contact during the 1930s. The first time they came together for frequent discussions on Pareto was in the fall of 1932 when Lawrence Henderson organized a seminar on Pareto's sociology. Henderson was a physiologist; by the time he was 48, he was well established in his field and had proven to be a tough-minded methodologist. At about this time he read Pareto's sociological writings and was extremely impressed with the Italian's efforts to apply scientific methods of analysis to social phenomena. From 1926 to 1932 Henderson read carefully the bulk of Pareto's work and became convinced that his approach was of momentous importance. He decided to conduct a seminar on Pareto, and early in the 1932 academic year he began recruiting faculty members to be participants. Schumpeter, DeVoto, Brinton and Parsons all came. Henderson's good friend Charles P. Curtis, Jr., a Boston lawyer and a Harvard Fellow, gave him some assistance in running the course. Henry Murray, Clyde Kluckhohn, and Robert Merton also attended, and Henderson personally invited Homans, who was in his first year of graduate school, to join the seminar.[12] Henderson conducted the seminar for two years, from 1932 to 1934.[13] Apparently, most of the same men attended regularly throughout its duration.

The seminar was particularly important for Homans. It was the beginning of his friendship with Henderson, and that friendship meant the beginning of his career as a sociologist. Homans had studied English literature as an undergraduate at Harvard with Bernard DeVoto as his tutor. DeVoto was a friend of Henderson and had assigned Homans a large portion of Pareto's sociological works to read in his senior year. DeVoto then introduced Homans to Henderson who later issued the invitation to join the seminar. As the first part of the seminar wore on, Curtis, who was not only a good friend of Henderson's but of Homans' family as well, asked Homans to collaborate with him in writing a book introducing Pareto's main theoretical framework to an English-speaking audience. Homans accepted, and the book appeared in 1934.[14] In the preface Homans and Curtis acknowledge the importance of the seminar and Henderson for their work.

This book is the first fruits of the Seminar on the Sociology of Pareto which Lawrence J. Henderson had conducted at Harvard for the last two years. Nearly everything it says which is not Pareto's and which is of any importance we have parroted from him.[15]

Parsons also attended the seminar regularly.[16] It is difficult to determine the impact of the seminar on Parsons, but it surely was important as his first intensive encounter with both Pareto and Henderson.[17] There is no mention of Pareto in Parsons' early writings, but after the seminar in 1932–34 references to Pareto appear frequently. The steady contact in the seminar fostered in Parsons and Henderson a concern for each other's work. When Henderson's book on Pareto appeared in 1935, the year after the seminar was completed, Parsons gave it a very complimentary, though brief, review in the *American Economic Review*.[18] Two years later, when Parsons published his *Structure of Social Action*, he acknowledged Henderson's generous help with regard to the analysis of Pareto's ideas.[19] Parsons had intensive contact with Henderson prior to the book's publication, primarily because Henderson had been asked to read the manuscript. Parsons recalls that Henderson took this with 'uncommon seriousness':

As a result we had a series of very long talks. These lasted for perhaps five or six months . . . It was quite a training in close thinking for me. I undertook major revisions after these talks.[20]

Crane Brinton attended the seminar, but not as steadily as Homans and Parsons. He presented at least one historical 'case' to the group each year. For Brinton the seminar may have been an important source of information about Pareto, as it was for the other two scholars. But unlike them the seminar did not serve as a steppingstone to friendship with Henderson – for Brinton already had such a friendship. That friendship dated back to 1918 when Henderson, as Chairman of the Bowdoin Prize Committee, had awarded the prize to Brinton in his junior year at Harvard. Henderson thought highly of Brinton's essay and invited the student to dinner. Brinton adds: 'I never lost touch with him after that.' Except for the years when Brinton was abroad – in 1919–23 and 1927–28 – he saw Henderson 'fairly often, dining with him, or visiting at the Henderson summer home in Morgan Center, Vt'.[21]

By 1936 the friendship with Henderson was not only still alive but manifested itself in Brinton's work. For in that year Brinton, in two of his books, referred to Henderson in prefatory remarks: one acknowledged that Henderson had introduced Brinton to Pareto's ideas[22]; and the second cited Brinton's debt to the biochemist for giving the historian not only the topic for historical study, but help in dealing with it. Brinton wrote:

I wish to thank . . . Professor L. J. Henderson for arousing my first interest in Talleyrand, and for constant critical attention to the details and to the generalizations of this study. . . .[23]

One year later Brinton became a member of the exclusive and prestigious Saturday Club of Boston. The Saturday Club had been founded in 1856 by Ralph Waldo Emerson and a group of important Bostonians; the function of the club was to bring together its members on the last Saturday of every month for a leisurely dinner in the afternoon.[24] Bliss Perry notes in his 'Recollections of the Saturday Club' that at his first dinner in 1903 – the same year Lawrence Lowell entered the Club – 'it was a gathering primarily of Harvard notables'.[25]

Henderson had been elected to the club in 1922, and within ten years of that date all the five founders of Harvard's Society of Fellows belonged, except Charles P. Curtis, Jr., who was elected in 1953.[26] The Saturday Club, then, was one place for a select group of Harvard scholars, from a variety of disciplines, to meet and converse. Brinton indicates that in the 1930s Henderson, Lowell, and a few others often talked about Pareto in discussions which lasted into late afternoon.[27] After Brinton joined the Club in 1937, he and Henderson probably saw each other regularly at the Club until the latter's death in 1942.[28]

It is obvious that this contact and sharing of ideas which existed before 1936 did not diminish as time passed. For in 1938 Brinton's *Anatomy of Revolution* appeared citing a debt to Henderson for the 'conceptual scheme' he employed as a framework for the entire book.[29] In fact, the friendship may well have grown stronger for in 1939 Brinton was named a Senior Fellow in the Society of Fellows, a move which had to involve Henderson, as he was the Chairman of the Society at the time.

The Society of Fellows is the third social–intellectual association effecting contact among the men under study, and it appears to have cemented the Henderson–Homans and Henderson–Brinton friendships. Henderson is the most important figure here for several reasons. He was *the* founder of the Society in the sense that it was his idea[30] and – as Homans and Bailey put it – 'Henderson was the first to begin thinking what might practically be done' about truly altering the usual graduate school experience.[31] He was a Senior Fellow and Chairman of the Society from its first days in 1933 until his death nine years later.[32] The senior Fellows are a set of nine men, professors or members of the Harvard Corporation. They select the Junior Fellows, a group of graduate men who work on projects of their own and who are free from the usual academic demands. One important aspect of the Society was designed specifically to encourage the sharing of ideas among Senior and Junior Fellows: all Society members dine together every Monday evening.[33]

In addition to their previous ties Brinton and Henderson were Senior Fellows together from 1939 to 1942. For those three years the two saw each other every week at the Monday dinners. In addition, the Senior Fellows met periodically as a group 'to administer the affairs of the Society'. They also interviewed the nominees for the new Junior Fellow positions and settled on their choices each year.[34] Brinton and Henderson were now hard-working colleagues. There was a twenty-year gap in their ages, and over twenty years had passed since they first became acquainted. Yet there were still close friends.

The first year the Society began accepting Junior Fellows was 1933–34. Homans had been nominated for a Junior Fellow position; he entered the competition as a poet, but was rejected.[35] But by the second year of the Society's existence Homans was on the road to becoming a sociologist. In 1934 Homans had completed two years of Henderson's seminar on Pareto and published his book with Curtis.[36] Homans describes his new situation.

... I had demonstrated my Paretan faith; I was Henderson's man, and Henderson was chairman of the Society of Fellows. I had worked with, and became a great friend of Curtis. ... If the society was what I wanted, I was in with the right crowd. ... Accordingly, when Henderson suggested that I should run for the Society a second time, and this time as a sociologist, I again agreed at once.[37]

As a Junior Fellow from 1934 to 1939, Homans was in weekly contact with Henderson all during this period. There can be no doubt of the enormous influence Henderson had on Homans. Homans describes himself as 'Henderson's man', and elsewhere in his autobiographical essay he calls Henderson 'my patron'.[38] Homans had agreed to run for the Society as a sociologist. But his undergraduate training had all been in English literature, and he relied on Henderson to help make him a sociologist. Homans recalls: 'to prepare for entering the Society, I asked Henderson, in effect, "Master, what shall I do to become a sociologist?"'[39] Henderson had some specific ideas of what Homans should do; and Homans followed the boss's orders.

In spite of the apparent master–slave relationship, respect did not flow all one way. Henderson thought highly of the work Homans had done with Curtis in 1934.[40] And he kept up with Homans' later writings; he read and criticized carefully the thick historical work which was the major product of Homans' years as a Junior Fellow.[41]

The friendships that developed between Henderson and Brinton and between Henderson and Homans seemed to bring Brinton and Homans together. But their contact was very slight compared with the close ties each had with Henderson. Both attended Henderson's seminar, but Brinton did so occasionally. It appears that no significant interaction occurred at that time, for Brinton marks his first contact with Homans as being in 1934 when Homans became a Junior Fellow. Though Brinton was not a Senior Fellow until 1939, he was a guest at the Society dinners three or four times a year all during the years Homans was with the Society.[42] Henderson, Chairman of the Society, may well have been Brinton's host at these times. At least by 1937 Homans was familiar with Brinton's writings and was using some of Brinton's historical materials and interpretations to support his own arguments.[43] By this time, however, Homan's friendship with Henderson was on solid footing, and Brinton's had long been so; whereas, there is no evidence for close contact between Homans and Brinton.

In 1939 Brinton came weekly to the Society dinners in anticipation of his appointment as a Senior Fellow, but his regular attendance overlapped with only the last three months of Homans' membership.[44] In 1940 Homans finished his *English Villagers of the Thirteenth Century*. By this time there had been some contact between the two men, for in his preface Homans expressed his warmest thanks to Brinton, who read and criticized part of the book.[45] Brinton was not, however, the major influence on Homans' work; this came instead from social anthropology, to which Homans had been introduced by Elton Mayo.[46]

Parsons writes that his contact with Homans came about 'essentially through the association with Henderson. Homans was a kind of protégé of Henderson from the beginning of his interest in Pareto.'[47] Parsons also indicated that he was in the Pareto seminar with Homans, but he makes no mention of any close relationship between them at this time. Parsons also states: 'With Brinton there was never any such intensive contact [as with Henderson], though it was a continuing relationship.'[48]

The contact the four scholars had with each other during the decade under study is summarized in the diagram below, indicating their participation in the three formal groups discussed above.

Diagram indicating formal association of 'Pareto Circle'

| | 1932 | 33 | 34 | 35 | 36 | 37 | 38 | 39 | 40 | 41 | 42 |

Henderson (1878-1942)

Brinton (1898-)

Homans (1910-)

Parsons (1902-)

Seminar ═══════════
Society of Fellows * * * * * * * * * * *
Saturday Club ÷ ÷ ÷ ÷ ÷ ÷ ÷ ÷ ÷ ÷ ÷ ÷

Although all four men participated in the seminar, the text above indicates that the interaction among Homans, Parsons and Brinton was much less intense than the contact each had with Henderson; this is because Brinton did not attend regularly but mostly because Henderson led the seminar. Not only was Henderson dynamic and stimulating in discussion, but he was *the* expert in the group on Pareto and scientific method. It is clear that in the seminar situation he would be the center of interaction.

The four scholars naturally belonged to other formal organizations which fostered interaction; e.g., beginning with 1939, Homans and Parsons were on the faculty in the same department. And informal contact, growing from friendship, common interests, or scholarly cooperation, was no doubt even more important in the sharing and influencing of ideas than the formal activities. But the point of the chart is that even with these three quite formal associations, a trend is apparent. Henderson was very active in the last decade of his life; and the other three men were drawn into groups of which Henderson was either a long-standing member (Saturday Club) or the founder (the seminar, the Society of Fellows). Furthermore, there appears to have been less contact among the other three than each had with Henderson.

Qualitative measures of friendship patterns (discussed earlier in this section) further testify to Henderson's central position within the group. Brinton's long-standing friendship with Henderson was reinforced by the steady contact near the end of Henderson's life; this, accompanied with his only occasional contact with Parsons and Homans, leads to the conclusion that he was influenced more by Henderson than by the other two. Homans' autobiographical essay makes clear that his relationship with Henderson was much more significant than with either Brinton or Parsons. And Parsons has recently stated that of the three – Homans, Henderson and Brinton – Henderson was his closest friend.[49]

II. Social system concepts
The central task of this study is to compare how the four scholars wrote about the concepts of social system and social equilibrium. But before a comparison is attempted, it seems wise to sketch some of the main ideas pertaining to these concepts presented by each of the authors in his major works during the 1930s and early 1940s. Any analogies or examples of the concepts of equilibrium and social system used by the authors will also be described here.

One of the first major works published in English on Pareto's writings was *An introduction to Pareto: His Sociology* (1934) by G. C. Homans and C. P. Curtis, Jr. As we have seen (see above, pp. 139–40), this book was an outgrowth of the seminar under Henderson. Homans and Curtis saw as their purpose to provide readers with a statement on scientific method which would help them understand Pareto's *Sociologie Générale.*[50] Only a few sections of the book are relevant to this study of social system concepts, because, as S. E. Finer states: 'This book is, for the most part, a set of everyday illustrations and examples of the subclasses of the residues.'[51] And a discussion of residues belongs to another sort of paper. Here we are concerned with the authors' ideas on the mutal dependence of variables and social equilibrium.

Homans and Curtis assert that most relationships between two variables are ones of mutual dependence rather than of cause and effect. The authors illustrate how changes in one variable will usually mean changes in other variables,[52] yet they do not delimit their statements about mutual dependence with a description of the system within which the elements interact. In fact, in this book the concept 'system' is not well developed at all, within either a physical or social framework.

The two authors discuss in detail the problems facing the social scientist – the analyst who wishes to determine the precise relationship between two or more mutually dependent variables. Mathematics, the authors assert, is the only means available for solving such problems, but in sociology we are unable to measure accurately enough to use such mathematical techniques.[53] The problem is not insuperable, however, since

> ... in few states of mutual dependence are all the variables equally important ... we could describe this state of mutual dependence very well ... if we first treated the variable of primary importance as the cause of the others and then took into account the reaction of the variables of secondary importance on the cause.[54]

Homans and Curtis use cause and effect reasoning while acknowledging that most related variables are, in fact, interdependent.

Though Homans and Curtis do not elaborate on the concept of 'system', they do present in detail their ideas on 'equilibrium'. Before they define the term, they give examples of equilibrium, the first of which is a marble in a bowl. 'If the bowl is jerked, the marble rolls up its side a little, but soon falls back to the bottom again. The marble is in a state of equilibrium.'[55] The second illustration is the recovery of weight in the human organism following sickness. If a baby contracts the measles and loses weight – but then gains weight again so that he weighs what he would have had he not been ill, then 'the baby, as far as its health is concerned, is in a state of equilibrium'.[56] Because growth is a process, this last example illustrates dynamic equilibrium, as opposed to the marble at rest which exemplifies static equilibrium. Since Pareto applies his notion of equilibrium to social processes, he is more concerned with the dynamic form of equilibrium than the static.[57] Finally, Homans and Curtis offer Pareto's own definition of equilibrium:

> This state is such ... that if a modification were artificially introduced in it unlike that which it in reality undergoes, immediately a reaction would be produced which would tend to bring it back to the real state.[58]

Under 'artificial changes' the old equilibrium is regained, but under more violent changes a *new* equilibrium may be reached.

Homans and Curtis note that the importance of the equilibrium concept is 'Pareto's belief ... that most societies at most times behave in a manner which indicates that they are in equilibrium. When a society suffers a disturbance, a reaction is set up which tends to bring it back to its original state'.[59] Examples which the authors give of equilibrium at work in society are, first, the manner in which a rich nation readjusts to the dislocations of a short war and, second, how a populous region, devasted by earthquake, gradually rebuilds, so that 'the region becomes more or less what it was before. Such a region is in equilibrium.'[60]

Henderson's book on Pareto appeared the year following that of the Homans and Curtis work. Early in the book Henderson launches into an analysis of systems – first 'the physico-chemical system' and then the social system. The physico-chemical system is based on the work of Willard Gibbs and is 'an isolated material aggregate' consisting of individual substances which are found in phases (liquid, solid, or gaseous), and the system is characterized by a certain temperature, pressure, and concentration of the substances.[61] When one of these factors changes, the others do too. 'Thus all the factors that characterize this system are seen to be mutually dependent. In this respect this system is typical of all systems.'[62]

Like Homans and Curtis, Henderson argues that cause and effect analysis is inadequate for describing systems with interdependent elements. Such analysis 'has to be replaced by some method ... involving the simultaneous variations of mutually dependent variables'.[63] But Henderson's solution is more sophisticated than that given by Homans and Curtis. In his notes at the end of the book Henderson devotes eight pages (twice the length of his chapter on the social system itself) to the forms the required method of analysis might take. Henderson – reflecting his physical science background – gives a most precise explication of the dynamics of mutual dependency.[64]

Henderson next describes Pareto's social system, which he finds analogous to Gibbs' physico-chemical system. In his notes Henderson assures the reader that Gibbs' work did not influence Pareto, but that Pareto's own thesis on the mathematical theory of equilibrium in elastic solids and his work in economics were major sources of his social system.[65] But Henderson cannot refrain from spelling out all the analogies between the physico-chemical system and the social system – individuals correspond to Gibbs' components, while sentiments, verbal elaborations and economic interests correspond to Gibbs' temperature, pressure and concentrations.[66] Thus, in describing Pareto's social system in detail, Henderson's chief analogy is that of a chemical system. Yet at various points he asserts that the social system is analogous to 'dynamical, thermodynamical, physiological, and economic systems',[67] to 'the solar system', as well as to the physico-chemical system.[68]

Of the four scholars under study Henderson alone truly analyzes the concept of system. He spells out the steps one would go through in 'building up the conceptual scheme of a new generalized system'. Briefly, one must recognize a certain set of phenomena which is analogous to those sets of phenomena designated by terms like dynamical system or physico-chemical system, then take into account a certain number of variables and discover the set of equations which 'completely describe' the system of variables.[69] 'The state defined by such a set of equations is frequently a state of equilibrium. . . .'[70] When Henderson illustrates this procedure, he uses a mechanical example.[71] But in the main sections of the book when he describes *social* equilibrium, his analogy is that of the self-healing power of the living organism.

The case of physiological equilibrium is similar [to social equilibrium]. In fact, it is logically identical. When recovery from disease is in question, the process is still often referred to as a result of the *vis medicatrix naturae*.[72]

Henderson then cites works in medicine and physiology in support of his statements.

Henderson appears to use the more formal and complicated analogies of mechanics and chemistry when explicating the scientific bases of or uses for the models of social system and equilibrium; this he does primarily in the notes. The body of the book contains more everyday analogies. Henderson, the physical scientist with an incredibly broad range of interests, was capable of writing for the educated layman or for the scientist. In *Pareto's General Sociology* he did both – the former in the first half of the book, the latter in the notes.

Parsons' *Structure of Social Action* (1937) was published two years after Henderson's work. It contains a precise analysis of Pareto's theories – an analysis which S. E. Finer calls a 'masterly and minute exegesis',[73] But, unfortunately, the exegesis touches only lightly on topics relevant to this paper. Parsons acknowledges in his preface that 'the primary aim of the study is not to determine and state in summary form what these writers [including Pareto] said . . .'.[74] Essentially, this had been the aim of the Homans and Curtis and Henderson books on Pareto. Parsons is concerned instead with analyzing the Paretan ideas in terms of the action frame of reference. Hence, he concentrates almost exclusively on what Pareto has to say about such concepts as sentiments, logical and non-logical actions, and values.

Because of Parsons' specific focus he never describes Pareto's concept of social system or equilibrium. When he uses the word *system*, even when referring to the content of Pareto's writings, he writes within the action framework. Parsons is concerned with finding in Pareto statements about action within society as a whole; that is, statements about the acts of all members of the society. Particularly useful to Parsons in this regard is Pareto's theory of social utility. He isolates Pareto's statement that there exists the 'end a society should pursue by means of logico-experimental reasoning', and rephrases it as a key point at which Pareto gives support to the action frame of reference.

This [Pareto's statement just quoted] may be restated to the effect that the actions of the members of a society are to a significant degree oriented to a single integrated system of ultimate ends common to these members. More generally the value element in the form both of ultimate ends and of value attitudes is in a significant degree common to the members of the society. This is one of the essential conditions of the equilibrium of social systems.[75]

This is a perfect example of how Parsons talks about social system and equilibrium in the 1930s. He has moved out of Pareto's social system, which is composed of residues, derivations and social heterogeneity, and into the social system of action, composed of means-ends schemes and values. And he has taken with him the concepts under study in this paper, undefined and never discussed *per se*, but used specifically to illuminate his own analysis. Not until fourteen years later did Parsons concentrate on the concept of social system for its own sake.[76]

Brinton's *Anatomy of Revolution* (1938) is an attempt to give a structural analysis of four revolutions. Brinton's aim is '. . . to establish, as the scientist might, certain first approximations of uniformities to be noted in the course of four successful revolutions in modern states. . . .'.[77] Brinton notes that his problem is to find a conceptual framework within which to fit the details of the revolutions, and that in the social sciences it is sometimes difficult to distinguish a conceptual scheme from a metaphor.[78] He discusses the metaphor of the storm as particularly appropriate for describing the course of revolutions. It becomes clear that Brinton seeks an analogy for something which rages and then subsides.

The conceptual scheme he then mentions is Pareto's 'social system of equilibrium'.[79] Brinton notes the equilibrium concept is useful in a variety of contemporary fields.

The concept of equilibrium helps us to understand, and sometimes to use and control, specific machines, chemicals, and even medicines. It may someday help us to understand, and within limits to mold, men in society.[80]

Brinton seizes upon this concept, applies it to society, and notes that no society is in perfect equilibrium and that under certain conditions of change 'a relative disequilibrium may arise, and what we call a revolution break out'.[81] As Brinton explicates the phenomenon of social equilibrium, he uses the exact same analogy and even the same terms as Henderson.

In social systems, as in the human organism, a kind of natural healing force, a *vis medicatrix naturae*, tends almost automatically to balance one kind of change with another and restorative change.[82]

Brinton feels that the conceptual framework of social equilibrium is the most useful one for sociologists who are analyzing revolutions. However, he indicates that he cannot use this framework because 'it ought to be formulated in terms more close to those of mathematics than we can honestly employ'.[83] Yet the scheme he does actually employ is the analogy itself: social equilibrium as a self-healing living organism. 'We shall regard revolutions . . . as a kind of fever.' It is the kind of fever from which the patient inevitably recovers.

Finally the fever is over, and the patient is himself again . . . societies which undergo the full cycle of revolution are perhaps in some respects the stronger for it; but they by no means emerge entirely remade.[84]

III. Patterns of influence

This last part of the paper compares the analogies used by the four scholars of the concepts of social system and equilibrium and relates this analysis to the interaction patterns within the group. The first section of the paper indicated that Henderson was the hub of friendships within this group. The links among the others were much weaker than those which tied each to Henderson. Brinton and Homans were in contact, but the evidence indicates that their relationship came about primarily as a result of their separate friendships with Henderson. Parsons had little contact with the other two, though a good friend of the physiologist. Given this situation and the fact that it was Henderson – particularly through his two-year seminar on Pareto – who introduced to the other three men the concepts of social system and social equilibrium, the discussion to follow compares Henderson's writings with those of each of the other three.

During the two years of Henderson's seminar Homans was in his two years of graduate school. He was 24 years old when he stopped graduate work and became a Junior Fellow in 1934, and Henderson – the red-bearded dynamo – was 56. The Homans and Curtis book appeared in 1934 as well, and since the authors indicate in their preface that all the core ideas came from Henderson, a strong similarity should exist between the analogies used by Homans and Curtis for the social system and equilibrium and those used by Henderson the next year in his book. And so it is. Henderson's explication of Pareto's concepts utilizes analogies from a whole range of fields – physics, chemistry, astronomy and economics. A parallel to Henderson's mechanical example to illustrate equilibrium[85] is Homans and Curtis' use of the marble in the bowl. For their examples of dynamic equilibrium both books employ analogies of a living organism recovering from disease. Henderson calls the homeostatic process of living bodies the result of *vis medicatrix naturae*.[86] The Homans and Curtis example is that of the baby regaining weight after the measles until

. . . its weight returns to what it would have been if it had never suffered the attack. . . . This effect of this sort of equilibrium used to be called the *vis medicatrix naturae*. We rely on it when we prescribe, as Hippocrates did: 'Let it alone and it will get well.'[87]

Both Homans and Henderson share still another way of looking at equilibrium. Henderson states about social equilibrium:

In the case of Pareto's social system the definition of equilibrium takes a form that closely resembles the theorem of Le Chatelier in physical chemistry, which expresses a property of physico-chemical equilibrium. . . .[88]

Homans and Curtis note the same fact:

Clearly his [Pareto's] definition of equilibrium as observed in society is closely analogous to the definition of equilibrium as observed, for instance, in chemistry.[89]

In Henderson's book the analogies for the social system and social equilibrium are drawn from a variety of fields: in each field the analogous system is carefully described in precise correspondence to the social system. In the case of physics and chemistry it is with words, diagrams and formulas; in physiology with examples from medicine; and in astronomy with a description of the solar system which exactly parallels a description of the social system. Henderson was the physical scientist with very broad interests and was at home in all these fields.

Homans' training in English literature and Curtis' in law did not allow them to elaborate on the scientific analogies. But as shown above, their description of the more everyday analogies is very similar to Henderson's. Another important point is that Homans and Curtis mention all the same fields as Henderson.

The curious thing is that the phenomena of equilibrium are observed over a wide range of fields. In physics, chemistry, biology, astronomy, and other sciences groups of mutually dependent factors are observed to form systems which have this property in common: the group as a whole acts so as to counteract any change in any of its components.[90]

It appears likely that the two authors learned the above fact not from their own experience – which was minimal in the physical sciences – but from Henderson.

Parsons participated in Henderson's seminar on Pareto, and it was no doubt Henderson who stimulated Parsons' interest in the Italian theorist. That interest first bore fruit when Parsons included Pareto in his *Structure of Social Action*. But the manner in which Parsons wrote about social system and equilibrium in the thirties did not closely resemble Henderson's treatment of the concepts or that of Homans and Curtis.

Parsons was no doubt not as impressionable a member of the seminar as Homans – not so much because Parsons is older, for Homans is only eight years his junior, but because Parsons had a position as Faculty Instructor in the Department of Sociology at Harvard.[91] He was teaching; Homans was a student. There is a difference in the frame of mind which accompanies the two occupations. One way a student can learn a great deal is by being open, impressionable before a competent teacher for a certain length of time; enthusiasm which springs from truly understanding the teacher's ideas can lead the student to much deeper study than is otherwise possible. But the teacher – in order to teach well – is thrown back on his own resources, his own creativity, his own ideas.

Parsons had been working out his own ideas on a number of topics which, though they were somewhat related to those pursued in the seminar, were definitely not Henderson's central concerns. Before he had had contact with Henderson, Parsons had become much interested in Max Weber's work and 'the problem of the nature of economic theory and its boundaries and limitations'; Parsons found Pareto's writings interesting because the Italian had explicitly addressed this problem.[92] Parsons notes that Henderson and Homans were not very concerned with the problem of economic theory and that 'Schumpeter's participation in the Seminar was entirely peripheral'.[93] Apparently, the seminar devoted little time to Pareto's ideas on economic theory. Parsons had also begun serious analysis of Durkheim's theory; Durkheim was discussed by members of the Pareto circle, but his ideas again were not one of their main interests.[94]

So before the seminar was organized Parsons' interests had

crystallized around different problems than those which stirred Henderson. Parsons was consequently less influenced by Henderson's views than was Homans, who entered the seminar with less knowledge of social theory or at least entered with interests more attuned to Henderson's.

However, Henderson did impress Parsons with the value of Pareto's concept of social system. *The Social System* appeared fourteen years after his *Structure of Social Action* with the following acknowledgement to Henderson:

The title, *The Social System*, goes back, more than to any other source, to the insistence of the late Professor L. J. Henderson on the extreme importance of the concept of system in scientific theory, and his clear realization that the attempt to delineate the social system as a system was the most important contribution of Pareto's great work.[95]

Henderson had introduced Parsons to the concept of social system; but between the 1930s and 1951, when *The Social System* was finished, Parsons worked with the concept and developed a somewhat different way of using it. Parsons' social system was based more on a physiological model of systems than was Henderson's which leaned heavily on a physical-chemical model.[96]

Brinton had written a great deal about revolutions[97] before he wrote *Anatomy of Revolution*, but the fact that he chose to put his knowledge of the historical data into a conceptual framework – and that he chose the particular framework he did – owes a great deal to Henderson. Some historians are content to describe. But Brinton felt that the scientific method could be applied to historical data and that 'the scientist cannot work without a conceptual scheme . . .'.[98] The main support for his belief came from Henderson's definition of fact – '"an empirically verifiable statement about phenomena in terms of a conceptual scheme".'[99] So Brinton needed a conceptual scheme to make his facts about revolution really facts, in a scientific sense.

Brinton's ideas about social equilibrium very much resemble Henderson's. Brinton points out that the concept of equilibrium is useful in the fields of mechanics, chemistry and physiology.[100] These are the three main fields from which Henderson drew his examples and analogies. Brinton next states:

The concepts of a physicochemical system in equilibrium, a social system in equilibrium, John Jones's body in equilibrium, do not in the least prejudice the immortality of anyone's soul, nor even the ultimate victory of Vitalists over Mechanists.[101]

The phrase 'physicochemical system' echoes a chapter title in Henderson's book. And Henderson makes a similar statement about the Vitalists.

It is instructive to note that these physiological phenomena [examples of equilibrium] have been used by philosophers as a foundation of the argument for vitalism. For they belong in the same class as those inorganic processes that arise from similar disturbances of physical and chemical equilibrium.[102]

Brinton uses the self-healing organism as his analogy for the process of social equilibrium, and fever is the analogy for disequilibrium (or revolution). Henderson does indeed use the same basic metaphors.[103] It is easy to see why, of all the other analogies Henderson uses, that the physiological one would appeal to Brinton. First of all, Brinton is not trained in the physical sciences; the chemical and mechanical examples would be difficult for him to make his own. But more importantly, Brinton insists:

The first job of the scientist is to be obvious. . . . We shall, then, hope that whatever uniformities we can detect . . . will turn out to be obvious, to be just what any sensible man already knew about revolution.[104]

With this belief Brinton cannot choose a highly technical analogy, or his conclusions will not be what any sensible man already knew. Everyone knows about fever. Thirdly, if Henderson says that social equilibrium works 'identically' to physiological equilibrium, as he certainly did,[105] he should be believed. For Henderson is not only the expert on social equilibrium, but is a physiologist as well. He alone among the four scholars had an MD; this was obviously Henderson's field – or rather one of his fields – of expertise.

IV. Conclusion

Henderson's impact on Homans, Brinton and Parsons was undeniably great – more extensive by far than the impact any one of the other three men had either on him or on one another. Under Henderson's influence a group of scholars in the 1930s embraced Pareto's theories and terminology. For Henderson, Pareto's system was the most sophisticated step yet taken toward building a science of society.

However, at this period in intellectual history sociologists are not enamored with Pareto. Despite Pareto's pretensions to the contrary, his conclusions are not the result of any experimental method; they are not facts about the social world, but rather high order abstractions. Pareto presents the reader with few techniques for studying sociological phenomena empirically. One does find in Pareto a conceptual scheme for categorizing non-logical aspects of human behavior. He filled over a thousand pages developing his system of residues and derivations; but it is not very useful at the heuristic level. Even Pareto found that only two of his residues were helpful when he applied his system to the sociopolitical world, the result of which was his theory of the 'circulation of the elites'. An important spinoff from this emphasis on residues and derivations was that his 'general sociology' ignored specifically sociological elements of human behavior – e.g. the importance of interaction networks and institutional roles.

There is another basis for today's critical view of Pareto's theories. Homans made explicit that one reason he accepted Pareto was that he could use him as a defense against the Marxists, since Pareto had demonstrated that Marx's theories, among others, contained rationalizations. Pareto at the same time set his 'debunking' of other social theories in a framework of such cold logic and scientific analysis that he convinced important readers like Henderson that he had kept *his* work primarily free of rationalization. Henderson's trust in Pareto was no doubt imparted to members of the Pareto circle in the 1930s. But in the 1960s Pareto has been 'justly celebrated as the greatest rationalizer of authoritarian conservatism in our time'.[106]

Our view from the 1960s enables us to see that the Pareto circle had difficulty doing sociology applying Pareto's framework and terminology. Brinton indicates that near the end of his life Henderson himself was revising Pareto in a general sociology of his own. In the same article, written in the 1950s, Brinton notes that Henderson's use of Pareto as a basis was perhaps a mistake.[107] Yet in the 1930s Brinton had himself relied heavily on Pareto's framework in two of his books: *Anatomy of Revolution* and *French Revolutionary Legislation on Illegitimacy*.

Homans has exhibited a particularly dramatic change of attitude regarding Paretan ideas. In 1936 he described the dynamics of social equilibrium as an inescapable fact of life.

[A] society is an organism and . . . like all organisms, if a threat be made to its mode of existence, a society will produce antibodies which tend to restore it to its original form.[108]

But in 1961 Homans completely reversed his view on the natural tendency of a social system to return to its original state if disturbed. He even changed the term to avoid confusion with his earlier meaning of the concept.

We speak of *practical equilibrium* instead of plain *equilibrium* in order to avoid the almost mystical arguments that have encrusted the latter word. . . . [We do not assume] that if a change from practical equilibrium does occur, behavior necessarily reacts so as to reduce or get rid of it. There is no homeostasis here: no belief that a group acts like an animal body shaking off an infection.[109]

Parsons' recent attitudes toward Pareto are somewhat more difficult to gauge. He did not embrace the Paretan social system and equilibrium concepts as immediately or as completely as did the others. But today he seems closer to Pareto than either Homans or Brinton. This is partly because the social system concepts are now more central to his thinking than to theirs, but also because his method of doing sociology – building large-scale, abstract theories – is similar to Pareto's. The latter is one primary source of the intellectual schism between Parsons and Homans. Homans insists that theories should be built from the psychological bottom up – from the empirically tested hypotheses to more general propositions – while Parsons, like Pareto, builds systems from the top down.

The members of the Pareto circle found themselves moving beyond the conceptual schemes Pareto had given them in order to solve their more immediate problems of studying social phenomena. They were aided in their move away from Pareto by the Second World War which made the Soviet Union our ally and altered the ideological situation within American sociology. Marxism seemed less of a threat and more of an historical antecedent to sociology. Pareto was no longer necessary as a defense against the Communist menace. In addition to these global events, Harvard was changing, too. The dynamic Henderson died in 1942. The Paretan framework seemed more and more remote and removed from the scene of battle.

Today we view Pareto and social system concepts through a different set of perceptual spectacles than did the Pareto circle. In the past Homans, Brinton and Parsons looked through lenses ground for them by Henderson. Time, however, has revealed the weaknesses in Pareto's sociology. The lenses ground by Lawrence Joseph Henderson have been discarded. And Pareto too has joined the long line of social theorists who form the corps of the classic tradition.

Source: Journal of The History of the Behavioral Sciences, IV (1968), No. 4: 316–34.

Notes

1. Vilfredo Pareto, *Sociological Writings*, selected and introduced by S. E. Finer. New York, Praeger, 1966: 31.
2. Ibid.: 28–9; also, George C. Homans, *Sentiments and Activities: Essays in Social Science*. New York, Free Press, 1962: 5; Elton Mayo's membership in the group was brought to my attention in letters from G. C. Homans (19 June 1967) and Talcott Parsons (18 July 1967).
3. Lawrence J. Henderson, *Pareto's General Sociology: A Physiologist's Interpretation.* Cambridge, Harvard University Press, 1935; George C. Homans and Charles P. Curtis, Jr., *An Introduction to Pareto: His Sociology.* New York, Alfred A. Knopf, 1934.
4. Talcott Parsons, *The Structure of Social Action*, 2nd edn. New York, Free Press, 1961.
5. Crane Brinton, *Anatomy of Revolution*, 2nd edn. New York, Vintage Books, 1952; Crane Brinton, *French Revolutionary Legislation on Illegitimacy 1789–1804.* Cambridge, Harvard University Press, 1936.
6. Talcott Parsons, 'Pareto's central analytical scheme', *Journal of Social Philosophy*, I (April, 1936): 244–62; Lawrence J. Henderson, 'Pareto's science of society', *Saturday Review of Literature*, XII (25 May 1935): 3–4 +; Crane Brinton, 'What's the matter with sociology?' *Saturday Review of Literature*, XX (6 May 1939): 3.
7. Although this paper deals primarily with the relationships *within* the Pareto circle in the 1930s, a brief discussion of the specific relationship between the Marxists and Paretans was required.
8. Homans, *Sentiments and Activities*: 4.
9. Letter from Crane Brinton, 17 Feb. 1967.
10. Crane Brinton (ed.), *The Society of Fellows.* Cambridge, The Society of Fellows of Harvard Univ., 1959: 3.

11. Crane Brinton, 'Lawrence Joseph Henderson', *Saturday Club: A Century Completed, 1920–1956*, E. W. Forbes and J. H. Finley, Jr. (eds). Boston, Houghton Mifflin, 1958: 213.

12. For biographical sketch of Henderson see ibid.; also Brinton (ed.), *Society of Fellows*: 1–37. On the seminar see Homans, *Sentiments and Activities*: 3, 5; Pareto, *Sociological Writings*: 28–9; Lloyd K. Garrison, 'Great-circle course', Review of *Lions Under the Throne* by Charles P. Curtis, Jr., *Saturday Review of Literature*, XXX (29 Mar. 1947): 9; also letters from Talcott Parsons, 10 Mar. 1967 and 18 July 1967.

13. Homans and Curtis, *Introduction to Pareto*, Preface; and Homans, *Sentiments and Activities*: 5.

14. On Homans, DeVoto, Henderson, and Curtis see Homans, *Sentiments and Activities*: 3–7.

15. Homans and Curtis, *Introduction to Pareto*, Preface.

16. Letter from Talcott Parsons, 10 Mar. 1967.

17. Prior to the seminar, Parsons had only occasional contact with Henderson, dating from 1927 when Parsons came to Harvard; ibid.

18. Talcott Parsons, 'Review of *Mind and Society* by V. Pareto and *Pareto's General Sociology* by L. J. Henderson', *American Economic Review*, XXV (1935): 502–8.

19. Parsons, *Structure of Social Action*: vii.

20. Interview with Talcott Parsons, 8 Mar. 1967.

21. All the information on the Brinton–Henderson friendship is from Brinton's letter to the writer, 17 Feb. 1967.

22. Brinton, *French Revolutionary Legislation*: xi.

23. Crane Brinton, *The Lives of Talleyrand*. New York, W. W. Norton, 1936: x.

24. *Saturday Club*, fold-out on back cover and p. 1.

25. Ibid.: 5.

26. From membership list on back cover of *Saturday Club*. The five founders of the Society of Fellows were Henderson, John Livingston Lowes, Alfred North Whitehead, Charles P. Curtis, Jr. and – important as President of Harvard and the original benefactor of the Society – A. Lawrence Lowell.

27. Letter from Crane Brinton, 17 Feb. 1967.

28. The membership list for the Saturday Club gives the dates each person was in the Club; it appears that individuals were elected to the Club for life.

29. Brinton, *Anatomy of Revolution*: 11.

30. Charles P. Curtis, Jr., 'Abbott Lawrence Lowell', *Saturday Club*: 128.

31. Brinton (ed.), *Society of Fellows*: 4–5.

32. Ibid.: 22, 76; letter from Crane Brinton, 17 Feb. 1967.

33. Brinton (ed.), *Society of Fellows*: 29–30.

34. Ibid.: 1–2, 24–5.

35. Homans, *Sentiments and Activities*: 6.

36. See text above: pp. 137–8.

37. Homans, *Sentiments and Activities*: 6.

38. Ibid.: 37.

39. Ibid.: 6.

40. Henderson, *Pareto's General Sociology*: 42, note 1.

41. George C. Homans, *English Villagers of the Thirteenth Century*. Cambridge, Harvard University Press, 1941, Preface.

42. Letter from Crane Brinton, 17 Feb. 1967.

43. George C. Homans, 'The futility of revolution', *Saturday Review of Literature*, XVI (15 May 1937): 14.

44. Letter from Crane Brinton, 17 Feb. 1967.
45. Homans, *English Villagers*, Preface.
46. Homans read under Mayo at Henderson's suggestion. Homans writes that Mayo and Henderson were great friends and had adjoining offices in the Harvard Business School. Mayo was influenced by Henderson and Paretan ideas, but Homans notes that 'Mayo was never a full-fledged Paretan' (letter from George C. Homans, 19 June 1967).
47. **Letter from Talcott Parsons, 10 Mar. 1967.**
48. Ibid.
49. Interview with Talcott Parsons, 8 Mar. 1967.
50. Homans and Curtis, *Introduction to Pareto*: 13–14.
51. Pareto, *Sociological Writings*: 84.
52. Homans and Curtis, *Introduction to Pareto*: 33, 80.
53. Ibid.: 35.
54. Ibid.: 36.
55. Ibid.: 271.
56. Ibid.: 272.
57. Ibid.
58. Ibid.: 273.
59. Ibid.: 276.
60. Ibid.: 272–3.
61. Henderson, *Pareto's General Sociology*: 10.
62. Ibid.: 12.
63. Ibid.: 13.
64. Ibid.: 13–14, 74–81.
65. Ibid.: 91–3.
66. Ibid.: 16–17.
67. Ibid.: 18.
68. Ibid.: 91.
69. Ibid.: 110–11, 114–15.
70. Ibid.: 114.
71. Ibid.: 113.
72. Ibid.: 46.
73. Pareto, *Sociological Writings*: 43.
74. Parsons, *Structure of Social Action*: v.
75. Ibid.: 707.
76. *The Social System* was first published in 1951.
77. Brinton, *Anatomy of Revolution*: 6.
78. Ibid.: 15.
79. Ibid.: 16.
80. Ibid.
81. Ibid.
82. Ibid.: 17; for Henderson's similar wording, see Henderson, *Pareto's General Sociology*: 46.
83. Brinton, *Anatomy of Revolution*: 17.
84. Ibid.: 18.
85. Henderson, *Pareto's General Sociology*: 113.
86. Ibid.: 46.
87. Homans and Curtis, *Introduction to Pareto*: 272.
88. Henderson, *Pareto's General Sociology*: 85–6.
89. Homans and Curtis, *Introduction to Pareto*: 283.
90. Ibid.

91. Pitirim A. Sorokin, *A Long Journey: The Autobiography of Pitirim A. Sorokin*. New Haven, College and University Press, 1963: 243–4.
92. Letter from Talcott Parsons, 18 July 1967.
93. Ibid.
94. Ibid.
95. Parsons, *Social System*: vii.
96. Letter from Talcott Parsons, 18 July 1967.
97. Crane Brinton, *A Decade of Revolution*. New York, Harper and Brothers, 1934; Brinton, *French Revolutionary Legislation on Illegitimacy*.
98. Brinton, *Anatomy of Revolution*: 6, 10.
99. Ibid.: 11.
100. Ibid.: 16.
101. Ibid.
102. Henderson, *Pareto's General Sociology*: 47.
103. Ibid.: 46.
104. Brinton, *Anatomy of Revolution*: 26–7.
105. Henderson, *Pareto's General Sociology*: 46.
106. H. Stuart Hughes, *Consciousness and Society*. New York, Vintage Books, 1961: 82.
107. Brinton, 'Lawrence Joseph Henderson', *Saturday Club*: 211.
108. George C. Homans, 'The making of a communist', *Saturday Review of Literature*, XV (31 Oct. 1936): 6.
109. George C. Homans, *Social Behavior: Its Elementary Forms*. New York, Harcourt, Brace & World, 1961: 113–14.

3.2 Robert M. Young, The Role of Psychology in the Nineteenth-century Evolutionary Debate

I

The history of psychology is a discipline whose relationship with psychology and with the history of science has yet to be defined. This paper is a case study in the relations between the history of psychology and an important issue in the mainstream of the history of science – the nineteenth-century debate on evolutionary theory. It was prepared as a talk for an audience of psychologists, and one of its aims is to suggest parallels between current debates on the relevance of psychology to society and an analogous debate that occurred in the eighteenth and nineteenth centuries . . .

I should like to provide a perspective on what I take to be some of the conceptual and methodological problems involved in the scientific investigation of human nature by considering earlier episodes in the debate on man's place in nature, debates that reveal an intimate union of psychological, biological, social, and ideological issues. My purpose in doing so is to suggest that there is an important *critical* function for the history of psychology.

I am an historian of science and am preoccupied with the history of attempts to apply the methods and assumptions of science to the study of man and society. I hasten to add that I do history because I believe that our present predicament might be eased if we could gain greater perspective on the assumptions we make and on the ways in which the heritage of the past constrains our present thinking. We could then act on the basis of what we see. Similarly, the structure and the conceptual affinities of past controversies can perhaps help us to take a broader view of our own situation. Historical studies thereby become an analytic tool, not an antiquarian quest in search of who did what first or whether or not A is buried in B's grave.

The writing of the history of a subject at any point in time is highly constrained by contemporary conceptions of the subject itself. This is particularly true of the history of psychology, because its small number of practitioners is on the defensive. They feel themselves to be under attack from colleagues who wonder why they aren't doing experiments, and they also feel shunned by professional historians of science who do not see the relevance of the history of psychology to the mainstream of the history of science. If I were giving this whole paper on the historiography of psychology, I would try to point out in some detail the fundamental philosophical issues behind these reactions. Instead, I shall only say that they reflect the methodological and metaphysical insecurity of all three groups – the historians of psychology and of science, and the practicing experimentalists. It is therefore not surprising that the history of psychology is in a very primitive state and that its practitioners have, until very recently, tended to write synoptic surveys – *The History of Psychology from Plato to NATO* – to show that they have culture, or they search (I think vainly and irrelevantly) for the first truly scientific treatment of this or that problem.

II

I should now like to address myself to the issues I have raised by directing attention to the evolutionary debate in the late-eighteenth and nineteenth centuries. In doing this I hope to provide some evidence for the claim that there are intimate relations between psychological, social, philosophical, and theological issues, and thereby to show the importance of wider issues for the history of psychology. More importantly, I shall try to show the crucial role of the history of psychology *in* the mainstream of the history of science, considered in its social and political context.

At first glance this seems to be an easy task. Every schoolboy knows that the furor over evolution was largely due to the wider implications of the theory, i.e. that the theory of organic evolution implied that man was descended from the apes. Put slightly more formally, it meant that the origin of man occurred by means of the continuous operation of natural laws and not by special creation. This, in turn, implied that it was no longer possible to separate mind and culture from the domain of scientific laws. Man and all his works – body and mind, society and culture – became, in principle, part of the science of biology. The continuity of types was based on the continuity of natural causes, and discontinuities between body and mind and between nature and culture became untenable. God did not act by isolated interpositions, and moral responsibility no longer had a separate, divinely ordained basis in the freedom of the will.

Notice, however, that I have not suggested that discoveries in psychology were central to this set of issues. On the contrary, it is usually argued that it was the theory of evolution which gave psychology – especially comparative, developmental, and physiological psychology – a sound conceptual basis. Charles Darwin's theory was not derived from such findings. Rather, it was derived primarily from studies in geology, paleontology, zoogeography, theory of classification, the study of domesticated animals, and the practices of breeders. Once established, the theory transformed psychology. It is true that Darwin indulged in some speculations about psychology and mental inheritance as he was working out his evolutionary theory, that he made numerous notes on instinct, and that he wrote a short study in child development as well as two books that dealt with issues recognizably psychological: *The Descent of Man*[1] and *The Expression of the Emotions in Man and Animals.*[2] However, it is clear that these were not his central concerns. He considered *The Descent of Man* to be an unoriginal work, and he handed his notes on instincts and comparative psychology to George J. Romanes, who set out to do for the evolution of the mind what Darwin had done for the evolution of the body.[3]

If we look closely at the *Descent of Man*, we find Darwin accepting the *principles* which follow from evolutionary continuity. However, his examples are excessively anecdotal, and his categories of analysis are drawn from a pre-evolutionary psychological tradition. Here is a sample of his approach to the issues:

It is, therefore, highly probable that with mankind the intellectual faculties have been mainly and gradually perfected through natural selection; and this conclusion is sufficient for our purpose. Undoubtedly it would be interesting to trace the development of each separate faculty from the state in which it exists in the lower animals to that which exists in man; but neither my ability nor knowledge permits the attempt.[4]

Darwin had taken the same, rather laconic, approach to the problem of instincts in *On the Origin of Species*.[5] This is slightly surprising, since he had been very aware of the importance of the problem of mind from the beginning of his researches. The following passages appear in his notebooks of 1837 and 1838, the period in which he was developing his conception of natural selection, but before he had the flash of insight which will be discussed below: 'My theory would give zest to recent and fossil comparative anatomy; it would lead to the study of instincts, heredity, mind-heredity, whole [of] metaphysics.'[6] In the same period he was also concerned with the origins and philosophical status of mind, and wrote: 'Why is thought being a secretion of brain more wonderful than gravity a property of matter?'[7]

However, as I have said, he gave his notes on instinct and mind to G. J. Romanes, whom some believe to be the most important pioneer in comparative psychology. When Romanes asked Darwin's advice about the question of the origins of mind, Darwin was again rather casual, and his reply did not reveal the intensity of interest and the penetrating reasoning which characterized, say, his correspondence about plants:

> I have been accustomed to looking at the coming of the sense of pleasure and pain as one of the most important steps in the development of mind, and I should think it ought to be prominent in your table. The sort of progress which I have imagined is that a stimulus produced some effect at the point affected, and that the effect radiated at first in all directions, and then that certain definite advantageous lines of transmission were acquired, inducing definite reaction in certain lines. Such transmission afterwards became associated in some unknown way with pleasure or pain. These sensations led at first to all sorts of violent action, such as the wriggling of a worm, which was of some use. All the organs of sense would be at the same time excited. Afterwards definite lines of action would be found to be the most useful, and so would be practiced. But it is no use my giving you my crude notions.[8]

Darwin was, therefore, very diffident about psychology and characteristically deferred to Spencer, Romanes, and Huxley on psychological topics. Historians of psychology have, nevertheless, habitually attributed a great deal to Darwin's influence on the subject. My point is that, whatever the *implications* of his work for psychology and whatever the long-term influence of evolutionism on psychology, it is clear that the main sources of Darwin's theory were derived from the studies of a field naturalist and from geology. This was where his real interests lay and where he made his own contributions to the *findings*, as opposed to the theories and assumptions of science. He wrote twenty books, four of which were major (*Origin, Variation of Plants and Animals, Descent of Man, Expression of the Emotions*), of which only the last was primarily psychological. And even that was really only intended as an essay to be appended to *The Descent of Man* and was published separately because of the excessive length of the *Descent.*[9] By comparison, eight of his books were strictly about plants.

Having denigrated the significance of psychology in Darwin's work, I want to make one important exception: the influence of Malthus' *Essay on Population* on Darwin's ideas. This was a crucial influence, but in order to understand it, I must make a large detour that will introduce my main theme, i.e. the extremely intricate ways in which psychological ideas influenced the evolutionary debate, both implicitly and explicitly, in its details as well as in its very wide implications. I want to address myself to this issue in two ways. The first is to stress the relative *isolation* of the internalist history of psychology in the nineteenth century from the main stream of the great debate on man's place in nature, and the second is to go back and look again to see that at another level psychological theories were at the very heart of the debate, providing its most fundamental conceptions and touching on its widest implications for human nature and society and indeed for the philosophy of nature itself.

III

First, let us consider the isolation of the evolutionary debate from the internal history of psychology. If we recall what I said about the significance of the evolutionary debate for views of human nature, one would expect that the findings of psychologists would provide the data for a central area of contention. What is the evidence for mental determinism? How closely comparable are the behaviors of men and lower organisms? Is there a perfect correlation between the mind's activities and the physiology of the nervous system? What are the grounds for believing that criminals and lunatics have no control over their actions? All of these issues were being assiduously investigated and debated in the late eighteenth century and throughout the nineteenth. Surely these debates should be closely integrated with the evolutionary debate.

One might object that the general public was not expert enough to consider the detailed findings of psychologists, physiologists, and psychiatrists. This is an initially plausible hypothesis until we look at the astonishing level of sophistication of the public debate in the nineteenth-century periodicals on such abstruse issues as the details of geological stratification and natural history. The most prestigious of the periodicals were the three main quarterlies, and from the beginning the *Edinburgh Review* (1802), the *Quarterly Review* (1809), and the *Westminster Review* (1824) contained extended critical essays on nearly every significant work in geology and biology, written by the leading scientists, philosophers, and theologians of the period. Works on the philosophy of science and on natural theology received the same exhaustive treatment. The complex interrelations among geological, biological, and theological issues were also discussed at great length.

In the same period three main sorts of investigation were occurring in psychology.[10] The first concerned the laws of mind and centered on the concept of the association of ideas. The tradition of associationist psychology goes back to the work of David Hartley (1749) (to which I shall revert in the next section) and can be traced ultimately to the mainstream of the Scientific Revolution in the works of Newton and Locke. The association of ideas was also a basic assumption of the epistemology and psychology of David Hume and had continental parallels in the work and influence of Condillac. In the nineteenth century there were extremely important writings in this tradition by Thomas Brown, James Mill, J. S. Mill, Alexander Bain, Herbert Spencer, and G. H. Lewes. The laws of mind, and their relations with the laws of nature, with society, and with education, were discussed at great length. These books were reviewed in the periodicals, but there is little sustained treatment of them in connection with the evolutionary debate until well after 1860. Also, the context in which they were considered was neither primarily psychological or biological.

Similarly, beginning as early as 1811, there were important experiments on the structure and function of the central nervous system. By the early 1820s, the functional division of the spinal nerve roots was being investigated experimentally in Britain, France, and Germany. In the same period there were extensive experiments on the functions of the brain which might have given strong support to the antideterminists and antimaterialists in the evolutionary debate. Between 1822 and 1870 numerous experimental tests provided *no* evidence for localization of functions in the brain or for the production of purposive movements by artificial stimulation of the cerebral cortex, and the interpretation placed on these experiments was that they supported the autonomy of an indivisible mental substance and belief in free will. There were extensive debates on these findings in the physiological and clinical journals, but the issues found almost no expression in the general debate. This is all the more curious, since the evolutionary debate focussed on the structure of the brain after 1860, and the people conducting the public debate about evolution were also intimately involved in the physiological and medical debates. I am thinking particularly of Richard Owen and T. H. Huxley.

There were also heated controversies on the concept of reflex and on how far up the neuraxis automatic, reflex functions prevailed. Eminent physiologists differed on issues that were perfectly parallel to those in the mainstream of the general evolutionary debate. Thus, while evolutionists argued about whether or not the activities of animals and men were entirely determined by unchanging laws, the physiologists differed on whether or not the thinking part of the brain – the cerebral cortex – obeyed the same laws as the automatic, reflex functions of the lower brain centers and the spinal cord. Once again, it was often the same people who took leading parts in both debates. William Carpenter and G. H. Lewes are notable examples.

Even when, in the 1840s, Thomas Laycock applied the reflex concept to all levels of the central nervous system and argued for complete continuity of function, he did not do so on the basis of the theory of evolution but, on the contrary, based his claims on the principle of continuity in the anti-evolutionary theory of the Great Chain of Being. He did this in spite of the fact that the general version of his theory was presented to the British Association in 1844, the year of publication of a widely read and hotly debated argument in favor of evolution, *Vestiges of the Natural History of Creation.*[11]

Twenty-five years later, when the tide had turned, and it was demonstrated by experiment that movements which had hitherto been attributed to free will could be produced by localized electrical stimulation of the cerebral cortex, these findings might have been taken up by the evolutionists and determinists. They might have told the general public that thought and action were entirely functions of brain centers and that free will was thereby proven to be a chimera. But they didn't. This work was based on inferences drawn from experiments on dogs and monkeys, whose relevance for man was based on evolutionary theory, but the implications of the findings for man were not driven home in the general debate.

An indication of this isolation can be found in Darwin's *Descent of Man*. There was a section on the brain added to the second edition, which Darwin persuaded Huxley to write. Even so, the two editions of that work – 1871 and 1874 – appeared in the period of greatest discovery in the physiology of the cortex, and no whisper of these developments occurs in the book: the argument is conducted entirely in terms of comparative anatomy, without reference to physiology. The issue of comparative anatomical structures in the brain had preoccupied Huxley in the 1860s, in his debates with Owen on the brain.[12] It could, of course, be argued that Huxley was not *au fait* with developments in cerebral physiology. If so, he must have learned it fairly soon thereafter, since the Royal Society turned to him to act as referee for some very delicate issues raised by papers which David Ferrier submitted on the subject in the mid-1870s.[13] Yet neither his contribution to Darwin's book nor his popular essays draw on the findings and theories of psychologists and neurophysiologists. The same can be said of William Carpenter, who was the chief expositor of experimental physiology in Britain at the same time that he was one of the most active and respected interpreters of evolutionary theory.[14]

Having discussed the curious sequestration of associationism and neurophysiology – two approaches with evolutionary and determinist implications – I want to increase our bewilderment by briefly discussing the example of phrenology. In this case our problem is made worse by the fact that phrenology was one of the most popular and publicly controverted theories in the nineteenth century. Its alleged determinist, materialist, and atheist implications were grasped at the outset, and its adherents were regularly reviled in the main periodicals – for example, in the third number of the *Edinburgh Review*, again in 1815, and again and again. I know of only one extensive, balanced treatment of it – by Richard Chevenix in the *Foreign Quarterly Review* in 1828.[15] The same people who attacked uniformitarian geology and evolutionism sallied forth against the putative atheism and degradation of man which lay in the phrenological doctrines of Gall, Spurzheim, and their Scottish exponent and popularizer, George Combe. Phrenological works were in the library of every Mechanics Institute, and by 1832 there were twenty-nine phrenological societies and numerous publications in Britain (with others in France and even more in America). Combe's *Constitution of Man* sold 50,000 copies between 1835 and 1838, 80,000 by 1847, and a total of over 100,000 by 1865. It is said that homes which contained only the Bible and *Pilgrim's Progress* chose Combe's *Constitution of Man* as their third book.[16] I shall argue below that phrenology played a central role in the development of evolutionary theories, but the close conceptual affinities between phrenology and the issues in the evolutionary debate were not a significant feature of the public debate. The condemnation of phrenology and that of uniformitarian geology and evolutionism went on side by side.

To complete our confusion, I should point out that these three themes in the history of psychology – associationism, neurophysiology, and phrenology – came together in the writings of Alexander Bain and that they were placed in the context of evolutionism by Herbert Spencer. In the crucial period of 1855–59, when they were providing perfect ammunition for the wider debate, the connection was not widely grasped. When psychophysiology was placed on an experimental basis in the early 1870s by the findings of Fritsch and Hitzig and of Ferrier in their research on cerebral localization, these findings were available for the debate surrounding *The Descent of Man*, but they were not taken up.[17] When they were extended by physiologists and neurologists – especially John Hughlings Jackson – the failure of the evolutionists and anti-evolutionists to exploit them becomes astonishing.

IV

Having borne with me this far, you will have anticipated that I have been setting up an elaborate rhetorical question which I shall now proceed to answer. If so, you will be half-disappointed, and the half that I cannot explain is the most puzzling one. That is, it is easy to point out that the anti-evolutionists understandably did not want to conduct the debate on grounds which their whole position required them to deny *in principle* as long as they could hope to defeat the enemy on safer territory. For example, it was preferable to defeat the uniformitarians on the battlefield of the history of the earth. If this could be shown to be unexplainable without recourse to divine intervention, then the question of the vulnerability of man's special place in nature need never arise. Even Lyell's uniformitarian geology represented an outwork, protecting the central citadel of man's special nature. He was uniformitarian about the history of the earth, but denied biological evolution. For safety's sake he added that even if evolution was required to explain the history of the animal kingdom, to apply it also to man would stretch analogy beyond all reasonable bounds.[18]

This is not to say that psychological, physiological, and phreno-logical issues never got mentioned in the general debate. However, when they did come up, they were isolated from the mainstream of the debate and/or treated polemically. Those who were upholding the traditional picture of the order of nature and society were unwilling to dignify the sciences of mind and brain by debating their findings, when their whole position required them to deny that man's mind lay within the domain of science. This interpretation is given support by the appearance of *Mind*, the first professional journal in psychology and philosophy in any country. By the time it was founded – we would say, very tardily – in 1876, psychologists were fed up with being fobbed off. The journal was financed by Bain, and the first editor was his protégé, George Croom

Robertson. In the prefatory remarks in the first issue, Robertson wrote:

> Even now the notion of a journal being founded to be taken up wholly with metaphysical subjects, as they are called, will little commend itself to those who are in the habit of declaring with great confidence that there can be no science in such matters, or to those who would only play with them now and again. . . . MIND intends to procure a decision on this question as to the scientific standing of psychology.[19]

This partially satisfactory explanation of why the defenders of man's special status were unwilling to debate the laws of mind and brain – because they denied the relevance of scientific methods and laws to higher functions – leaves us with the problem of why the *proponents* of evolutionism and of a science of man did not draw on the detailed findings of psychologists and neurophysiologists to bolster their own case. Since I simply don't know the answer to this at the straightforward level, I will spare you my paltry speculations on the subject while inviting you to look at the much more significant role which psychological theories played at a deeper level in the theories of the evolutionists.

Before doing so, however, I should add that psychology was in some respects in good company. While I cannot explain its isolation to my own satisfaction, I can remind you that the relevance of findings in one discipline to those in another can take time to dawn on very clever men – even to men whom we see as most directly concerned with the integration of those sets of findings. In the nineteenth-century debate on man's place in nature there were relatively few articles which, as they say nowadays, got it together. I have found only one essay in the pre-1859 period which closely integrates the question of evolution with the question of the natural history of man, and that was written by a medical psychologist, Henry Holland, in the *Quarterly Review* in 1850.[20] For the most part, however, the debate on man's place in nature was not an integrated debate. It was, rather, a network, involving partial overlaps between issues that seem obviously to be related from the anachronistic vantage point of a current observer. There was uniformitarian geology, but its chief preoccupation was to combat catastrophism. Its use of paleontology was in that context.[21] Viewing paleontology from another perspective, it was closely integrated with the study of comparative anatomy under the inspiration of Cuvier. Richard Owen was called 'the English Cuvier', but it was not until the 1860s that debate centered on comparative anatomy as applied to man, a debate in which the advocate of evolution, Huxley, engaged the opponent, Owen. The quarrel was refereed by Lyell, who was still six years away from accepting evolution.[22] Physical anthropology was related to the problem of the antiquity of man, while its preoccupation with skulls was strongly influenced by issues that had been raised by phrenology.[23] But anthropology in general was concerned with issues which it had inherited from speculative history and Utilitarianism. The foundation of an Anthropological Society in 1863 was not a consequence of evolutionism but a reflection of earlier preoccupations.[24] When anthropologists drew ideas from evolutionism, it was not a new inspiration, but an alliance which served to obscure other, deeper problems in their work.[25] The study of the geological evidences of the antiquity of man was done for a variety of reasons, but the concentration on the issue in the 1860s was not in any simple sense seen as a consequence of Darwin's theory. Finally, students of social development, like those in anthropology, had preoccupations which owed more to questions that had faced the Scottish Enlightenment and the Utilitarians, rather than to evolutionism.[26] My point in parading these disciplines before you is to apply to the past a truism of current intellectual life. Whatever an outsider may consider to be the relevant conceptual affinities and common interests, different disciplines have their own preoccupations, and the relations between these form a loose network of issues and interests, which does not conform to any abstract pattern. Professionals, students, and laymen differed as much then as they do now about the relevance of given issues. It could be argued that our ability to see this clearly in examples drawn from the past might help us to loosen our categories and perhaps recover some of our lost curiosity.

The example of the geological evidence of the antiquity of man is revealing.[27] What could be more central to the question of man's place in nature? Many alleged fossil remains of man had been found in the eighteenth century and again in the 1820s by William Buckland, the main exponent in that period of the recent and special origin of man. These were found in strata which also contained the remains of extinct animals, but they were all explained away, and Buckland wrote confidently in 1836 that there was no convincing evidence of fossil human remains which were laid down with the bones of extinct animals. Charles Lyell – Buckland's geological opponent but a believer in the special creation of man – saw some very convincing evidence in 1833 and wrote that the circumstances of the remains were 'far more difficult to get over than I have previously heard of'. As late as the 1855 edition of his *Principles of Geology*, he remained unconvinced. However, as early as 1849, and again in 1857, Boucher de Perthes had reported unequivocal evidence of fossil men with extinct animals, but nobody was convinced, and I have seen no report of these findings in the general debate at the time. Even Darwin was forced to admit in 1863, 'I am ashamed to think that I concluded the whole was rubbish.'

However, in 1858, In Brixham Cave in South Devon near Torquay, some flint implements were found in direct association with bones of extinct animals from the Pleistocene. This was reported to the British Association in 1858, but there is no reference to these findings in Darwin's *Life and Letters*. There was another paper read to the Royal Society in May 1859, after the joint Darwin–Wallace paper to the Linnean Society in July 1858 in which they announced their theory of evolution by natural selection. This paper was concerned with additional findings in the Somme Valley. The debate was largely confined, however, to geologists and antiquaries. Lyell went to France to see for himself and confirmed the findings in his Presidential Address to the British Association in 1859. Once again, however, these startling confirmations of the worst fears of the anti-evolutionists did not appear in the reviews of Darwin's *Origin of Species* in 1860. Indeed, they were hardly mentioned at all in the public debate, until Lyell forcibly drew attention to them in 1863 in *The Geological Evidences of the Antiquity of Man;* and even he remained – as Darwin saw it – maddeningly uncommitted to their implications in support of evolution.[28] By 1871, the connection between the question of man's antiquity and acceptance of evolution was much clearer, and Darwin could allude to them rather complacently in the introduction to *The Descent of Man:* 'The high antiquity of man has recently been demonstrated by the labours of a host of eminent men, beginning with M. Boucher de Perthes; and this is the indispensable basis for understanding his origin. I shall, therefore, take this conclusion for granted, and may refer my readers to the admirable treatises of Sir Charles Lyell, Sir John Lubbock, and others.'[29] My point in giving this example is that, as with the study of psychology, neurophysiology, and phrenology, we can understand the failure of opponents of evolution to give publicity to such findings, but the tardy integration by the proponents of evolution is odd. In any given period, intellectual life is fragmented in ways that appear bizarre to those who have the benefit of hindsight.

V

I appreciate that the structure of this paper is maddeningly parenthetical and even recursive. I now want finally to make my positive case. I want to begin with a truism that is often forgotten by those who view the past through spectacles crafted in the workshops of current disciplinary boundaries: the river of nineteenth-century naturalism was fed by many streams.[30] The part of that river which interests us most is that which led to the interpretation of man in naturalistic terms. I have mentioned that evolutionary theory drew on geological, paleontological, natural historical, and breeders' interests and issues. But it also drew on Utilitarianism, associationism, phrenology, mesmerism, Owenite Socialism, Positivism, and scientific historical criticism of the Bible. For present purposes I want to draw attention to the number of streams in that second list which are psychological. Indeed, we could extend it to include classical economics and the development of sociological and anthropological theory. I want to point out the specific role of psychology in evolutionary and social theories, but in order to do so we must look below the surface. Before Bain, psychology was not seen as a discipline in its own right. It was playing a broader social and intellectual role, and in the development of evolutionary theory its influence was more abstract. Thus, for example, important psychological works were seen in social, theological, ethical, educational, colonialist, and logical contexts. This is particularly true of association-ist psychology in the writings of the Utilitarians or Philosophical Radicals.[31] Similarly, phrenology was propagated and well received as a platform for social, educational, and public health reforms. It was offered as the key to all philosophical and social problems – a panacea for all social ills. If we are to understand the role of psychology in the evolutionary debate, we must stop looking for it at the level which has been reviewed thus far in the argument. My point for the remainder of this paper is that psychological theories lay at the very basis of much of evolutionary theory. Indeed, it can be argued that in many respects evolutionary theory (as well as theories of progress, Utilitarianism, and social science) was *applied psychology*.

In order to see this, however, we must look again at certain key figures, not all of whom are usually regarded as psychologists. It is generally acknowledged that Descartes and Locke were figures in whose work there were important and intertwined strands of ontology, epistemology, physical science, theology, psychology, and (implicitly in Locke, explicitly in Descartes) physiology. But the effective beginning of the modern tradition in empiricist psychology is usually attributed to David Hartley's *Observations on Man, His Frame, His Duty and His Expectations*, which appeared in 1749.[32] Hume's work also contained very influential arguments about the association of ideas, and although he spelled out in detail some of the laws of association, he eschewed any speculation on the physical basis of the associative process. The association of ideas had been an afterthought in Locke's *Essay*, and although it was central to Hume's argument, it is still the case that he was preoccupied with epistemological and ethical issues. The same argument can be made about Hartley's *Observations*, since he was really most concerned with questions of natural theology, morality, and the afterlife. Nevertheless, in integrating arguments drawn from Locke, John Gay, and Newton, he went into great detail about the associative process in the primarily psychological context of learning and constantly related this to a theoretical framework based on vibrations in the brain. For present purposes, it was Hartley's formulation which was most influential in providing a mechanism for changing utilities and adaptations. His detailed theory allowed others to speculate on ordered change through experience in both the psychological and, by analogy, the somatic structural realms. His ideas were the fountainhead for the development of the associationist tradition in psychology, but they were also used in theories of progress and evolution. Thus, for example, Erasmus Darwin introduced the section on 'Generation', in which he puts forth his theory of evolution in *Zoonomia* with the following passage:

The ingenious Dr Hartley in his work on man, and some other philosophers, have been of the opinion, that our immortal part acquires during this life certain habits of action or of sentiment, which become for ever indissoluble, continuing after death in a future state of existence; and add, that if these habits are of the malevolent kind, they must render the possessor miserable even in heaven. I would apply this ingenious idea to the generation or production of the embryon, or new animal, which partakes so much of the form and propensities of the parent.

Owing to the imperfection of language, the offspring is termed a *new* animal, but is in truth a branch or elongation of the parent; since a part of the embryon-animal is, or was, a part of the parent; and therefore in strict language it cannot be said to be entirely *new* at the time of its production; and therefore it may retain some of the habits of the parent-system.

At the earliest period of its existence the embryon, as secreted from the blood of the male, would seem to consist of a living filament with certain capabilities of irritation, sensation, volition, and association; and also with some acquired habits or propensities peculiar to the parent: the former of these are in common with other animals; the latter seem to distinguish or produce the kind of animal, whether man or quadruped, with the familiarity of feature or form of the parent.[33]

Erasmus Darwin has here adapted Hartley's argument and employed it as the basis for a theory of evolution. First, he has taken principles which Hartley had used to refer to the afterlife; he has secularized them and treated inheritance as an extended form of learning. In so doing, he treats the offspring, its inherited habits, and its bodily features as a prolongation of the acquired experiences of the parent. Associationist psychology, suitably extrapolated, becomes evolution. The other examples I shall cite are similar in using psychological theories as the basis for other sorts of theories, although there is space only to mention them.

Joseph Priestley also drew on Hartley's psychological theories in support of his necessitarianism and abrogated Hartley's vestigial mind–body dualism in support of his Unitarian materialist progressive philosophy of nature. This served as the foundation for his work in such diverse fields as theology, chemistry, and politics.[34] In the sphere of social theory, William Godwin, one of the founders of modern anarchism, based his theory of inevitable and indefinite human progress on Hartlean psychological mechanisms.[35] The writings of James and J. S. Mill, and the logical, educational, social, and political theories which they espoused, were also based on associationist principles.[36]

Looking across the Channel, the epistemological and psychological writings of Condillac also employed a sensationalist epistemology and an associationist mechanism.[37] These were taken up by Condorcet as the basis for his theory of human progress in the *Sketch for a Historical Picture of the Human Mind* (1795) in a way that parallels Godwin's use of associationism in England.[38] Paralleling Erasmus Darwin's evolutionism, one finds one aspect of Lamarck's evolutionary theory also dependent on the inheritance of characteristics that were acquired through the repeated strivings of individuals. This aspect was secondary in his theory to an inherent tendency to progress, but, as we shall see, some of his interpreters made it the primary factor.[39]

I have provided these examples of the influence of the psychological writings in the period before the nineteenth-century debate to show their fecundity in generating biological and social theories. It is an important task – one to which historians of psychology and of the other sciences have not begun to address themselves – to consider how many fundamental aspects of the so-called Scientific Revolution can be reinterpreted from the point of view of psychological theories. After all, the fundamental ontological, epistemological, and methodological shifts during the sixteenth and seventeenth centuries were, in large measure, concerned with problems of perception, purposiveness, and objectivity – topics that are central concerns of current psychology.[40]

But our main concern is with the nineteenth-century debate, and I want to address myself to the writings of the three main evolutionary theorists in Britain: Darwin, Wallace, and Spencer. In the crucial period in which he was formulating his theory, Charles Darwin needed a basis for making an analogy from the artificial selection of breeders to a natural process whose directionality did not depend on the intentions of a conscious selecting agent. How, to put it crudely, could *nature* select?[41] In providing an answer to this, Darwin drew on a theory which was rooted in associationist psychology but which was at two removes from the basic Hartlean doctrine. As a social doctrine, Utilitarianism depended on the associationist pleasure–pain theory of learning. In its economic form it required that men act in their rational self-interest, seeking the pleasures that flow from employment and avoiding the pains of poverty. The equilibrium and wealth of society depended on this mechanism. Pleasure and pain became the rewards and punishments of rational social and economic behavior. The system worked if there was enough to go around or if enough wealth could be created – if nature was bountiful enough and man industrious enough. Adam Smith tended to feel that they were.[42] Twenty years later, T. R. Malthus took the opposite view, putting a different face on social equilibrium and pointing out the checks on progress which, you will recall, was being advocated on the basis of parallel extrapolations from sensationalism and associationism by Godwin and Condorcet.[43] Malthus pointed out that nature was not bountiful enough, human industry was not inventive enough, and the sexual appetite was too strong. The conflict between the limits of nature and industry, on the one hand, and population growth, on the other, produced laws of struggle. He interpreted struggle as a benevolently designed cosmic learning theory:

I should be inclined . . . to consider the world and this life as the mighty process of God, not for the trial, but for the creation and formation of mind, a process necessary to awaken inert, chaotic matter into spirit, to sublimate the dust of the earth into soul, to elicit an etherial spark from the clod of clay. And in this view of the subject the various impressions and excitements which man receives through life may be considered as the forming hand of his Creator, acting by general laws, and awakening his sluggish existence, by the animated touches of the Divinity, into a capacity of superior enjoyment. The original sin of man is the torpor and corruption of the chaotic matter in which he may be said to be born.[44]

He continues:

The first great awakeners of the mind seem to be the wants of the body. They are the first stimulants that rouse the brain of infant man into sentient activity, and such seems to be the sluggishness of original matter that unless by a peculiar course of excitements other wants equally powerful, are generated, these stimulants seem, even afterwards, to be necessary to continue that activity which was first awakened. . . . From all that experience has taught us concerning the structure of the human mind, if those stimulants to exertion, which arise from the wants of the body, were removed from the mass of mankind, we have much more reason to think that they would be sunk to the level of brutes, from a deficiency of excitements, than that they would be raised to the rank of philosophers by the possession of leisure. . . . Necessity has been with great truth called the mother of invention. . . .[45]

To furnish the most unremitting excitements of this kind, and to urge man to further the gracious designs of Providence, by the full cultivation of the earth, it has been ordained that population should increase much faster than food. . . . Strong and constantly operative as this stimulus is on man to urge him to the cultivation of the earth, if we still see that cultivation proceeds very slowly, we may fairly conclude that a less stimulus would have been insufficient. . . . Had population and food increased in the same ratio, it is probable that man might never have emerged from the savage state.[46]

The mechanism had produced progress, but the means were painful. The only hope of mitigating the resulting suffering was 'moral restraint' from premature marriage, and Malthus did not put much faith in this partial palliative. I have quoted the rationale for his mechanism at length in order to offer convincing evidence that his explanations of progress and the restraints on it were generalizations of the pleasure-pain theory of learning: it caused progress, just as the survival or death of individuals constituted the ultimate sanctions of the law of population. As I said above, Darwin read Malthus' *Essay on Population* (in the sixth edition, where the arguments I have quoted are not offered in a summary form) at the crucial time when he was looking for a basis for an analogy between artificial and natural selection. Both his working notebooks and his retrospective accounts make it clear that the Malthusian population theory, when applied to plants and animals (and, secondarily, to man), provided the concepts of law and of natural pressure which Darwin required in order to formulate the theory of evolution by natural selection.[47] He wrote to Wallace in 1859:

You are right, that I came to the conclusion that selection was the principle of change from the study of domesticated productions; and then, reading Malthus, I saw at once how to apply this principle. Geographical distribution and geological relations of extinct to recent inhabitants of South America first led me to the subject: especially the case of the Galapagos Islands.[48]

The case of A. R. Wallace, the co-discoverer of the theory of evolution by natural selection, was different in important respects. I only want to make a few observations about it. He did not reach the theory by analogy to artificial selection. Indeed, he denied the analogy.[49] However, Malthus' conception of struggle also provided him with the key to the theory of natural selection. In the cases of both Darwin and Wallace, the concepts of species survival and extinction were explicit generalizations of the Malthusian concepts of population survival and death, and these were based on utilitarian concepts which, in turn, were derived from the association psychology. It is ironic that Wallace later claimed that the principle of utility was a corollary of the concept of natural selection: 'The utilitarian hypothesis . . . is the theory of natural selection applied to the mind. . . .'[50] I should perhaps add that in addition to the influence of Malthus and Lyell on him, Wallace's naturalistic approach to man was derived from three main sources: Robert Owen's socialism, Robert Chambers' *Vestiges of Creation*, and George Combe's *Constitution of Man*. All three of these drew on the principles of phrenology to support their biological approach to human nature.[51]

If we turn to the writings of the third main evolutionist, Herbert Spencer, we find a complicated situation which begins, for our purposes, with his first book, *Social Statics* (1851).[52] His central purpose in that book was to rebut moral and social theories based on Utilitarianism, theories which, of course, depend on associationism. Spencer considered the psychology of the Utilitarians to be too abstract, thereby paying insufficient attention to individual differences. Individualism was a central belief of Spencer's, and in answering the Utilitarians he turned to a theory which at once provided a sufficient number of variables to account for individual differences and made the context for the study of man that of biological adaptations rather than mind in general. That theory was phrenology. But Spencer's individualism was coupled with another basic belief in social progress, and the faculty psychology of phrenology was a static one that allowed for only partial modifications as a result of experience. After all, phrenology had been derived by Gall in explicit reaction against naive sensationalism, and Gall had postulated innately given instincts as the basis for his faculties. Thus, Spencer's adaptive view of man and his organic view of society gave no promise of social progress. In search of this he turned to another psychological theory on biological form: Lamarckian evolution. Lamarck's theory, you will recall, had two aspects: an inherent tendency to progress in life, and perturbations of this due to the recalcitrance of the environment. The secondary factor led organisms to acquire structural modifications as a result of striving, and these were passed on to the next generation. I want to stress these two aspects because what most of us mean by 'Lamarckian' evolution is the version of the theory which became popular once Spencer got through with it. He conflated the two aspects of Lamarck's doctrine and made the inheritance of acquired characteristics the mechanism of inevitable biological and human progress.

Notice that the 'Lamarckian' aspect of Lamarckianism was derived from continental expressions of the associationist psychology – the tradition inspired by Condillac and united with biology and physiology by the French Idéologues. Spencer set aside his former belief in the faculty psychology of phrenology and adopted the associationist psychology as expounded by J. S. Mill in his *Logic*. He had developed a renewed interest in psychology through his friendship with G. W. Lewes, and when he decided to write a book on the subject, he drew on Mill's expressions of the Hartlean doctrine. The heart of Spencer's *Principles of Psychology* (1855)[53] was Part III, in which he extended learning by association from the experience of the individual to that of the race, and made this the basis for biological evolution. Two years later he presented a general theory of progress based on evolution, which was, in turn, an extrapolation from the inheritance of functionally produced modifications according to the mechanisms of associationist psychology. After publishing this essay, entitled 'Progress: Its Law and Cause',[54] he further generalized his theory in *First Principles* (1862), and this contained the foundations for his synthetic philosophy as applied to psychology, education, sociology, and ethics.[55] It also provided the basis for much of later functionalist theory in psychology, sociology, and anthropology as well as for so-called Social Darwinism in political theory.[56] (It is worth adding parenthetically that it had important direct influences on neurology through the work of John Hughlings Jackson who, in turn, had an important influence on certain of the assumptions of Freud's psychoanalytic theory.)[57] My reason for reiterating the history of Spencer's intellectual development and some aspects of his influence is to draw your attention to its roots in two psychological theories: phrenology and associationism, placed in an evolutionary context and then reapplied to society in the form of the organic analogies which have been central to functionalist thought in the behavioural and social sciences. If there were space to pursue these issues further, we might explore the role of psychological theories in political and social thought – for example, in Walter Bagehot's conservative *Physics and Politics* and in John Dewey's liberal theories of industrial democracy.[58] But instead of pursuing the influence of psychological conceptions further outward into society, I would like to look briefly at a still deeper level of scientific thought.

VI

There is still another way in which psychological conceptions played an important part in the nineteenth-century debate. Once again, the issue is not straightforward. The domain of this influence was the philosophy of nature, with particular effect on the concepts of 'cause' and 'force'. This aspect of the debate is very elusive, and most scholars have only seen one bit of it in the epistemological debate between William Whewell and J. S. Mill on induction.[59] This controversy, however, was only the tip of a very large iceberg that has been investigated by Dr Roger Smith of the University of Lancaster. In an extremely interesting doctoral dissertation on 'Physiological Psychology and the Philosophy of Nature in Mid-Nineteenth Century Britain',[60] he discusses a network of ideas and influences which involved debates on the role of touch in learning (going back to Berkeley), the organic sense (or 'muscle sense'), the principle of the conservation of energy, and the concept of force itself. I am not competent to summarize his very illuminating findings, but it is clear from his research that a number of ideas were exploited in psychology, physiology, epistemology, and the philosophy of nature in an effort to overcome the problems raised by Cartesian mind–body dualism in the course of the period of the evolutionary debate. Putting the issues very crudely, there seem to have been two camps, one of which was phenomenalist in its approach and used a Humean conception of the association of ideas to argue that science can only be concerned with the constant conjunctions of phenomena. Phenomenalists argued that the concepts of cause and force were, as far as science was concerned, not amenable to further analysis. The other group, represented by some of the arguments of the Scottish School, by William Whewell, and by the later thought of A. R. Wallace, wanted to be anthropomorphic about nature and to project a dynamic view. Thus, they argued that the concept of cause referred to a power behind the phenomena, while the concept of force was not only derived by analogy from the concept of human intention but also implied will in nature. As Wallace said, 'All Force is probably Will-Force.'[61] Proponents of this view were very much in sympathy with spiritualism and other manifestations of alleged forces behind the phenomena of nature which might bridge the gaps between mind and body, human will and mechanism, dynamic nature and dead nature. I do not want to pursue this issue further except to say that Dr Smith seems to me to have opened up a whole new dimension of research which promises to help to show the real connections among psychological, philosophical, scientific, social, and popular theories in the period. He has shown that below the surface of the histories of psychology, biology, and related disciplines, as hitherto written, lies a much more potentially illuminating field of research in which psychological ideas were fundamental.

I have tried to shed light on three roles for mind in the evolutionary debate. My first pass at the problem produced a puzzling set of negative findings. Issues that appear central from the point of view of hindsight play a surprisingly small role in the great debate on man's place in nature. The opponents of scientific naturalism about man were not prepared to discuss in detail what they were not prepared to concede in principle, while the proponents were working with a set of overlapping conceptions which only came into focus after evolutionism had gained the center of the debate and the burden of proof had shifted to the opponents. A second view of the same debate reveals that psychological conceptions played a central role in social theory and in the theoretical assumptions of the evolutionists and those who influenced them most. Thirdly, I briefly reported that at the level of the philosophies of nature which underlay the scientific ideas not only of psychologists but also of physiologists, evolutionists, and philosophers, psychological conceptions appear to be fundamental. It seems clear that in the nineteenth-century debate there was an intimate mixture of psychological, social-philosophical, biological, and theological issues. These were linked with basic beliefs about man, nature, and society which were themselves playing an important role in the period. Those of us who find ourselves cut off as professional psychologists or historians of psychology from the contemporary mainstream of social and political issues can perhaps learn something from the earlier debate. It is only by taking a narrow and superficial view of that debate that the isolation of psychology can be made to appear real. It wasn't actually so isolated. It was engaging in scientific, philosophical, theological, political, and ideological work.

It seems to me that the task of a critical history of psychology is to use historical research to help us to consider the work that our own research and theoretical conceptions are doing, to evaluate that work, and then to debate among ourselves and with our students and colleagues what work we think we ought to be doing. That, of course, is a political question, but if we ponder both the negative and positive roles of psychology in the evolutionary debate, we may begin to see that it has always been political. Conceptions of psychology lie at the center of debates on man's place in nature. Such debates, however, are at the same time fundamentally concerned with man's place in society and – overtly or covertly – with the putative desirability and possibility of alternative social structures.

Psychology is never sequestered. It is only that its students, teachers, and researchers are more or less self-conscious and critical about the actual roles which their theoretical and applied scientific activities are playing in the maintenance or transformation of conceptions of nature, human nature, and society.

Source: In Mary Henle, Julian Jaynes, John J. Sullivan (eds), *Historical Conceptions of Psychology*. New York, Springer, 1973: 180–204.

Notes

1. Charles Darwin, *The Descent of Man, and Selection in Relation to Sex* (1871), 2nd edn. London, Murray, 1874.
2. Charles Darwin, *The Expression of the Emotions in Man and Animals.* London, Murray, 1872.
3. George J. Romanes, *Mental Evolution in Animals, with a Posthumous Essay on Instinct by Charles Darwin.* London, Kegan Paul, Trench, 1883.
4. Darwin, *Descent of Man*: 128–9.
5. Charles Darwin, *On the Origin of Species by Means of Natural Selection, or the Preservation of Favoured Races in the Struggle for Life.* London, Murray, 1859; facsimile reprint: New York, Atheneum paperbacks, 1967, Ch. 7.
6. Francis Darwin (ed.), *The Life and Letters of Charles Darwin*, 3 vols. London, Murray, 1887, 2: 8.
7. Quoted in Sir Gavin de Beer, *Charles Darwin, Evolution by Natural Selection.* London, Nelson, 1963; also paperback: 108.

8. Francis Darwin (ed.), *More Letters of Charles Darwin*, 2 vols. London, Murray, 1903, 2: 51–2.

9. Darwin, *Descent of Man*: 3–4.

10. I have discussed the issues which are here considered only briefly in *Mind, Brain and Adaptation in the Nineteenth Century* (Oxford, Clarendon, 1970) and in outline in The functions of the brain: 'Gall to Ferrier (1808–1886)', *Isis*, **59** (1968): 251–68.

11. Thomas Laycock, 'On the reflex functions of the brain' [read at York before the Medical Section of the British Association for the Advancement of Science on 28 Sept. 1844], *British & Foreign Medical Review*, **19** (1845): 298–311; cf. his systematic treatise, *Mind and Brain: Or the Correlations of Consciousness and Organization*, 2 vols. Edinburgh, Sutherland & Knox, 1860; [Robert Chambers], *Vestiges of the Natural History of Creation*. London, Churchill, 1844; facsimile reprint: New York, Humanities Press, 1969. [Names in square brackets denote anonymous authorship.]

12. Thomas H. Huxley, *Man's Place in Nature* (1863; reprint: Ann Arbor, Michigan, Ann Arbor paperback, 1959): see especially: 133–8 for 'A succinct history of the controversy respecting the cerebral structure of man and the apes', which is not reprinted in the standard edition of Huxley's *Collected Essays*.

13. George Rolleston *et al.*, Referees' reports on Ferrier, 1874, Archives of the Royal Society, RR. 7: 299–305, RR. 12: 103.

14. William B. Carpenter, *Nature and Man. Essays Scientific and Philosophical*. London, Kegan Paul, Trench, 1888.

15. [Thomas Brown], 'Villers *sur une Nouvelle Theorie du Cerveau* . . . by Dr Gall of Vienna', *Edinburgh Review*, **2** (1803): 147–60; [John Gordon], 'The doctrines of Gall and Spurzheim', ibid., **25** (1815): 227–68; [Francis Jeffrey], 'Phrenology', ibid., **44** (1826): 253–318, 515, and **45** (1826): 248–53; [Richard Chevenix], 'Gall and Spurzheim – phrenology', *Foreign Quarterly Review*, **2** (1828): 1–59. Anyone who wishes to investigate the fine texture of the Victorian debate as understood in its contemporary context must rely on the indispensable key to the authorship of the (usually anonymous) essays in the main periodicals provided by the research of Walter E. Houghton (ed.), *The Wellesley Index to Victorian Periodicals, 1824–1900*, of which vol. I has so far appeared. London, Routledge, 1966.

16. [Chevenix], op. cit.: 17–20; A. Macalister, 'Phrenology', in *The Encyclopedia Britannica*, 9th edn. Edinburgh, Black, 1885, 17: 844; advertisement in *The Leader*, **1** (1850): 24; George Combe, *The Constitution of Man Considered in Relation to External Objects* (1828); 9th edn (the Henderson Edition). Edinburgh, Maclachlan & Stewart, 1866: viii; J. D. Y. Peel, *Herbert Spencer. The Evolution of a Sociologist*. London, Heinemann, 1971: 11; Charles Gibbon, *The Life of George Combe*, 2 vols. London, Macmillan, 1878.

17. [W. Boyd Dawkins], 'Darwin on the descent of man', *Edinburgh Review*, **134** (1871): 195–235; [St George J. Mivart], 'Darwin's *Descent of Man*', *Quarterly Review*, **131** (1871): 47–90; Alvar Ellegard, *Darwin and the General Reader. The Reception of Darwin's Theory of Evolution in the British Periodical Press, 1859–1872*. Göteborg, Acta Universitatis Gothoburgensis 1958, Ch. 14.

18. Charles Lyell, *Principles of Geology, Being an Attempt to Explain the Former Changes of the Earth's Surface by Reference to Causes now in Operation*, 3 vols. London, Murray, 1830–3, i: 156.

The Role of Psychology   175

19. John Croom Robertson, 'Prefatory words', *Mind*, **1** (1876); cf. Alexander Bain & T. Whittaker (eds), *Philosophical Remains of George Croom Robertson*. London, Williams & Norgate, 1894. The only document I have encountered in the mainstream of the evolutionary debate which makes anything like a serious attempt to integrate the implications of psychology with those of the general evolutionary theory is John Tyndall's notorious 'Belfast Address' to the British Association in 1874, reprinted in his *Fragments of Science*, 7th edn, 2 vols. London, Longmans, 1889, 2: 135–201; cf.: 202–23. He draws primarily on the theories of Spencer, and although his address is a central document in the debate, it is usually considered to represent the final victory statement of the evolutionists. As such, it is consistent with my general thesis about the sequestration of psychology until after evolution had won the day.

20. [Henry Holland], 'Natural history of man', *Quarterly Review,* **86** (1850): 1–40.

21. M. J. S. Rudwick, 'The strategy of Lyell's *Principles of Geology, Isis,* **61** (1970): 5–33.

22. Charles Lyell, *The Geological Evidences of the Antiquity of Man With Remarks on the Origin of Species by Variation.* London, Murray, 1863; Leonard G. Wilson (ed.), *Sir Charles Lyell's Scientific Journals on the Species Question.* New Haven, Yale, 1970; cf. above, note 12.

23. James Hunt, 'On the localization of functions in the brain, with special reference to the faculty of language'. *Anthropological Review,* **6** (1868): 329–45, and **7** (1869): 100–16, 201–14.

24. John W. Burrow, 'Evolution and anthropology in the 1860s: The Anthropological Society of London, 1863–71', *Victorian Studies,* **7** (1963): 137–54.

25. John W. Burrow, *Evolution and Society. A Study in Victorian Social Theory.* Cambridge, 1966; also paperback reprint.

26. Ibid.

27. I have drawn this example from Jacob W. Gruber, 'Brixham Cave and the antiquity of man', in M. E. Spiro (ed.), *Context and Meaning in Cultural Anthropology, in Honor of A. I. Hallowell.* New York, Free Press, 1965: 373–402.

28. *Life and Letters of Darwin*, op. cit., 3: 8–14.

29. Ibid.: 15–16; *Descent of Man*, op. cit.: 2.

30. For interpretation of the general shape of the evolutionary debate, see Walter F. Cannon, 'The basis of Darwin's achievement: A revaluation', *Victorian Studies,* **5** (1961): 109–34; R. M. Young, 'The impact of Darwin on conventional thought', in Anthony Symondson (ed.), *The Victorian Crisis of Faith.* London, SPCK, 1970: 13–35.

31. Elie Halévy, *The Growth of Philosophic Radicalism*, trans. M. Morris, 2nd edn, corrected. London, Faber & Faber, 1952; Ernest Albee, *A History of English Utilitarianism.* 1901; reprint: New York, Collier paperback, 1962.

32. David Hartley, *Observations on Man, His Frame, His Duty and His Expectations*, 2 vols. London, Leake & Frederick, 1749; reprint: Gainesville, Florida: Scholars' Facsimiles & Reprints, 1966; cf. excerpts in Richard J. Herrnstein and E. G. Boring (eds), *A Source Book in the History of Psychology.* Cambridge, Harvard, 1965: 279–83, 348–55. There are passages in this collection from the writings of nearly all the figures mentioned in this paper. See also R. M. Young, 'Association of ideas', in P. P. Wiener (ed.), *Dictionary of the History of Ideas* (New York, Scribner's, in press) and

'David Hartley', in C. C. Gillispie (ed.), *Dictionary of Scientific Biography* (New York, Scribner's, in press).

33. Erasmus Darwin, *Zoonomia; or the Laws of Organic Life*, 2 vols. Dublin, Byrne & Jones, 1794–6), 1: 524–5.

34. Joseph Priestley, *Hartley's Theory of the Human Mind, on the Principle of the Association of Ideas, with Essays Relating to the Subject of it*. London, Johnson, 1775; 2nd edn, 1790; *The Conclusions of . . . Dr Hartley's Observations on the Nature, Powers, and Expectations of Man; Strikingly Illustrated in the Events of the Present Times, with Notes and Illustrations by the Editor*. London, 1794; John A. Passmore, *Priestley's Writings on Philosophy, Science and Politics*. New York, Collier paperback, 1965: 9–10.

35. William Godwin, *Enquiry Concerning Political Justice, and its Influence on Morals and Happiness* (1793): 4th edn, 2 vols. London, Watson, 1842, 1, Ch. 9, especially p. 190n. This explicit acknowledgement to Hartley does not appear in the more readily available edition edited and abridged by K. Codell Carter (Oxford, Clarendon paperback, 1971); George Woodcock, *Anarchism. A History of Libertarian Ideas and Movements* (New York, World, 1962; reprint: Harmondsworth, Penguin paperback, 1963 and 1970), Ch. 3, in which the author points out that Godwin's influence was most strongly felt in literature and in the development of socialist theory.

36. See above, note 31; [J. S. Mill], 'Bain's psychology', *Edinburgh Review,* **110** (1859): 287–321.

37. Etienne Bonnot de Condillac, *Condillac's Treatise on Sensations*, trans. G. Carr. London, Favil, 1930.

38. Antoine-Nicolas de Condorcet, *Sketch for a Historical Picture of the Progress of the Human Mind* (1795), trans. June Barraclough. London, Weidenfeld & Nicolson, 1955; Alexandre Koyré, 'Condorcet', *Journal of the History of Ideas,* **9** (1948): 131–52.

39. Jean-Baptiste de Lamarck, *Zoological Philosophy. An Exposition With Regard to the Natural History of Animals* (1809), trans. H. Elliot. London, Macmillan, 1914; reprint: New York, Hafner, 1963; W. M. Wheeler & T. Barbour (eds), *The Lamarck Manuscripts at Harvard*. Cambridge, Harvard, 1933, showing Lamarck's abiding interest in psychology.

40. I have suggested other aspects of the desirability of integrating the history of psychology with the history of science in 'Scholarship and the history of the behavioural sciences', *History of Science*, **5** (1966): 1–51, at: 18–25, and 'Animal soul', in P. Edwards (ed.), *The Encyclopedia of Philosophy*. New York, Macmillan, 1967, 1: 122–7.

41. R. M. Young, 'Darwin's metaphor: Does nature select?' *The Monist*, **55** (1971).

42. Adam Smith, *An Inquiry into the Nature and Causes of the Wealth of Nations* (1776; reprint: 2 vols. London, Dent, 1910). For analyses relating his arguments to associationism, see Leslie Stephen, *History of English Thought in the Eighteenth Century* (1876); reprint: 2 vols. London, Hart-Davis, Harbinger paperback, 1962), 2: 53–68; Halévy, op. cit. (note 32: 16–18, 88–120); Sir Alexander Gray, *Adam Smith*. London: Historical Association, 1948 and 1968; cf. the editor's introduction by Andrew Skinner to the Pelican (abridged) edition of *The Wealth of Nations*. Harmondsworth, Penguin, 1970.

43. Thomas R. Malthus, *An Essay on the Principle of Population as it Affects the Future Improvement of Society with Remarks on the Speculations of Mr Godwin, M. Condorcet, and Other Writers*. London, Johnson, 1798; reprint:

Ann Arbor, Michigan, Ann Arbor paperback, 1959, and Harmondsworth, Penguin paperback, 1970, with an excellent introduction by Anthony Flew.

44. Ibid. (Ann Arbor edn): 123–4.

45. Ibid.: 124–5.

46. Ibid.: 126–7.

47. R. M. Young, 'Malthus and the evolutionists: The common context of biological and social theory', *Past & Present*, No. 43 (1969): 109–45.

48. *More Letters of Darwin*, op. cit. (note 8), 1: 118–19.

49. Alfred R. Wallace, 'On the tendency of varieties to depart indefinitely from the original type' (1858), reprinted in Sir Gavin de Beer (ed.), *Charles Darwin and Alfred Russel Wallace, Evolution by Natural Selection*. Cambridge, 1958: 275–7; A. R. Wallace, *Darwinism*. London, Macmillan, 1889: vi; cf. above, note 41.

50. A. R. Wallace, *Natural Selection and Tropical Nature. Essays on Descriptive and Theoretical Biology*. London, Macmillan, 1891: 199–200.

51. On the influence of Owenite Socialism, see A. R. Wallace, *My Life. A Record of Events and Opinions*, 2 vols. London: Chapman & Hall, 1905, 1: 87, 91–105; 2, Ch. 24, and the appendix to *The Wonderful Century. Its Successes and Failures*. London, Swan Sonnenschein, 1898; on the influence of Chambers' *Vestiges*, see *My Life*, 1: 254–5; on the influence of Combe and phrenology, see *My Life*, 1: 234–5, and *The Wonderful Century*, Ch. 16. Wallace's beliefs in phrenology and socialism played important parts in his retreat from belief in the adequacy of natural selection to account for fundamental features of man's body and mind. See R. M. Young, '"Non-scientific" factors in the Darwinian debate', in *Actes du XIIe Congrès International d'Histoire des Sciences*. Paris, Blanchard, 1971, tome 8: 221–6. The influence of phrenology on Owenism is not straightforward. The phrenologists contributed the general orientation that men should be seen in intimate relationship with their environments, and this point of view was used to support innumerable movements and schemes for social improvement, including Owenism. On the other hand, even as modified by Spurzheim and Combe, Gall's theory did not allow sufficient scope for alteration of character as a result of alteration of external conditions. Thus, although the phrenologists provided a general warrant for seeing man in adaptive, biological terms, they clashed with the Owenites over the degree of improvement which changed social conditions could achieve. (George Combe visited New Lanark in 1820, and his brother, Abram, founded an Owenite community. Owenism was, of course, in full flower before phrenology became a popular movement in Britain, and George Combe was certainly influenced by Owen's environmentalism.) There is a systematic ambiguity between the fundamental assumptions of phrenology and the meliorist uses to which it was put. See J. F. C. Harrison, *Robert Owen and the Owenites in Britain and America*. London, Routledge & Kegan Paul, 1969): 86–7, 239–40, where the uneasy relations are mentioned, although Harrison stresses the conflict at the expense of the common assumptions. Chambers' debt to phrenology is unequivocal. See *Vestiges*, op. cit. (note 11): 322–3, 324–60. Chambers' acknowledgment of his debt to Combe's *Constitution of Man* is reprinted in the 12th edn of *Vestiges*, in which the authorship of the book is first acknowledged. Edinburgh, Chambers, 1884: xxx, xvi.

52. The development and affiliations of Spencer's evolutionary theory are considered in detail in *Mind, Brain and Adaptation*, op. cit. (note

10), Chs 5–6; 'Malthus and the evolutionists', op. cit. (note 48): 134–7; R. M. Young, 'The development of Herbert Spencer's concept of evolution', *Actes du XIe Congrès International d'Histoire des Sciences.* Warsaw: Ossolineum, 1967, 2: 273–8.

53. Herbert Spencer, *Principles of Psychology.* London, Longmans, 1855.
54. [Herbert Spencer], 'Progress: Its law and cause', *Westminster Review,* **11** (1857): 445–85.
55. Herbert Spencer, *First Principles.* London, Williams & Norgate, 1862.
56. J. D. Y. Peel's *Herbert Spencer. The Evolution of a Sociologist,* op. cit. (note 16), provides an excellent analysis of Spencer's work in its social and intellectual context, with an assessment of his influence. This should be complemented by the account given by Donald Macrae in his introduction to the Pelican edition of Spencer's *The man versus the state with four essays on politics and society* (Harmondsworth: Penguin, 1969), in which the essay, 'The social organism', is particularly relevant to the points being stressed here.
57. E. A. Stengel, 'A re-evaluation of Freud's book "On aphasia": its significance for psycho-analysis', *International Journal of Psycho-Analysis,* **35** (1954): 85–9; 'Hughlings Jackson's influence on psychiatry', *British Journal of Psychiatry,* **109** (1963): 348–55.
58. Walter Bagehot, *Physics and Politics, or Thoughts on the Application of the Principles of 'Natural Selection' and 'Inheritance' to Political Society.* London, King, 1869; C. H. Driver, 'Walter Bagehot and the social psychologists'; and 'The development of a psychological approach to politics in English speculation before 1869', in F. J. C. Hearnshaw (ed.), *The Social & Political Ideas of Some Representative Thinkers of the Victorian Age.* London, Harrap, 1933; reprint: London, Dawson, 1967: 194–221, 251–71; John Dewey, *Human Nature and Conduct. An Introduction to Social Psychology.* New York, Holt, 1922; reprint with new introduction, New York: Random House, 1930; George R. Geiger, 'Dewey's social and political philosophy', in Paul A. Schilpp' (ed.), *The Philosophy of John Dewey* (1939; 2nd edn. New York, Tudor, 1951): 337–68.
59. E. W. Strong, 'William Whewell and John Stuart Mill: Their controversy about scientific knowledge', *Journal of the History of Ideas,* **16** (1955): 209–231; A. Ellegard, 'The Darwinian theory and nineteenth-century philosophies of science', ibid., **18** (1957): 362–93; Edward H. Madden (ed.), *Theories of Scientific Method: The Renaissance Through the Nineteenth Century.* Seattle, Washington, 1960, Chs 10–11.
60. Doctoral dissertation, University of Cambridge, 1970.
61. A. R. Wallace, *Natural Selection and Tropical Nature,* op. cit. (note 50): 211; cf. *Darwinism,* op. cit. (note 49), Ch. 15.

Chapter 4

4.0 Introduction

This chapter echoes earlier concerns: the history of the human sciences and the Victorian debates over evolution. The theme that unites the book extract and the two articles which follow might be expressed by resurrecting the term 'palaetiology', a word coined by William Whewell to connote the application of the methods of science to the study of the past (see Reading 2.2). But these studies of the history of geology, evolutionary biology, palaeontology and historical anthropology also raise a common set of questions, not always explicit, about the relations between the substance of science and the beliefs, expectations and values of its practititioners.

The extract from Joe D. Burchfield's *Lord Kelvin and the Age of The Earth* culminates in an account of Kelvin's debate with T. H. Huxley in 1869, over the former's assertion that the principle of uniformity in geology ran counter to the recently discovered laws of thermodynamics. In the argument of the *Origin*, Darwin had relied on the unlimited amount of time for the operation of natural selection provided by Lyell's uniformitarian geology, and was considerably discomfited by the 'odious spectre' of Kelvin and his greatly reduced estimate of the age of the earth. Darwin's adherence to natural selection in the face of an apparently overwhelming rebuttal, and his anxious search for a less damaging revision of Kelvin's argument, are revealing, as are the reasons, apart from his physical objections, for Kelvin's antagonism to natural selection: in particular, his belief that its emphasis on chance was incompatible with design in nature.

Elizabeth Fee's article extends the period covered in the Burchfield extract, and considers how a number of anthropologists in the second half of the nineteenth century, in the face of evidence from alien cultures, reformulated the earlier view that patriarchal monogamy was a *natural* state of affairs. Certain characteristics of wealthy middle-class Victorian families (sexual control, marital fidelity, an emotional life based on home and children, and an exaggerated respect for wives and mothers) were now held to represent the *civilized* evolutionary crowning-point of a process originating in repugnant primitive promiscuity.

Fee is largely content to allow her subjects to speak for themselves in support of her belief that they constructed a past instructive to their own social order. Stephen Jay Gould advances a similar motivation – within disciplinary rather than cultural bounds – as one explanation for the eager acceptance by leading British palaeontologists of the authenticity of the Piltdown remains. His short case study of this significant episode in palaeontological history supplies an outline of the evidence for the dependence of scientific judgement in part on transient beliefs and practices. Speculating variously on the role of national rivalry with the French, cultural biases towards a brain-led evolutionary scheme for man or indeed towards white racial supremacy, he also notes how certain 'facts' about the remains conformed to expectation, and concludes that the view of science as an enterprise based on objective facts must here too give way to its conception as a human activity modified within a specific culture, albeit one that stumbles towards 'a better understanding of nature'.

4.1 Joe D. Burchfield, Kelvin and the Age of the Earth

Kelvin, Darwin, and Wallace
Charles Darwin's (1809–82) theory of evolution by natural selection was the most famous, the most popularly discussed, and perhaps the most important scientific hypothesis of the nineteenth century. It explained the progressive modification of species in terms of minute chance variations in individuals, reinforced by slow degrees over vast periods of time. The limitation of geological time was consequently of great importance to evolution. If Kelvin's arguments proved valid, they denied the possibility that the earth could have existed long enough to meet the demands of natural selection, and no one was better aware of the problem than Darwin himself. Indeed, unable to cope with the physicist's mathematics, Darwin found himself forced into an awkward and reluctant retreat.

Darwin had been profoundly influenced by Lyell's *Principles of Geology*, and when he proposed his theory of evolution by natural selection in 1859, he was supremely confident that he could count on the world having endured for as long as his theory demanded. Lyell had taught him the importance of time in accounting for geological change – and by analogy for biological change – and had convinced him that the time available for those changes had been inconceivably vast. 'He who can read Sir Charles Lyell's grand work on the Principles of Geology', Darwin challenged, 'and yet does not admit how incomprehensibly vast have been the past periods of time, may at once close this volume.' [1] This challenge appeared in every edition of *On the Origin of Species*, but although few accepted the invitation to close the book, not a few took the opportunity to answer the challenge.

The opening for the debate came in Darwin's single quantitative statement about geological time. Confident in his faith in Lyell's doctrine, he abandoned his usual caution and instead of the meticulous amassing of evidence which characterized the *Origin* as a whole, he chose to illustrate the vastness of time by a single example drawn from a familiar locale, the denudation of the Weald. His approach was simplicity itself. Looking at the great width and depth of the Weald, the great eroded valley stretching between the North and South Downs across the south of England, he compared the total volume of material which must have been eroded away during its formation with a rough estimate of the rate at which marine denudation would wash it away. He concluded that the process must have required about 300 million years.[2] It seems clear that Darwin did not think it necessary to prove the vastness of time. Lyell, after all, had already provided such arguments in abundance. Thus, rather than a careful calculation based on the available quantitative evidence, he contented himself with a naïvely conceived, hasty calculation purely for illustration. It was a mistake he came quickly to regret.

The reaction was immediate. On December 24, 1859, less than a month after the official appearance of the *Origin*, the *Saturday Review* carried a critical review which singled out the calculation of the denudation of the Weald as the focus for attack.[3] Shortly afterwards John Phillips, who had frequently disagreed with Lyell and Darwin in the past, launched his challenge before the Geological Society of London and later included it in his book, *Life on Earth*.[4] But Darwin had already become disconcerted by his unfortunate lapse of caution. Indeed when the second edition of the *Origin* appeared only two days after the criticism in the *Saturday Review*, it already contained a statement that the calculation of the denudation of the Weald might need to be reduced by a factor of two or three.[5] Nonetheless, the reviews stung, and by early January 1860 the calculation had become 'those confounded millions of years (not that I think it is probably wrong)'.[6] Even the parenthetical reservation did not last the year, however, and in November Darwin wrote Lyell of the changes to be made in the third edition: 'The confounded Wealden Calculation to be struck out, and a note to be inserted to the effect that I am convinced of its inaccuracy from a review in the Saturday Review and from Phillips, as I see in his Table of Contents that he alludes to it.'[7] The concession hurt bitterly, for a few days later he again wrote Lyell: 'Having burned my fingers so consumedly with the Wealden, I am fearful for you . . . for heaven's sake take care of your fingers: to burn them severely, as I have done, is very unpleasant.'[8] When the third edition of the *Origin* appeared in early 1861 the Wealden calculation was conspicuously missing, but it was too late. Darwin had set up a straw man and his opponents attacked it joyously.

It was almost a year after the appearance of the third edition of the *Origin* that Kelvin first attacked Darwin's chronology. In all probability he was unaware of Darwin's retreat; but even had he known, it is unlikely to have made any difference in his attitude. The Wealden calculation was gone, but not the necessity for vast periods of time to accommodate natural selection. And Kelvin was opposed to natural selection, as he was to uniformitarianism, because he believed it to be directly contrary to the fundamental principles of nature. His objections were actually threefold: natural selection could not account for the origin of life; it required far more time for its operation than the laws of physics would allow; and its emphasis upon chance did not allow for the evidence of design in nature. Of these objections, Kelvin considered the third to be the most important, but only the second was concrete and subject to scientific verification. It was consequently this objection that he stressed in his attacks upon Darwin's theory.

It should be noted that Kelvin did not object to evolution *per se*, but to the mechanism of natural selection. He was a firm believer that science should seek recourse in a first cause (or divine creation) only after every possibility for natural explanation had been exhausted, and he recognized that evolution could account for the diversity of life by natural means. He was evolutionist enough, in fact, to propose publicly that the primitive seeds of life on earth may have first arrived upon the meteoric fragments of some previously inhabited world and subsequently evolved into forms now existing.[9] Kelvin seems to have feared, however, that the Darwinians would attempt to push their theory to the origins of life itself, and this he was anxious to deny them. Indeed, he denied the validity of such speculations on any specific grounds whatever.

I need scarcely say that the beginning and maintenance of life on earth is absolutely and infinitely beyond the range of sound speculation in dynamical science. The only contribution of dynamics to theoretical biology is the absolute negation of automatic commencement or automatic maintenance of life.[10]

On this point, at least, Kelvin could have had Darwin himself as an ally, and thus his strongest objections were directed at the emphasis placed upon chance in natural selection and the implicit denial of providential guidance. Life may have originated in a simple primitive germ, but Kelvin was convinced that the complexity of life bore witness to the work of a Creative Intelligence. He was equally convinced that whereas natural selection would require almost endless time, divine guidance would enable evolution to produce the diversity of life in a relatively short period. Thus as far as evolution was concerned, his arguments for limiting the earth's age were also proofs of design in nature.

The first important reference to the basic incompatibility between Darwin's theory and Kelvin's geochronology appeared in Fleeming Jenkin's famous review of the *Origin* in 1867.[11] Jenkin (1833–85) had much in common with Kelvin. He was a Scot, a physicist, an engineer, and an inventor, and at the time of his views of the *Origin*, the two men were working in close association. They had served together on several transatlantic cable laying enterprises, and as early as 1860 had created a partnership to market a patent. In 1865, the partnership was extended to include the joint marketing of all their patents, and a second was formed as consulting engineers.[12] The partners were more than successful business men, however: they were close personal friends. Thus Jenkin, who clearly looked upon Kelvin as the senior partner (he once referred to himself as a 'great worshipper' of Kelvin,[13] was in a position to be thoroughly familiar with the older man's views on geological time, and it is hardly surprising that he used them with such telling effect in his review of the *Origin*.

Jenkin's criticism of the Darwinian theory hinged upon his contention that it was mathematically impossible for a single fortuitous variation in an individual organism to be perpetuated. Any individual variation, he believed, would be swamped in succeeding generations by the preponderance of normal individuals. The inadequacy of geological time was a secondary point in his criticism, but he stressed the value of Kelvin's work as strong reinforcement for his primary argument. Jenkin knew and candidly admitted that Kelvin's results were mere approximations. He conceded that some of the information upon which they were based was totally wrong, and that new data might materially alter Kelvin's quantitative conclusions. Nonetheless, he accepted the basic argument as valid and insisted that it must therefore limit the uniformitarian demands for time. He was convinced, moreover, that whatever quantitative changes had to be made in Kelvin's results, the maximum time available since the formation of the earth's crust would prove inadequate for Darwin's theory.[14]

In his discussion of geological time, Jenkin's views were very similar to Kelvin's. He saw the principle of the dissipation of energy as absolute proof against the possibility of geological uniformity. The geological activity of the earth must diminish over long periods of time, and thus the present rates of geological processes cannot in themselves provide an accurate measure of past time. Jenkin did not argue for catastrophes in the old sense any more than Kelvin had, but because the rates of geological change could not be constant some other measure of geological time had to be adopted.

So far as the world is concerned, past ages are far from countless; the ages to come are numbered; no one age has resembled its predecessor, nor will any future time repeat the past. The estimates of geologists must yield before more accurate methods of computation, and these show that our world cannot have been habitable for more than an infinitely insufficient period for the execution of the Darwinian transmutation.[15]

Also like Kelvin, he chose Darwin's calculation of the denudation of the Weald as an example of the errors which could arise from an indiscriminate reliance upon the principle of uniformity. The erosion of the Weald, he said, might have taken place a thousand times faster or a million times slower than Darwin had estimated. The available data was simply too meagre for judgment. 'The whole calculation', he concluded with biting precision, 'savours a good deal of that known among engineers as guess at the half and multiply by two.'[16]

Of the other critics who used Kelvin's chronology to refute Darwin, perhaps the most notable was St George Mivart (1827–1900). By the time Mivart's *On the Genesis of Species* appeared in 1871 both Huxley and Wallace had yielded to Kelvin's arguments without abandoning Darwinism. But Mivart refused to accept the possibility of such a compromise. A significant part of his criticism of the *Origin* hinged upon the absence of any trace of the transitional forms that must have existed if one species evolved from another. This was an objection that Darwin had attempted to answer by emphasizing the imperfection of the geological record. If the time available were much reduced, however, Darwin's defense would be severely weakened, and this was the point Mivart hoped to make with the help of Kelvin's physics. He made no pretext of being able to evaluate the validity of the physicist's arguments, but argued that the fact that they had not been refuted spoke loudly in their favour. He also contended that since neither biology nor geology alone could resolve the question of how imperfect the geological record really was, the results of physics had to be considered. In Mivart's opinion there was no question but that natural selection had to give way, and to that end he strongly endorsed Kelvin's conclusions.[17]

Darwin was greatly troubled by the use of Kelvin's arguments, and in his letters after 1868 he frequently alluded to his doubts and uncertainties. Early in 1869, for example, he complained to Wallace that 'Thomson's views on the recent age of the world have been for some time one of my sorest troubles.'[18] He had reason to be troubled, for Kelvin's most sustained attack upon natural selection and the uniformitarian view of time came between 1868 and 1871, just when the last two editions of the *Origin* were being prepared. And neither Darwin nor his supporters could find adequate refutations for his arguments.[19] The only ray of hope seemed to come from Croll, and Darwin grasped it eagerly.

Like Lyell, Darwin was first attracted by Croll's glacial hypothesis. For years he had hoped to find evidence that the ice ages had not occurred simultaneously throughout the world, but until he read Croll had begun to doubt that the hope would ever be realized.[20] He consequently greeted Croll's work enthusiastically, even the parts dealing with geological time. As he wrote in September 1868:

I have never, I think, in my life been so deeply interested by any geological discussion. I now first begin to see what a million means, and I feel quite ashamed of myself at the silly way in which I have spoken of millions of years.[21]

As he prepared the fifth edition of the *Origin*, Darwin asked for further elaboration on the question of time, and was greatly pleased to learn of Croll's efforts to extend the physical estimate of the sun's age.[22] Nonetheless he remained disturbed by the fear that even Croll's most generous estimate of solar heat would still allow too little time. Thus in January 1869 he wrote again:

Notwithstanding your excellent remarks on the work which can be effected within a million years, I am greatly troubled at the short duration of the world according to Sir W. Thomson, for I require for my theoretical views a very long period before the Cambrian formation.[23]

There appeared to be no way out. Whatever time might be added to the age of accumulated strata had to be bought at the expense of the Precambrian. The sun limited the age of both, and natural selection demanded ample time in both. The fifth edition of the *Origin* bears witness to Darwin's discomfort and his reluctant attempts to compromise. The Lamarckian reliance upon the direct effects of environment and the use and disuse of parts, never entirely absent from the *Origin*, became more pronounced as Darwin attempted to speed up the process of evolution. And where the denudation of the Weald had once illustrated the immensity of time, he substituted a detailed discussion of Croll's views on the inability of man to conceive of millions of years and on the consequent errors of geologists in attempting to express geological time in years. He also endorsed Croll's assessment of how rapidly denudation actually takes place, and though clearly still hoping that the 60 million year estimate of time since the beginning of the Cambrian might be extended, conceded that great changes could have taken place during the time allotted.[24]

Compromise was not total retreat, and there were occasions when Darwin felt a surge of optimism. One such occasion came early in 1869 when he read an anonymous review of Kelvin's opinions and of the recent controversy between Kelvin and Huxley. The reviewer, who turned out to be P. G. Tait, Kelvin's long-term collaborator, was obviously a physicist and obviously strongly partisan in Kelvin's favor. But paradoxically, he took an excessively hard line in insisting that Kelvin had been over generous in his estimate of the sun's age. The true limit of time, he asserted, must be closer to 10 million years than to Kelvin's 100 million. Darwin saw a faint ray of hope, however, in the fact that the physicists disagreed among themselves. As he gleefully remarked to Hooker, geologists should be well warned and be careful in accepting the physicists' views. Physics, after all, had been wrong before. 'Nevertheless,' he continued, 'all uniformitarians had better at once cry "peccavi" – not but what I feel a conviction that the world will be found rather older than Thomson makes it, and far older than the reviewer makes it. I am glad I have faced and admitted the difficulty in the last [fifth] edition of the *Origin*. . . .'[25]

Darwin's concern over the limitation of the earth's age was not shared by his co-discoverer of natural selection, Alfred Russel Wallace (1823–1913). Although initially cautious about accepting Kelvin's results, the arguments of Geikie and Croll had quickly convinced Wallace that the uniformitarians had greatly overestimated the time necessary for the actions of denudation and sedimentation.[26] By late 1869 he was willing to accept Kelvin's arguments, and before the end of the year enthusiastically wrote Darwin that he had found a way out of the dilemma of time[27]. His solution appeared in a fascinating short paper early the following year.[28]

The basis for Wallace's new hypothesis was Croll's theory of the cause of the ice ages. A year earlier he had still leaned toward Lyell's geological explanation of widespread glaciation and had given Croll's astronomical hypothesis only secondary consideration, but thanks to Darwin's enthusiasm he completely reversed his views.[29] Not only did he accept Croll's hypothesis as the best key to determining geological time, he also saw in it a previously overlooked factor which he believed must influence the operation of natural selection. Referring to Croll's calculation of the changes in the earth's orbital eccentricity during the last three million years Wallace observed that during most of that time the eccentricity was considerably higher than it is at present. In fact, for the last 60,000 years the earth has been in a period of uncommonly low orbital eccentricity. According to Croll's hypothesis, therefore, the earth must now be experiencing an abnormally prolonged period of climatic stability. Applying this conclusion to the problem of species change through natural selection, Wallace drew an amazing yet logical conclusion.

Croll's hypothesis asserted that during periods of high orbital eccentricity radical changes in climate would occur every 10,500 years (half the period of a complete precession) as glaciers covered first one hemisphere and then the other. Wallace argued that these changes would be particularly pronounced in the extratropical regions where they would result in vast migrations of both plants and animals. This almost constant migration, he continued, would intensify the competition between allied species and force many to extinction. And finally, in a notable example of continued Lamarckian influence, he suggested that these 'altered physical conditions would induce variation' among the individual organisms and thus accelerate the process of change. All in all, he concluded:

. . . we should have all the elements for natural selection and the struggle for life, to work upon and develop new races. High eccentricity would therefore lead to a rapid change of species, low eccentricity to a persistence of the same forms; and as we are now, and have been for 60,000 years, in a period of low eccentricity, *the rate of change of species during that time may be no measure of the rate that has generally obtained in past geological epochs.*[30]

Turning next to the problem of the earth's age itself, Wallace again looked to the arguments of Croll and Lyell. He had already been convinced that Lyell had greatly underestimated the rates of geological activity, and thus not surprisingly, he favored Croll's estimate of 80,000 years as the time since the last ice age rather than Lyell's choice of nearly one million years. Nonetheless, he believed that Lyell's attempt to measure the duration of geological epochs from the changes in the species of marine mollusca would still prove valuable. Lyell had not only chosen the earlier period of eccentricity for the glacial epoch, however, he had also overlooked the 60,000 years of relative stability immediately preceding the present age. Taking all of these factors into account Wallace asserted that 100,000 years rather than one million years would probably be a better (although perhaps still excessive) estimate for time required for the extinction of 5% of the existing species of marine mollusca. This substitution would reduce Lyell's results by a factor of **ten and would fix the age of the Cambrian formation at only 24 million** years. Thus, Wallace's hypothesis could easily accommodate Darwin's demands for a long Precambrian period for the development of life and still fit within Kelvin's 100 million years for the total age of the earth. The Precambrian would in fact be three times longer than all post-Cambrian time; and the rate of biological change during that and subsequent periods would on the whole be faster than the relatively slow change in present species.

Darwin was not convinced. Returning a draft of the article he commented: 'your argument would be somewhat strengthened about organic changes having been more rapid, if Sir W. Thomson is correct that physical changes were more violent and abrupt'. He was not yet ready to make such a concession. The real rub, however, was simply that: 'I have not yet been able to digest the fundamental notion of the shortened age of the sun and earth.'[31]

Despite mounting pressures from the seemingly irrefutable arguments of the physicists, Darwin never really accepted Wallace's compromise. He neglected it even in his letters to Wallace, as for example, in 1871 when he was searching for a reply to Mivart: 'I can say nothing more about missing links than I have said. I should rely much **on pre-Silurian times; but then comes Sir W. Thomson like an odious spectre.[32]** Wallace's solution was clearly not the one he sought, and when he turned to the question of time in the sixth edition of the *Origin*, it was Croll's calculation that he discussed rather than Wallace's. There had been one significant effect, however, for after discussing both Kelvin and Croll and the difficulty in reconciling natural selection with even the greatest period that their combined hypotheses would allow for the Precambrian, he continued:

It is, however, probable, as Sir William Thompson [sic] insists, that the world at a very early period was subjected to more rapid and violent changes in its physical conditions than those now occurring; and such changes would have **tended to induce changes at a corresponding rate in the organisms which then existed.[33]**

This was, of course, the very concession that he had refused to make to Wallace two years earlier.

Like all of Darwin's compromises, his concessions to Wallace and Kelvin included a hedge. Thus his final published statement on the subject of time attempted to retrieve with one hand what he felt obliged to concede with the other:

> With respect to the lapse of time not having been sufficient since our planet was consolidated for the assumed amount of organic change, and this objection, as urged by Sir William Thomson, is probably one of the gravest as yet advanced, I can only say, firstly that we do not know at what rate species change as measured in years, and secondly that many philosophers are not as yet willing to admit that we know enough of the constitution of the universe and of the interior of our globe to speculate with safety on its past duration.[34]

In effect, Darwin asked for a suspension of judgment because he still could not accept the idea that Kelvin's allowance of time would be adequate for what he saw as the necessarily slow progress of Precambrian evolution. But he stood against the tide. Geikie, Croll, and Wallace were heralds of the changed geological opinions that were to dominate the next few decades, and even Huxley, rising to defend geology and natural selection from Kelvin's attack, found that he could do little more than lead a graceful retreat.

The Kelvin–Huxley debate

That Thomas Henry Huxley (1825–95) should respond to Kelvin's challenge to the uniformitarian and Darwinian theories was almost inevitable. He was both Darwin's self-appointed 'bulldog' and the president of the Geological Society of London at the time when Kelvin's influence first came to be strongly felt; and he was certainly not averse to a good fight. He therefore took the occasion of his presidential address in 1869 to reply to Kelvin's assertion of a year earlier that 'a great reform in geological speculation seems now to have become necessary'.[35] Kelvin's rebuttal, delivered like his original challenge before the Geological Society of Glasgow, came within the month, and there, uncharacteristically for both, the matter was allowed to rest. Neither Huxley nor Kelvin had attended the other's lectures, and neither was primarily a geologist; but the nature of their exchange left no doubt that a debate had in fact taken place, and its effects influenced geological theory for decades.

When Kelvin initiated the debate in February 1868, the revisions for the fifth edition of the *Origin* were yet to be begun, Geikie's paper on denudation was yet to be delivered, and the full account of Croll's theory was still at the printers. And though Phillips, Croll, and a few others had publicly endorsed the physical limits on the earth's age, the recent edition of Lyell's *Principles* had again raised the spectre of the earth as a great perpetual motion machine. Kelvin consequently felt, with some justification, that his arguments, had been ignored and that geologists in general were still neglecting the basic principles of natural philosophy.

Nonetheless, the attack began mildly. At the end of the eighteenth century, he conceded, geology clearly needed more time than the dogmatic diluvian hypotheses of the period allowed, and the geologists had justifiably sought to break free from the restrictions of Mosaic chronologies. He also praised the efforts made since then to make geology into a truly exact science. Some modern geologists, however, had gone too far in their reaction against the old hypotheses, and the time had long passed for them to reassess the principle of uniformity and its demands for unlimited time. Curiously, after his ringing call for reform, Kelvin chose to resurrect John Playfair, the spokesman for a much earlier generation, as the focus for attack. Yet the implication was plain; geology had not kept up with the advances in physical science.

Kelvin's criticism centered on the accusation that Playfair had confused *present order* or *present system* with the *laws now existing*. The laws of nature had not changed, Kelvin agreed, but the condition of the earth had changed continuously. Indeed, it was just because the second law of thermodynamics was unvarying, that one could be certain that the energy available for geological activity must have decreased constantly since the world began. Friction alone must have cost the earth a measurable part of its store of energy; and Kelvin used Playfair's neglect of this point, and the neglect of others including the mathematicians Laplace and Lagrange, as a springboard for proposing his theory of tidal retardation. The calculation of time which this argument permitted was but a small part of his concern, however. His other arguments had already provided much more explicit results than it could supply. His real aim was to show, once and for all, that the laws of thermodynamics utterly negated the possibly that either continuously uniform or perfectly cyclic processes could exist in a finite world, and therefore that geological uniformity could not be a law of nature. Kelvin was convinced, moreover, that so long as geologists continued to cling to uniformitarianism, geology would never take its place as a truly exact science. Thus it was his contemporaries rather than Playfair or Hutton whom he challenged with the words: 'It is quite certain that a great mistake has been made – that British popular geology at present is in direct opposition to the principles of natural philosophy.'[36]

Huxley, a formidable champion in the arena of public debate, rose to answer the challenge. In this instance, however, his efforts must be praised more for their vigor than their strength. His address was characteristically lucid and elegant, but it seldom came to grips with the fundamentals of Kelvin's argument. To use his own metaphor, he sought by 'mother-wit' to act as an attorney for the defense of geology.[37] But though the jury before whom he tried his case – the Geological Society of London – was friendly, mother-wit alone could not combat the carefully reasoned, physically and mathematically correct arguments of his opponent.

Kelvin had attacked British 'popular geology' which to him meant uniformitarianism, but in choosing to focus his attack on Playfair rather than on Lyell or another contemporary geologist, he provided an excellent opening for rebuttal. Huxley seized on it immediately. If

Playfair and Hutton had improperly assumed that the intensity of the earth's activity has never changed, then point out the fallacy, he challenged, but do not assume that a valid attack upon them is also a valid attack upon modern geology. In effect, he accused Kelvin of attacking a straw man as he continued:

I do not suppose that, at the present day any geologists would be found to maintain absolute Uniformitarianism, to deny that the rapidity of the rotation of earth *may* be diminishing, that the sun *may* be waxing dim, or that the earth itself *may* be cooling. Most of us, I suspect, are Gallios, 'who care for none of these things', being of the opinion that, true or fictitious, they have made no practical difference to the earth, during the period of which a record is preserved in stratified deposits.[38]

Certainly Huxley's assertion here was not entirely unjustified, but neither was it particularly accurate. In the year since Kelvin's challenge had been issued both Geikie and Croll had come out strongly in support of the belief that geological theory could be reconciled with his limited time scale. A few other geologists had also abandoned strict Lyellian uniformity, and some indeed had never really accepted it. But the majority of their fellows seldom confronted the question directly. They were content to assume uniformity for the sake of description and explanation, and to accept the idea that geological time was virtually limitless if not actually infinite. Huxley's remark had thus sacrificed accuracy for rhetoric and yet had not avoided conceding the substance of Kelvin's argument.

The weakness of Huxley's position was still more apparent in his treatment of Kelvin's quantitative estimates of time. He began by asking whether it had ever been denied that 100 million years *may* have been long enough for the purposes of geology; a question that could have been answered emphatically yes had he chosen to answer. But instead he reversed his ground completely to assert that two, three, or four hundred million years made 'all the difference' in Kelvin's theory. He had missed the point entirely – or had carefully avoided it – for Kelvin's position had been that if any limit whatever could be placed upon the earth's age, it would refute uniformitarianism. It was sheer sophistry, moreover, to criticize Kelvin's broad allowances for error as vagueness, while in the same breath asserting that a change of a factor of two or three in his results would negate his argument. Huxley had sidestepped the fundamentals of Kelvin's objections to uniformitarianism, and yet in conceding the possibility of limiting the earth's age, he essentially admitted his willingness to abandon the indefinite time of Lyell. Darwin too must have felt abandoned when he heard:

Biology takes her time from geology. The only reason we have for believing in the slow rate of the change in living forms is the fact that they persist through a series of deposits which, geology informs us, have taken a long while to make. If the geological clock is wrong, all the naturalist will have to do is to modify his notions of the rapidity of change accordingly.[39]

Certainly Darwin could take little comfort in these remarks since Fleeming Jenkin had already shown that natural selection alone could not adequately account for species change within Kelvin's limited time.

Huxley's address was not all mere sound and fury, however. Even his disregard for time was based on his belief that geology could and should accommodate the laws of physics. Thus the most significant part of his address concerned a proposed reform in geological theory that, while not exactly what Kelvin had in mind, would alleviate some of the major difficulties. Kelvin had attacked only uniformitarianism as British 'popular geology', and Huxley questioned whether this was either just or accurate since catastrophism still claimed many able adherents. Reform might be necessary, but Huxley asserted that it should involve eliminating the weaknesses of both uniformitarianism and catastrophism while preserving the strengths of both. After all, he argued, there was actually no basic antagonism between catastrophism's practically unlimited bank of force and uniformitarianism's practically unlimited bank of time. 'On the contrary, it is very conceivable that catastrophes may be part and parcel of uniformity.'[40] The real disagreement, Huxley believed, arose from the antiprogressional bias of the uniformitarians, and from the catastrophists' neglect of the present as the only reliable guide to the past. Both schools had failed to approach geology in what he termed the aetiological sense; that is, neither had attempted 'to deduce the history of the world as a whole from both the known properties of the matter of the earth and the conditions in which the earth has been placed'.[41] Huxley's new approach, which he dubbed *Evolutionism*, would address itself to just this problem. It would allow for geological forces in the past that were of greater power or of a different nature than those in the present – in other words, it would allow for catastrophes in a natural but not a supernatural sense – but would stress the necessity of studying present geological processes as the guide to understanding the record of the past. Furthermore, by taking progression rather than uniformity as the key to the earth's history, evolutionism would recognize that geological change was a function of time. It would if necessary accommodate even the limited time scale of Kelvin. Clearly, Huxley's evolutionism bore a marked resemblance to the kind of uniformitarianism proposed by Geikie, and under either name it had a profound effect.

A final point in the address raised the question of mathematical certainty, and here he confronted Kelvin most directly. In looking at the physical arguments, Huxley granted Kelvin the accuracy of his mathematics and the probable validity of his principles, but not the necessary validity of his assumptions. Repeatedly he pointed out how Kelvin had required his readers to suppose this or assume that, while he stressed the hypothetical nature of these assumptions and argued that other suppositions had an equal right to consideration. Singling out Kelvin's most recent argument, the case of the tidal retardation of the earth, Huxley assailed it as an example of how 'the admitted accuracy of mathematical process' had been allowed 'to throw a wholly inadmissible appearance of authority over the results obtained by them.' Kelvin's mathematics were no doubt beyond question, but:

Mathematics may be compared to a mill of exquisite workmanship, which grinds you stuff of any degree of fineness; but, nevertheless, what you get out depends upon what you put in; and as the grandest mill in the world will not extract wheat-flour from peascod, so pages of formulae will not get a definite result out of loose data.[42]

Huxley's analogy was frequently quoted, but the full weight of his cautionary remarks was not immediately appreciated. Although in retrospect it seems his strongest argument by far, Kelvin's opponents failed to capitalize on it until late in the century. Thus for nearly two decades the most influential part of the address was not Huxley's warning about the uncertaties of Kelvin's assumptions, but his assertion that biology could take its time from geology.

Kelvin responded eagerly. His rebuttal, delivered only two weeks later, was obviously hurried but it contained a point by point refutation of Huxley's position.[43] Lacking the gift of prophecy, Kelvin could not appreciate the value of his opponent's *evolutionism*. He dismissed it facilely, but with some justice, as essentially the geology that he had learned at Glasgow University thirty years before. Kelvin's dispute, after all, was with uniformitarianism, and he emphatically denied Huxley's contention that few geologists were true uniformitarians. He also vigorously condemned the flippancy of Huxley's appeal to mother wit and the know-nothing attitude implied in the assertion that geology cared nothing about whether the sun were cooling or the earth's rotation slowing down. This was precisely the attitude that he believed had led geology into conflict with natural philosophy and that he sought to combat. As for the assertion that the limitation in geological time made no difference to biology, he replied:

The limitations of geological periods, imposed by physical science, cannot, of course, disprove the hypothesis of transmutation of species; but it does seem sufficient to disprove the doctrine that transmutation has taken place through 'descent with modification by natural selection'.[44]

Darwin himself could only reluctantly concur in this judgment.

Kelvin's address contained nothing original. In fact, it consisted largely of extended quotations to illustrate his points, but the quotations were effectively used. He turned Huxley's own words upon him, bombarded his listeners with quotations from geologists proclaiming the virtual limitlessness of the earth's age, and cited Geikie and Phillips to show how the geological evidence could agree with natural philosophy when 'properly interpreted'. With regard to Huxley's more substantial argument, Kelvin admitted that his calculations involved many assumptions, but vehemently denied the accusation that his mathematics gave too much authority to results from 'loose' data. His position remained firm that the wide limits which he had allowed in his results were quite sufficient to account for any errors in either his data or assumptions. One hundred million years was the maximum probable duration of life on earth that his calculations would allow, but he remained convinced that even 400 million years would not satisfy the demands of the uniformitarians.

Although the exchange between Huxley and Kelvin contributed nothing new to the solution of the problem of the earth's age, it inaugurated a new era in geological speculation that went far beyond the single question of time. Huxley's defense of uniformitarianism had failed to materialize. Instead he had championed a new approach to the problems of physical geology, an approach which could be made compatible with Kelvin's conclusions about the age of the earth and which was already being followed by a few geologists like Geikie and Croll. The debate succeeded, however, in bringing the question of geological time into the public forum and in generating considerable public interest. No longer could Kelvin complain that his arguments were ignored by the geologists. The apparent strength of his position had been demonstrated, and the weight of both scientific and public opinion had begun to swing decidedly in his direction.

Source: Taken from Joe D. Burchfield, *Lord Kelvin and the Age of the Earth.* London, Macmillan 1975, Ch. 3.

Notes

1.   Charles Darwin, *On The Origin of Species*, 1st edn, London, Murray, 1859,: 282.

2.   Ibid., : 285–7. This estimate is excessive even by today's standards which put the origin of the Weald at about 135 million years.

3.   Anon., 'Darwin's *Origin of Species*', *Saturday Review* (London), (24 Dec. 1859) **8** 775–6.

4.   John Phillips, 'Presidential address', *Quar. Jour. Geol. Soc. Lon.,* **16** (1860): lii–lv; *Life on Earth: Its Origin and Succession.* London, Macmillan, 1860.: 130. Phillips had crossed swords with Lyell and Darwin as early as 1838, shortly after the latter's return on the *Beagle.* (See [Kathrine] Lyell, *Life, Letters and Journal of Sir Charles Lyell, Bart*, 2 vols London, Murray 1881 II: 39–41. This letter also appears misdated as 1858, ibid., II, : 281–2.) It is apparent from Darwin's correspondence with Lyell in 1859–60 that he regarded Phillips' views with reluctant respect. (See: Francis Darwin (ed.), *Life and Letters of Charles Darwin*, 3 vols London, Murray, 1888, II, : 309, 349; F. Darwin and A. C. Seward (eds), *More Letters of Charles Darwin: A Record of His Work in a Series of Hitherto Unpublished Letters*, 2 vols New York, Appleton, 1903, I, : 127, 130, 141).

5.   C. Darwin, *The Origin of Species by Charles Darwin: A Variorum Text.* Philadelphia, Univ. of Penn., 1959, : 484.

6.   F. Darwin (ed.) Life of Darwin, II, : 264.

7.   Ibid., II, : 350. This statement was made before Darwin read *Life on Earth* (see ibid., II, : 349) which he subsequently criticized quite harshly. (See letters to J. D. Hooker. 15 Jan. 1861, and A. Gray, 5 June 1861. Ibid., II, : 358, 373–4). Thus, although he left the calculation out of the third edition, he did not include the proposed footnote.

8. F. Darwin and Seward, *More Letters of Darwin*, II, : 139.
9. William Thomson, First Baron Kelvin, *Popular Lectures and Addresses*, 3 vols. London, Macmillan, 1891–94, II, : 197–205 ('Presidential Address', *Brit. Assoc. Report*, Edinburgh, 1871.) Huxley referred to the theory as 'Thomson's 'creation by cockshy' – God Almighty sitting like an idle boy at the seaside and shying aerolites (with germs), mostly missing, but sometimes hitting a planet!' (See Leonard Huxley, *Life and Letters of Sir Joseph Dalton Hooker*, 2 vols. New York, Appleton, 1918, II, : 126).
10. Kelvin, *Popular Lectures*, I, : 422 (On the Sun's Heat, read 21 Jan. 1887).
11. Fleeming Jenkin, '*The Origin of Species*', *North British Review*, (June 1867): 277–318.
12. Silvanus P. Thompson, *The Life of William Thomson, Baron Kelvin of Largs*, 2 vols. London, Macmillan, 1910, I, : 408–9, 552–3.
13. Robert Louis Stevenson, 'Memoir of Fleeming Jenkin', in S. Colvin and J. A. Ewing (eds) *Papers of Fleeming Jenkin*, 2 vols. London, Longmans, 1887, I, : lxi.
14. Jenkin, '*The Origin of Species*', : 294–305.
15. Ibid., : 295.
16. Loc. cit.
17. St George Mivart, *On The Genesis of Species*. New York, Appleton, 1871, : 142–57.
18. **James Marchant, *Alfred Russel Wallace, Letters and Reminiscences*, :** London, Cassell, 1916, I, : 242. Letter dated 14 Apr. 1869.
19. For some indication of the magnitude of Kelvin's influence at this time see Alvar Ellegård, *Darwin and the General Reader: The Reception of Darwin's Theory of Evolution in the British Periodical Press, 1859–1872*. Göteborg, 1958, : 237–8.
20. F. Darwin and Seward, *More Letters of Darwin*, I, : 460–5. Letters to Lyell, Hooker, Huxley, and H. W. Bates between 1 Nov. 1860 and 26 Mar. 1861.
21. Ibid., II, : 211. Letter dated 19 Sept. 1868.
22. James Campbell Irons, *Autobiographical Sketch of James Croll with Memoir of his Life and Work*. London, Edward Stanford, 1896, : 200–3.
23. F. Darwin and Seward, *More Letters of Darwin*, II, : 163. Letter dated 31 Jan. 1869.
24. C. Darwin, *Origin, Variorum*, : 482–6.
25. F. Darwin and Seward, *More Letters of Darwin*, I, : 313–14. Letter dated 24 July 1869.
26. Alfred Russel Wallace, 'Geological climates and the origin of species', *Quar. Review*, **126** (1869) : 375–6.
27. Marchant, *Letters of Wallace*, I, : 246. Letter dated 4 Dec. 1869.
28. A. R. Wallace, The measurement of geological time', *Nature*, I (17 Feb. 1870) : 399–401 and 452–5.
29. Wallace sent a prepublication copy of 'Geological climates', which was a review of the new editions of Lyell's *Principles and Elements*, to Darwin early in 1869. Darwin immediately responded with the opinion that Wallace had not placed as much confidence in Croll's work as he, Darwin, was willing to do. (See Marchant, *Letters of Wallace*, I, :242. Letter dated 14 Apr. 1869.) There can be little doubt that this opinion was in some degree influential in Wallace's subsequent about-face and his adoption of Croll's hypothesis in 'Geological time'.
30. Wallace, 'Geological time', : 454.
31. Marchant, *Letters of Wallace*, I, : 250–1. Letter dated 26 Jan. 1870.

Actually, Darwin was confusing Wallace's argument for rapid climatic change due to alterations of astronomical conditions with Kelvin's argument for greater past meteorological and plutonic activity due to higher temperatures in the earth and sun. The two arguments have little in common, but I have found no record of Wallace correcting Darwin.

32. Ibid., I, : 268. Letter dated 12 July 1871.
33. Darwin, *Origin, Variorum*, : 513.
34. Ibid., : 728.
35. Kelvin, *Popular Lectures*, II, : 10 ('On Geological Time', read 27 Feb. 1868).
36. Ibid., II, : 44.
37. T. H. Huxley, *Discourses, Biological and Geological Essays. New York, Appleton, 1909*, : 308–42.
38. Ibid., : 329. Huxley's allusion was to the Bible, Acts 18:17. Gallio, the Proconsul of Achaea, refused to try Paul under Roman law for breaking Jewish law. Jewish law *was not his province* and he 'cared for none of those things'. Huxley's reference was perhaps more pointed than it appears at first.
39. Ibid., : 331.
40. Ibid., : 327.
41. Ibid., : 322.
42. Ibid., : 335–6.
43. Kelvin, *Popular Lectures*, II, : 73–127 ('Of Geological Dynamics', read 5 Apr. 1869).
44. Ibid., : 89–90.

4.2 Elizabeth Fee, The Sexual Politics of Victorian Social Anthropology

Some modern anthropologists have characterized the work of their nineteenth-century forebears as politics masquerading as science. The systems of evolutionary anthropology, riddled as they were with blatant racism, are now written off as the pseudo-scientific apologetics of early imperialism: *modern* anthropology is assumed to be value-free. But historical anthropology – indeed history itself – is inevitably, dialectically intertwined with politics. The analysis of the past is shaped by the present, our choice of questions, our selection of evidence, our analyses, all are influenced by contemporary concerns. The reconstruction of the past in turn serves present needs, as it clarifies or justifies the contours of present reality.

In the past fifty years, social anthropology has not escaped from politics; it has merely succeeded in gaining the appearance of political neutrality at the cost of a retreat into cautious microanalysis. But earlier anthropologists, attempting to come to grips with a changing political world, boldly sought for answers in grand syntheses of the past. They aimed at nothing less than an understanding of the entire history of the human race; and if in the end they succeeded in telling us more about themselves than about human pre-history or the cultures they ostensibly studied, we may still admire their daring, their forthrightness, and their intellectual ambition.

The reassessment of the past becomes particularly important when social conventions are threatened or the future appears uncertain. When, in the late nineteenth century, feminists challenged complacent assumptions about the timeless quality of women's role, the theoretical underpinnings of the status quo were reexamined. They were found partially unsatisfactory and had to be revised and recast in a modern form. Until about 1860, marriage, the family, and sexual roles were assumed to belong to the natural condition of man, institutions beyond and above any mere geographical or historical accident. Between 1860 and 1890, however, social anthropology demonstrated that the idealized family of the Victorian middle class was dictated by no law of nature, that monogamous marriage was only one of various human sexual possibilities and that women were not necessarily born only to domestic and decorative functions. Yet the scholarly work of the anthropologists also supported the conventional vision of perfection in family life; it was not, perhaps, a natural institution, but it was the result of a long and painful evolutionary struggle away from nature. Current social arrangements should be seen as the final culmination, the glorious end-product of man's whole social, sexual, and moral evolution from savagery to civilization. Other forms of marriage and the family were still surviving remnants of worn-out cultural patterns that had long been superseded. By thus presenting 'civilized' marriage as the end-point of social evolution, these men provided a solid, historical, evolutionary justification for the role of women in their own culture. To illustrate this process of theoretical reformulation, I will examine the work of six men whose theories were critically important: Henry Maine, Johann J. Bachofen, John McLennan, John Lubbock, Lewis Henry Morgan, and Herbert Spencer.

One of the main traditional theoretical justifications for English political authority had been a body of thought known as patriarchalism. In the seventeenth century, men like James I and Sir Robert Filmer had advanced the concept that royal authority could be legitimated by arguing, analogically, its paternal character. The vast bulk of the populace, it was suggested were like the women and children of a patriarchal family. In that institution, the father possessed natural authority over his wife, children and slaves. So, too, the argument went, the king held a natural authority over the subjects of his realm. This political model, of course, accepted, indeed drew strength from the widespread assumption that the patriarchal family was a natural, i.e. God-given, and a just institution.

By the nineteenth century, patriarchal theory had been profoundly challenged by emergent liberal concepts of democracy and individualism. While some conservatives clung to an attenuated patriarchalism as a means of defending the status quo, the doctrine had been seriously eroded as a prop to political authority by Whig contractualists and

Benthamite liberals. But the old tradition maintained force in other arenas. Imperialists delighted in legitimating, indeed applauding their sway over non-whites by comparing themselves to benevolent parents seeking the well-being of infantile natives. And patriarchalism still seemed of use in justifying the subordination of women.

But there were problems with patriarchalism. For one thing, in the 1860s the feminist challenge raised severe questions about the nature of the 'natural family' itself – most specifically about the subordinate position of women within it. The elemental source of authority in the tradition was thus questioned: if the power of father-husbands were proved not to be a natural, God-given, timeless fact of life, then might not the basis of sexual, imperial, and, to some extent, even political authority be undermined?

Another difficulty arose from the archaic, sweeping quality of the claims of patriarchal theory. Most of the anthropological community (and all of my sample with the exception of Maine) were liberals; they repudiated the claims of arbitrary, naturally inherent authority in political leaders; to rest the case for the subordinate position of women on such a harsh doctrine – to imply that women and children were virtual slaves or extensions of the head of the household – seemed reactionary and crude. The scientists were great believers in contract as the basis for human and political interrelationships, and so patriarchalism made them uneasy; it did not square with their economic, political, or social outlook and self-image as progressives. Yet it seemed somehow improper for slogans such as 'Individualism' and 'The Rights of Man' (so useful in combating governmental interference with free trade or capitalist enterprise) to be taken up by women or by non-whites. Certainly there could be no merit to the feminist charge that existing social arrangements were both unjust and unnecessary, the product of arbitrary, essentially political decisions, that could in turn be reversed. Somehow the essence of patriarchal theory still seemed particularly relevant to the woman question: if only it could be shorn of its archaic harshness. A generation of social anthropologists turned to a consideration of these fascinating intellectual and political problems. Perhaps they might discern the proper role of women by an examination of the past. Social anthropology, then, turned to a study of the role of women in the history of mankind [*sic*].

In 1861, Sir Henry Maine, a successful lawyer and Corpus Professor of Jurisprudence at Oxford University, had summed up the state of conventional wisdom concerning the patriarchal family. His *Ancient Law* attempted to demonstrate that the power of the father had always been the basis of law and of society; as Filmer had before him, he denounced liberal assumptions about the State of Nature as ahistorical nonsense: patriarchs had *always* ruled. The patriarchal father had always held absolute authority over his wife, children, servants and slaves: only in modern times had the family begun to disintegrate, with yet unrealized consequences for social order. The French Revolution had been only the first result of the new-fangled disregard for history and tradition; long political experience could not be so easily tossed aside.

In order to prove the timeless and inherent nature of patriarchal authority, Maine, a legal historian, turned to written records, using as his sources the Old Testament, the Hindu Laws of Manu, and the Twelve Tables of Rome. These were, he claimed, the earliest reliable records of man's history; the sources usually used by ethnologists he discounted as 'the slippery testimony concerning savages which is gathered from travelers tales'.[1] From a brilliant reconstruction of fragmented ancient histories and legal codes, Maine built a case for the foundation of all known societies from the same, patriarchal basis: 'The effect of the evidence of comparative jurisprudence is to establish the view of the primeval condition of the human race which is known as the Patriarchal Theory.'[2]

This accurate summation of the basis of patriarchal theory worried the anthropological community. From their perspective, Maine's work demonstrated severe methodological inadequacies. Maine had made no use of the available ethnological data: he had ignored field work. His formulation of patriarchal theory reflected the narrow range of his source material. For one thing, he had reconstructed only a fair case for a very small geographical area, and a very restricted historical period, and then blithely extrapolated to the rest of human history. Then, too, the anthropologists preferred the direct or indirect observation of tribal society to Maine's reliance on written sources. As Spencer explained: 'I see no reason to ascribe to the second-hand statements of modern explorers.'[3] It seemed that Maine had unintentionally demonstrated the weakness of the patriarchal theory's evidential base; the very inadequacies of his work, at least from an anthropological perspective, led others to an awareness of its problematic character. As Donald McLennan later explained: 'The patriarchal theory . . . was most simple, and agreeable to current prejudices. It used to be accepted as palpably true, like the fact of the sun moving daily round the earth. No one thought of proving it, and but few of seriously doubting it.'[4]

The second problem with Maine's theory was that it was a static, non-evolutionary view in an era of Darwinism, an age when evolutionary thinking permeated all of the social sciences. In considering the history of 'civilized' man. Maine had assumed that his conclusions could be extrapolated to man's entire history. Yet those attuned to work in primitive societies were aware that a great dichotomy existed between 'savage' and 'civilized' man. It seemed to go almost without saying that unless the two types had developed in a parallel manner, one was in reality the remnant of an 'earlier stage'; and Victorians had no doubt about the direction in which history was flowing. Every contact of British imperialists with 'primitive' cultures proved how far Victorian England had risen above the savage state: much of anthropology in this period was given over to documenting and glorifying the triumph of man over nature. Rousseau's savage was no longer noble, rather a living testament to civilized man's rude origins; Victorians gloried in the fact that economic, technological and cultural progress in the West had all but vanquished the remnants of natural man. But civilized man had to go beyond self-congratulatory affirmations of his superiority. It was the business of the anthropologists particularly to develop a theory that would tell him how that superiority had been achieved.

The consideration of remaining 'primitive' social orders, then, was of utmost importance, and when anthropologists turned to consider the primitive family, they found material that both raised problems for Maine's simplistic, static model, and provided the means to transcend it. The experience of 'primitive' societies made it clear that patriarchalism was *not* an immutable fact of nature. It became apparent that sexual control, marital fidelity, the centering of the emotional life around the home and children, the exaggerated respect shown to wives and mothers – these were *not* natural to man. Did this mean that patriarchal theory rested on false foundations? Not at all. One need only suggest that the contrast between savage and civilized family and sexual life could be accounted for by postulating that the former had evolved into the latter. That, in other words, if the male and female roles of the nineteenth-century middle class were not natural, they were something better: they were the final culmination of a millennia-long development – the very point to which all recorded history had been heading. They were, in short, not natural, they were *civilized*.

The theorist who opened up this promising line of analysis was J. J. Bachofen. In his most notorious work, *Das Mutterrecht* (published in 1861), Bachofen provided patriarchal theory with a novel, indeed a romantic account of social evolution. Bachofen saw the continuing struggle between male and female as the central theme in social evolution; each stage of human society could be described in terms of the balance of power between the sexes. He believed human history to have consisted of three main stages: the hetaerist-aphroditic, the matri-archal, and the patriarchal. During the hetaerist-aphroditic stage, marriage was unknown. Sexual relationships were unregulated: women were at the mercy of male lust; promiscuity and sexual exploitation triumphed. Eventually, however, the women had rebelled and had staged a worldwide Amazonian revolt. Following their military success, they established the second stage of human history, the matriarchal stage. Women, as mothers, dominated social and cultural institutions; female sexuality triumphed: now they could force marriage and monogamy onto the reluctant males. In the final stage, men rebelled in their turn, triumphed, and replaced matriarchy with the patriarchal system. Women were dethroned and male supremacy everywhere recognized. Each stage had its religious or ideological basis. In the matriarchal period, the female fertility principle had been glorified, while the transition to patriarchalism represented the emancipation of man from material nature, 'the sublimation of earthly existence to the purity of the divine father principle'.[5] Echoing much of mythological thought, woman to Bachofen represented materialism, and man, spirituality.

Indeed, Bachofen was a student of mythology, and argued that it was a valuable source of historical reconstruction: 'Multiform and shifting in its outward manifestation, myth nevertheless follows fixed laws, and can provide as definite and secure results as any other source of historical knowledge.'[6] Bachofen also made extensive use of other forms of symbolism from painting and sculpture to architecture, all freely interpreted. His approach was intuitive; he discounted the laborious approach of English scientists with their piling up of minute pieces of data in favor of a creative use of the imagination: 'Aroused by direct contact with the ancient remains, the imagination grasps the truth at one stroke, without intermediary links.'[7]

Neither Bachofen's methods nor his sources held any appeal for most English anthropologists. Like Maine, Bachofen had ignored the necessary hard work in the field; instead he constructed what seemed to be fairy tales. Worse, his mechanism of the transition to matriarchy seemed founded on obvious absurdities, given what every Victorian knew about the eternal nature of woman. His armed Amazonian rebellion seemed preposterous – women (as everyone knew) would never take up arms in open warfare with their men; and if they did, they could – it was patently clear – never have been successful. Nor was it plausible that within the structure of monogamous marriage women would have absolute authority over their children. Natural authority was represented by the father; only under extraordinary conditions could the mother attempt to take over his role. Women – it was well known – were the gentle sex, physically weak, and constitutionally timid; men *must* always have held social and political power.

Bachofen's theory, then, was badly flawed by unorthodox sources, an alien methodology, and absurd notions of ancient feminine superiority; still, there were many facets of this thought that held great appeal. Bachofen had, after all, argued that the modern Victorian family had to be understood as the end result of a long historical struggle against the crude desires of nature. This dovetailed with the contemporary identification of sexuality with animality; brute passions struggled with man's rationality – the sexual instinct was the 'animal' part of man which had to be kept under control by the higher dictates of conscience and the active striving of the will. The struggle that took place in the individual had once taken place in society; as Freud in *Totem and Taboo* postulated an actual Oedipal struggle in the history of man, so many anthropologists were to see the struggle against sexual instincts as the earliest triumph of man against nature. The evolution of moral feeling had led from sexual anarchy to the rigid mores of the Victorian middle class.

Anthropologists, then, could find congenial Bachofen's story which began with a period of promiscuity, when man's sexual desires were unhampered by social restraints. Women were the possessions of men, human property was communally shared. But then, just as a child learned sexual control, so did primitive man. According to Bachofen, the restraints were instituted by the women themselves, through the power of woman as mother. Matriarchy was a necessary stage in maturation, later to be supplanted by patriarchy as the more abstract 'masculine' values and activities supplanted the concrete 'feminine' ones.

Bachofen's thesis, then, clearly resolved some of the problems of Maine's work. Yet surely the evolutionary model could be freed from the assumption that the wheel of progress had been largely turned by women. It was, after all, well known that woman was a passive being whose role was to complement the aggressive creativity and achievements of the male through her altruism and selfless devotion to the cares of the home. In the end, those aspects of Bachofen's theory that accorded with Victorian sexual and racial mythology were accepted. In the next twenty-five years, John McLennan, Lewis Henry Morgan, John Lubbock, and Herbert Spencer reconstructed his model to reflect more closely their social and intellectual needs and those of their middle-class audience.

Unlike Maine or Bachofen, these men used the accounts of primitive societies brought back by travelers, colonial administrators, and missionaries as their main sources of evidence. Of the four, only Morgan had direct contact with the peoples about whom he was writing; his theories and data were correspondingly more sophisticated than those of men who worked primarily in their studies and in libraries. Yet all saw themselves as empirical scientists, carefully weighing the evidence at hand, comparing and evaluating contradictory reports and applying sound judgment to the results; they were often painstakingly careful in their scholarship as in their 'science'. They were also aware that many of their informants viewed other cultures in a highly prejudiced manner; they worked as best they could with the data available. And they all worked toward a similar end: the construction of a refurbished, scientific patriarchalism, based now on the evolutionary perspective suggested by Bachofen.

John McLennan, like Maine, a lawyer, devoted his three major works *Primitive Marriage* (1865), *Studies in Ancient History* (1876), and *The Patriarchal Theory* (1885) to demolishing the static version of patriarchalism. He constructed in its place a dynamic three-stage evolution that developed from primitive promiscuity through group marriage with descent in the female line, to monogamous marriage with male descent; at the same time he avoided Bachofen's absurdities.

McLennan had three types of evidence that suggested three distinct forms of social organization. There was, first of all, the evidence of the nature of contemporary primitive cultures. Ethnological studies revealed one very vivid fact about primitive societies: they were steeped in sex and sin. A recurrent theme of the missionaries, explorers and administrators was the lax or non-existent moral standards of the savages. Their lack of clothing, their apparent sexual freedom offended: specific customs such as wife-lending and the public defloration of virgins seemed clear evidence of a degraded moral sensitivity. The evident lack of female modesty shocked: as McLennan stated: 'Women are usually depraved and inured to scenes of depravity from their earliest infancy.'[8]

Secondly, kinship studies certainly demonstrated that the reckoning of descent through the female, or matrilineality, was a common enough custom to be accepted as evidence of a distinct type of social organization. (McLennan distinguished this from matriarchy, however; it was surely explicable without the improbable assumption that it stemmed from the actual social power or authority of women themselves, as Bachofen had assumed.) Thirdly, there was the monogamous, patrilineal, patriarchal, civilized family of McLennan's own culture.

Since the aims of historical anthropology were first, to isolate stages of development, and then, to arrange them in a plausible evolutionary sequence, McLennan postulated promiscuity as the earliest stage, and placed the monogamous family at the apex of evolutionary development.

He had, however, to account for the existence of matrilineal succession without the aid of Amazonian warriors. Bachofen's theory in that regard seemed even more improbable than matriarchy itself: 'That the children of a man and a woman living together as husband and wife should be subject to the mother's authority and not the father's, be named after her and not after the father, be her heirs and not the father's is simply incredible; and it is surely not rendered credible by the statement that these singularities were the direct consequences of women having been victorious in a war with men.'[9]

Kinship through women could have occurred only if kinship through men was impossible, that is, if paternity were uncertain. McLennan tied matrilineal succession directly to the prevalence of promiscuity. Savages, he reminded his reader, were 'unrestrained by any sense of delicacy from a copartner in sexual enjoyments'.[10] If tribal groups of men had held their wives in common, the children produced would belong to the horde rather than to the individual. In the absence of individual paternity, they would have been forced to reckon descent through the mother. Thus McLennan cleansed Bachofen's theory of the unseemly notion of matriarchal power.

But McLennan had still many identifiable social forms and customs which had to be fitted into his evolutionary sequence somewhere between the promiscuous horde and the patriarchal family. He had to explain exogamy (the custom of marrying outside one's own tribe or totem), polyandry and polygamy as transitional stages in the development of human marriage, and to show how the transitions from one stage to the other might have occurred.

Where Bachofen had made women responsible for a sudden shift to monogamy. McLennan postulated a gradual transition, created and engineered by male activity. His mechanism is worth following in some detail. First, he 'demonstrated' that women were less useful to society than were men. This was easy, for as all bourgeois Victorians knew, women could not support themselves, nor could they contribute by their labor to 'the common good'. In McLennan's theoretical system, women seem completely passive social units of property who may be either individually or collectively owned by men, but who initiate no action of their own.[11] Since women were of relatively little use to society. McLennan concluded that in times of scarcity, female infanticide would be common. Unfortunately, this in turn would result in a scarcity of wives, leaving the men only two alternatives: either several men could peaceably agree to share one woman (polyandry) or they would have to steal wives from other tribes (exogamy).[12] Through the single assumption of female infanticide, which he assumed to have been a generally universal practice. McLennan had explained both polyandry and exogamy, thus satisfying the scientific desire for simplicity of explanation. As long as polyandry was the most common form of marriage, descent would of course continue in the female line. To make the transition to male descent, either the polygamous or monogamous form of the family must have come into existence.

McLennan explained this transition through the mechanism of the capture of wives. He had compiled an enormous quantity of evidence of the practice of marriage-by-capture; marriage ceremonies which symbolized the forcible abduction of a woman by her husband were assumed to be survivals of a once universal system of the actual capture of women. This practice, thought McLennan, must have been responsible for the fragmentation of the horde into individual families. A man who had captured his wife would treat her as an individual possession. Neither the woman not her children could now belong to her matrilineal group; they must now become the individual property of their captor. Once the practice of capturing wives became well established, polyandry would vanish, since each man would want to venture into the jungle to find a wife of his own. When women were individually owned, paternity would no longer be uncertain and patriarchal succession would follow as a matter of course. As McLennan says, '. . . the superiority of the male sex must everywhere have tended to establish that system'.[13]

At this stage, McLennan introduced a new factor into his social history: an economic motive. The transition from matrilineal to patrilineal succession must have been aided, he thought, by the growth of private property. Fathers would for the first time have a real motive for desiring to identify their sons, since they would, of course, want them to inherit the family property, lest it be returned to the matrilineal line. Even if wife capture later became impractical, men would be sufficiently eager for male descent that they would be prepared to acquire their wives by purchase from the matrilineal group. The purchase price would cover the wife and her unborn children; today this is symbolized by 'giving away' the bride in the marriage ceremony.

As had Bachofen, McLennan had incorporated Maine's patriarchal family as the last of a succession of historical forms; yet Bachofen's theory had now been rewritten to exclude the Amazons and matriarchs who had been such unwelcome intruders in the history of man.

Sir John Lubbock, one of the most widely known pre-historians of his day, specialist in both anthropology and archeology, a prominent member of many scientific societies and the author of numerous articles and books, added further refinements to McLennan's theory. In his work – particularly *The Origin of Civilization and the Primitive Condition of Man* (1870) – Lubbock emphasized the emerging theme: the patriarchal family was not *natural*, but the proud product of millennia of development. The system whereby males ruled their families was *better* than natural (i.e. primitive, i.e. savage, i.e. disgusting); it was *civilized*.

In *Primitive Times* (1865) Lubbock had graphically described the immorality of savages and the horrors of their lives, dwelling particularly on their treatment of women. This degradation of women was best demonstrated by the fact that some tribes had no word for 'love'. Love, apparently, was a by-product of civilization, and respect for women a modern invention. Since savages knew nothing of 'love', they could not know the pleasures of monogamous marriage; instead, they practiced various objectionable customs which Lubbock delicately summarized in the term 'communal marriage'. He did not care for the careful distinctions between polygamy, polyandry, serial pairing, and the like; all these came under the general rubric of 'lax morality'. He strongly objected to the central importance that McLennan had given to polyandry. This he saw as an odd and exceptional marriage form. 'I cannot', he stated, 'regard polyandry as having been a general and necessary stage in human development', though he nowhere gave reasons for this feeling.[14] The prevalence of matrilineal succession simply showed, he said, how little faith was placed in the virtue of women in those barbarian days. Besides, the men had so many wives and children that they could have cared little about any of them. Matriarchal power he thought particularly implausible since – it was well known – women do not assert themselves, and savage women would be particularly unlikely 'to uphold their dignity'.[15] Matrilineal succession was only the consequence of the indifference of savage men to their progeny, certainly no evidence of female supremacy. In 'communal marriage', savage men shared their wives; individual marriage could have arisen only when individual men took wives by force: 'Capture and capture alone, could give a man the right to monopolise a woman, to the exclusion of his fellow clansmen.'[16] The sheer physical effort involved in wife-capture must have made the men proud of their new possessions. This pride and consequent possessiveness led to the strengthening of the bond between man and wife and to the beginning of marital 'love'. At the same time, paternal 'love' also became possible, because men would now be able to identify their own children. Marriage was born in 'brutal violence and unwilling submission', yet the very control of man over woman opened the possibility of love between the sexes.[17] The male loved the female not in spite of her weakness and vulnerability, but because of it; possessiveness was not an outgrowth of love, but love a consequence of possession. For Lubbock and doubtless many of his contemporaries, the essence and beauty of the male–female relationship resided in inequality.

Finally, Lubbock elaborated on the connection that McLennan had perceived between the monogamous family and private property. The first socially important consequence of the blossoming paternal affections was the father's desire to transmit his property to his sons. Apparently, these paternal feelings were not extended to daughters. In view of the fact that the association of monogamous marriage and male descent with the transmission of property is usually attributed to Engels, it is interesting to see how widely this idea was held by the most conservative anthropologists. None of them considered the possibility that property might have been held by women, although instances of female ownership are often cited (with surprise) in the ethnological monographs.

Lewis Henry Morgan is now best known as the immediate authority for Engels's *Origin of the Family, Private Property, and the State* (1884). Perhaps because he was an American, Morgan enjoyed little esteem among English anthropologists, though his work now seems more sophisticated than that of either Lubbock or McLennan. Morgan had been trained as a lawyer and had made a modest fortune in the practice of law, but his avocation was the study of Indian culture and system of kinship. Of his three major works, only two concern us here: *Systems of Consanguinity and Affinity* (1871) and his more developed thesis, *Ancient Society* (1877). Morgan's evolutionary scheme (consisting of fifteen or sixteen distinct stages) was considerably more complicated than those already sketched. His first stage is now the familiar one, 'primitive promiscuity'; the final one is monogamy. The story he relates is one of the gradual emplacement of restrictions on the natural passions of man: the evolution is a moral one, each stage an 'unconscious reformatory movement' testifying to the 'growth of the moral idea'. Morgan did not hide his consciousness of the superiority of monogamy which he saw as one of the higher cultural achievements of mankind: '. . . upon the family, as now constituted, modern civilized society is organized and reposes. The whole previous experience and progress of mankind culminated and crystallised in this one great institution.'[18] Morgan was not immune to the widespread moral distaste for 'primitive' man, yet he clearly admitted that the evidence for primitive promiscuity was, at best, indirect. In no existing society was there completely unregulated sexual conduct; the assumption of a promiscuous stage was however a logical deduction from knowledge of the consanguine family, in which intermarriage between brothers and sisters was permitted. 'Promiscuity may be deduced theoretically as a necessary condition antecedent to the consanguine family; but it lies concealed in the misty antiquity of mankind beyond the reach of positive knowledge.'[19]

According to Morgan, the very first organization of society was made on the basis of sex; the division between male and female was the first class division. Through a long succession of subdivisions, too complex to detail here, tribal organizations and various primitive forms of the family evolved. Morgan's comments on the systems of polygamy and polyandry are, however, amusing and instructive. Despite his admiration for monogamy, Morgan still had some sympathy for an institution in which each man might have several wives. Polygamy he considered 'essentially modern', a 'reformatory' and not a 'retrograde' movement. Polygamy, after all, implied a settled condition and a plentiful means of subsistence since a single male was able to support many individuals. With the polygamous family came the emergence of a single powerful family authority and therefore the rudiments of a system of government.

If polygamy constituted an advanced stage on the road to civilization, polyandry was worthy only of deep scorn. 'An excrescence of polygamy, and its repulsive converse', it was 'scarcely entitled to the rank of a domestic institution'.[20] Polyandry could only be the result of economic scarcity; men were unable to feed and maintain the standard of one wife and one set of children and so were forced to share. Morgan, like others, assumed that women had no productive function; large numbers of women and children existed as the result of surplus production. When the man was unable to support a wife and children he was dishonored; the state of polyandry was both economically and sexually 'repulsive'.

Morgan was no foe to the emancipation of women; he left a large sum of money to the University of Rochester for the furthering of female education and he optimistically predicted that equality of the sexes would be the result of the next great change in the history of the family. Since Morgan welcomed sexual equality in the future, he did not react with the usual derision to Bachofen's postulate of matriarchy in the past. Gyneocracy, the social power of women, would be a predictable consequence of matrilineal succession and the social arrangements of the matrilineal household: '. . . gyneocracy seems to require for its creation descent in the female line. Women thus entrenched in large households, supplied with common stores, in which their own gens so largely predominated in numbers, would produce the phenomena of mother-right and gyneocracy, which Bachofen has detected and traced with the aid of fragments of history and tradition.'[21] As far as Morgan was concerned, the transition to patrilineality had had a very unfavorable influence on the position of women – thus, Engels could derive from Morgan the world historical defeat of the female sex.

Morgan was not impressed with Lubbock's attempt to document the first appearance of paternal love in human history, but he was deeply convinced of the vast importance of private property to civilized man. He emphasized its role as a crucial one in the transition from mother-right to father-right; through this concern with the inheritance of property, men came to the realization of the advantages of monogamy: '. . . property, as it increased in variety and amount, exercised a steady and constantly augmenting influence in the direction of monogamy'.[22] The family, said Morgan, was a creation of the social and economic system; as society advanced, the family would continue to change and evolve to new forms.

Unlike other anthropologists, then, who tended to see their own form of the family as the final culmination of evolution, the most perfect possible form, Morgan left his evolutionary scheme open-ended, but moving toward more perfect monogamy and greater sexual equality.

Herbert Spencer held no such hopes or expectations; he was quite happy with things the way they were. Spencer, perhaps England's most revered nineteenth-century philosopher, polymath, and staunch supporter of the harsher bourgeois values, clearly believed that since evolution had placed women in the home, there they should stay. Spencer gathered the full weight of anthropological theory behind this opinion. A dilettante in social anthropology, he brilliantly summarized the findings of others, and integrated them within his own system of synthetic philosophy, thus carrying anthropological theories to a wider intellectual audience.

Spencer did more. In his *Principles of Sociology* (1876) he provided the now dynamic version of patriarchal theory with new support. Not only was the patriarchal Victorian family supported by and supportive of private property, decency, order, civilization, and love; it was now scientifically shown to be the inevitable product of natural selection.

Forms of marriage and the family, Spencer argued, like biological species, had to prove their superiority in the struggle for survival. A form that might have been adapted to the environment in the past could have become maladaptive when environmental conditions changed; in the long run, the fittest survived. Certainly, the patriarchal family based on monogamous marriage had survived the test and been proved most worthy. The development of rigidly defined sex roles (with males dominant) and the tight controls clamped on sexuality (particularly female sexuality) were not historical accidents but adaptations to social survival.

Civilization, Spencer would argue, rested on the control of sensuality. And there were few more fit than he to be the chronicler of the culture of repression, for he was, by all reports, its epitome. He was a man whose emotions and 'passions' appeared to have been completely subjugated to his intellect. As Francis Gribble said, in reviewing his autobiography for the *Fortnightly Review*. 'What is almost uncanny about Herbert Spencer is his triumphant superiority to natural instincts.'[23] Spencer, we are told, never unbent; he never loved or hated, he merely judged; his only expressed emotion was intellectual pride.

Spencer's survey of the evolution of the family in the *Principles of Sociology* began in the now standard fashion, positing promiscuity as the initial condition of humanity. Through 'logic' alone, Spencer decided that 'union of the sexes must have preceded all social laws'. This was a most unsatisfactory state of affairs. For one thing, the lower forms of domestic life were characterized by 'the abject condition of women' and an 'entire absence of the higher sentiments that accompany the relations of the sexes'.[24] Another difficulty with the 'unregulated relations of the sexes' had been the fact that it led to no settled political control: the establishment of subordination was hindered. In addition, Spencer thought that children must suffer from the lack of individual fathers because mothers would be unable to give them adequate support and protection. Women must also suffer since, after maturity, they would decline and die without male protection. Promiscuity was evidently a highly inefficient form of social organization; clearly things had to change.

Spencer saw three possibilities open to social evolution: polyandry, polygamy, or monogamy. Humanity experimented with these various familial forms: the evolutionary tree branched out in a variety of ways, yet finally only monogamy proved successful in the competitive struggle.

Polyandry, he found sure, would have fared poorly in the race for survival. With several fathers in one family, the resulting authority conflicts must have created serious evils. The first result would be poor family cohesion; poor family cohesion would have inhibited the development of social cohesion: poor social cohesion meant a disordered society and such societies would have lost out in warfare or economic competition with more disciplined cultures. Spencer also disputed McLennan's suggestion that the practice of wife-capture developed from an acute shortage of women (a demographic state conducive to polyandry). No, said Spencer, wife-capture arose because women *had* to be captured to overcome their feminine modesty and coyness: 'Coyness, either real or affected for reputation's sake, causes resistance of the woman herself.'[25] Natural man may have been at the mercy of his passions, but natural woman zealously guarded her reputation. She too, perhaps, was repulsed by polyandry, for certainly the idea of one woman having several husbands was – from the English perspective – a highly unnatural one.

Polygamy, says Spencer, does not create as much 'surprise and repugnance'; we are more used to the idea of one man with many women. And it seemed a more logical outcome of the earlier stage: the men who were strongest simply appropriated as many women as they desired. An additional motive for polygamy (surprising in that it clearly admitted the important economic function of women) was suggested by several polygamous chieftains whom Spencer quoted. They explained that many wives were desirable because the more women, the more food; their wives did all the agricultural labor. The emergence of polygamy, then, seemed dictated by the unrestrained sexual instincts of savage men, and by economic advantages.

Once established, polygamy had many advantages in the race for survival. Under polygamy, there was but one head of the household; order was more easily maintained. The single head of the household, this polygamous patriarch, would establish family cohesion and eventually patriarchal inheritance. With patriarchal inheritance would come economic stability and social control. The reverence due to the lord of the family would, after his death, generate ancestor worship, and with the development of religion, law and order would become firmly grounded. The logical progression here is fairly convincing if the following assumptions are granted: that the only authorities are male authorities, that men care only for their own children, that economic power has always been, and can only be, in the hands of males, and that only males could have been the subjects of ancestor worship and sacred authority. Polygamy, argued Spencer, was admirable for biological as well as social and political reasons: it maximized human fertility. Sons lost in battle could be easily replaced. Indeed, polygamy would be ideally suited to a military social organization: if many men were engaged in warfare, the surplus women could still be easily absorbed. Polygamy, however, could never have raised society to the stage of civilization, since for civilization to be attained, (as his colleagues had already agreed), marital love and respect for women were necessary.

Monogamy emerged from polygamy when the husband of many wives began to make one the favorite. He subsequently demoted the others to the status of slaves or concubines; a survival of this custom may be seen, Spencer said, where married men support a mistress or patronize a prostitute. Alternatively, since the state of having two wives must always be preceded by the state of having one, monogamy may sometimes have resulted from the unavailability of second wives. True monogamy, however, required not only that a man should have one wife, but that he should desire only one. Probably, men had come to place a higher value on their wives when they were forced to purchase them: the concept of wife as valuable property would also lead to a constraint on divorce.

When monogamy did appear, its superiority would become evident: a variety of indices showed the superiority of monogamous relationships. For one thing, when stable couples did form, the children would be well looked after and would be much healthier than the sickly offspring of unmarried mothers. Through the survival of the fittest, the healthy children would then in turn transmit, through the inheritance of acquired characteristics, the tendency toward forming stable couples to their own children. Moreover, the children with protective fathers would grow up healthier and better soldiers, so that societies with monogamous marriage must triumph over societies of 'lower domestic arrangements'. Then again, all of the benefits accruing to polygamy would follow even more strongly from monogamy: stronger ancestor worship, increased social cohesion and greater political control.

The greatest boon of the monogamous family, however, was that it alone created the conditions for the strict definition of sex roles. This was, to be sure, of value to the children: with a diminished family size and a thoroughly domesticated mother, the father could devote his attention to breadwinning, and the mother to child rearing. But still more was it of value to the adults: it allowed for the replacement of sexual instinct by 'the sentiments which characterise the relations of the sexes in civilised peoples'.

In the monogamous family, women could finally be placed on the pedestal on which they belonged. They were emancipated from physical labor and from the burden of male 'passions'; with settled family life, women lived longer because their husbands continued to support them even after their sexual attractiveness and general usefulness were long dead; in monogamous marriage, women came to be protected, respected, and insulated from the necessity of production and the evils of the marketplace. In primitive societies, Spencer reminded his readers, women had received brutal and cruel treatment; eye-witnesses testified to women's fishing, carrying and pitching tents, digging up roots, planting, plowing and reaping, building houses, climbing trees for small animals, and even hunting and going to war. We should, Spencer argued, congratulate ourselves on 'the improvement of women's status implied by limitation of their labors to the lighter kinds'.[26] The less a woman worked, the higher her status, and the less all women worked, the better for society, since the health of the future generation would then be improved.

At least in the wealthier middle-class circles of Victorian society, the pinnacle of female evolution had been reached. Women had ascended the pedestal to become 'the angels of the home': no longer sexually or economically exploited, the women were safe in their drawing rooms, far removed from the frightful realities of 'natural' life. Men, of course, were firmly in control. Though the demands of marital fidelity might sometimes conflict with their own 'natural passions', they, too, felt far removed from the savage state. And the rewards of civilized life compensated for any lingering sympathy for the primitive freedoms.

Spencer proceeded to draw the moral and political conclusions that social anthropology dictated. In the light of these evolutionary achievements, feminism appeared distinctly reactionary. The course of domestic evolution in the future, he cautioned, must be determined by the best interests of children. Besides, if 'in some directions the emancipation of women has to be carried further, we may suspect that in other directions their claims have already been pushed too far'.[27] Absolute equality with men could not be achieved; the law must continue to give supremacy to the husband, since he is more 'judicially-minded' than his wife. Besides, thought Spencer, if women would only perceive the glory of the domestic sphere, they would wish for none other. Probably all the fuss was being raised by spinsters who had no opportunity to enjoy the feminine role; normal women should be pleased to have no worries other than their domestic duties.

With Spencer the anthropological invigoration of a moribund patriarchalism reached a peak; here we should take note of the nature of the final achievement. The anthropologists had jettisoned a key provision of the old theory that had appeared increasingly untenable in the light of investigations of alien cultures: no longer did they posit that patriarchal monogamy was the natural, eternal human family order. Quite the reverse: in the beginning there was natural, and despicable licentiousness, and, if not female power, certainly female independence from male control. But they argued that the course of history and evolution flowed inexorably *away* from such an unnatural nature in the direction of patriarchal monogamy. Indeed patriarchalism was now inextricably linked with the progress of civilization; Victorian culture and its attendant social relations represented the capstone of all evolution. Male superiority, then, was sanctified not by nature, but by civilization.

More than that: the anthropologists had managed to harness a whole series of sacred concepts to help pull the patriarchal vehicle. Male dominance was now bound up with decency, the refinement and control of the passions, private property, and natural selection. But perhaps the strongest underpinning of male domination was to be found in the quintessential characteristic of the civilization it had produced – the thorough domestication of women (that is to say, male dominance). Admittedly not all women were angels; some were still slaves to economic production and others allowed themselves to be sexually used. The real vicissitudes of the lives of working women caused no great perturbation however; the lives of these women were pronounced anthropological survivals of an earlier stage. Like the savages, the lower classes had simply not advanced very far in evolutionary terms. But if Victorian society had not attained perfection, at least its civilized members could reassure themselves that they were aware of the evils yet lurking in the recesses of the culture, whereas savages had not sufficient moral sensitivity to know that their activities could be the cause of righteous indignation: as Lubbock observed, 'that which is with us the exception, is with them the rule: that which with us is condemned by the general verdict of society, and is confined to the uneducated and the vicious, is among savages passed over almost without condemnation, and treated as a matter of course'.[28]

This Victorian morality drama was, of course, a massive exercise in circular reasoning – despite the technical nature of many of the discussions and the impressive scholarly apparatus used. Their own culture provided the model by which all others had to be compared: the more divergent another culture from the 'civilized' ideal, then the more 'primitive' it must be. And 'primitive' was often synonomous with 'evil'; for if the Victorians had produced the most perfect of all social systems, they had also evolved the most sensitive perception for unveiling moral laxity in others. 'Primitive promiscuity' was more projection than fact, as Morgan himself almost admitted. In fact projections abounded throughout the scholarly studies. In their speculations about primitive man, Victorians projected fantasies of his enormous physical and sexual power: the untamed brute often seemed fashioned out of their own repressed psyches. Primitive woman emerged as an odd mixture of blushing bride and lascivious whore; since she was both 'primitive' (sensual) and 'woman' (pure and innocent). The stage was now set: it remained only to turn the psychic dichotomy between primitive and civilized into a evolutionary tale by filling in the intermediate links. In doing so, cultural anthropologists utilized the prevailing worship of progress and brilliantly confirmed their own social order by constructing an appropriate past.

Source: In Mary S. Hartman and Lois W. Banner, (eds), Clio's *Consciousness Raised: New Perspectives on the History of Women.* New York, Harper and Row, 1974, : 86–102.

Notes

1. Sir Henry Summer Maine, *Village Communities in the East and West* (orig. 1841) London, John Murray, 1887, : 17.
2. Maine, *Ancient Law: Its Connection with the Early History of Society, and its Relations to Modern Ideas*, 4th edn., London, J. Murray, 1870, : 123.
3. Herbert Spencer, *The Principles of Sociology, Works.* Osnabrück, Otto Zeller, 1966, reprint edn, : 683n.
4. Donald McLennan (ed), *The Patriarchal Theory: Based on the Papers of the Late John Ferguson McLennan.* London, Macmillan, 1885, : 3.
5. J. J. Bachofen, *Myth, Religion and Mother Right.* Trans. R. Manheim, Princeton, Princeton University Press, 1967, : 119.
6. Ibid., : 76.
7. Ibid., : 12.
8. John McLennan, *Primitive Marriage: An Inquiry into the Origin of the Form of Capture in Marriage Ceremonies,* Edinburgh: A. & C. Black, 1865.
9. John Ferguson McLennan, *Studies in Ancient History.* London, Macmillan, 1886, : 324.
10. McLennan. *Primitive Marriage,* : 167.
11. Much attention was paid to the psychology of natural man, but natural woman is a strangely shadowy figure in evolutionary anthropology, remarkable only for her ubiquitous 'maternal instinct'.
12. Note the conventional descriptive form: in polyandry several men agree to share one woman; polygamy results when one man possesses a number of women. The reverse description would have been inconceivable: in polyandry one woman possesses a number of men: polygamy results when a number of women agree to share one man.
13. McLennan, *Primitive Marriage,* : 228.
14. Sir John Lubbock (Lord Avebury), 'Review of McLennan's *Studies in Ancient History*', *Nature,* **15** (1876): 133.
15. Sir John Lubbock, *The Origin of Civilization and the Primitive Condition of Man.* London, Longmans, 1870, : 68.

16. Ibid., : 72.
17. Ibid., : 73.
18. Lewis Henry Morgan, *Systems of Consanguinity and Affinity of the Human Family*, Smithsonian Contributions to Knowledge, vol. 17. Washington, Smithsonian Institution, 1871. : 493.
19. Morgan, *Ancient Society*, Leslie A. White (ed.) 1877–78. Reprint edn, Cambridge, Mass, Harvard University Press, 1964, : 424.
20. Morgan, *Systems*, : 477.
21. Morgan, *Ancient Society*, : 297–8.
22. Ibid., : 426.
23. Francis Gribble, 'Herbert Spencer, his autobiography and his philosophy', *Fortnightly Review,* N.S. **75** (1904) : 987.
24. Spencer, *Works*, 6, : 629.
25. Ibid., : 623.
26. Ibid., : 724.
27. Ibid., : 755.
28. Lubbock, *Prehistoric Times*. London, Williams and Norgate, : 561.

4.3 Stephen Jay Gould, Smith Woodward's Folly

Nothing is quite so fascinating as a well-aged mystery. Many connoisseurs regard Josephine Tey's *The Daughter of Time* as the greatest detective story ever written because its protagonist is Richard III, not the modern and insignificant murderer of Roger Ackroyd. The old chestnuts are perennial sources for impassioned and fruitless debate: Who was Jack the Ripper? Was Shakespeare Shakespeare?

My profession of palaeontology offered its entry to the first rank of historical conundrums a quarter-century ago. In 1953, Piltdown man was exposed as a certain fraud perpetrated by a very uncertain hoaxer. Since then, interest has never flagged. People who cannot tell *Tyrannosaurus* from *Allosaurus* have firm opinions about the identity of Piltdown's forger. Rather than simply ask 'whodunnit?' this column treats what I regard as an intellectually more interesting issue: why did anyone ever accept Piltdown man in the first place?

In 1912, Charles Dawson, a lawyer and amateur archaeologist from Sussex, brought several cranial fragments to Arthur Smith Woodward, keeper of geology at the British Museum (Natural History). The first, he said, had been unearthed by workmen from a gravel pit in 1908. Since then, he had searched the spoil heaps and found a few more fragments. The bones, worn and deeply stained, seemed indigenous to the ancient gravel; they were not the remains of a more recent interment. Yet the skull appeared remarkably modern in form, although the bones were unusually thick.

Smith Woodward, excited as such a measured man could be accompanied Dawson to Piltdown and there, with Father Teilhard de Chardin, looked for further evidence in the spoil heaps. (Yes, believe it or not, the same Teilhard who, as a mature scientist and theologian, became such a cult figure some 15 years ago with his attempt to reconcile evolution, nature, and God in *The Phenomenon of Man*. Teilhard had come to England in 1908 to study at the Jesuit college in Hastings. He met Dawson in a Sussex quarry on 31 May 1909; the mature solicitor and the young French Jesuit soon became warm friends, colleagues, and co-explorers.)

On one of their joint expeditions, Dawson found the famous mandible, or lower jaw. Like the skull fragments, the jaw was deeply stained, but it seemed to be as apish in form as the cranium was human. Nonetheless, it contained two molar teeth, worn flat in a manner not rare in humans but never encountered in apes. Unfortunately, the jaw was broken in just the two places that might have settled its relationship to the skull: the chin region, with all its marks of distinction between ape and human, was gone, and so was the articulation with the cranium.

Armed with skull fragments, the lower jaw, and an associated collection of worked flints and bone, plus a number of mammalian fossils to fix the age as ancient, Smith Woodward and Dawson made their splash before the Geological Society of London on 18 December 1912. Their reception was mixed, although on the whole favourable. No one smelled fraud, although the association of such a human cranium with such an apish jaw indicated to some critics that remains of two separate animals might have been mixed together in the quarry.

Over the next three years, Dawson and Smith Woodward countered with a series of further discoveries that, in retrospect, could not have been better programmed to dispel doubt. In 1913, Father Teilhard found the all important lower canine tooth. It, too, was apish in form but strongly worn in a human manner. Then, in 1915, Dawson convinced most of his detractors by finding the same association of two thick-skulled human cranial fragments with an apish tooth worn in a human manner at a second site two miles from the original finds.

Henry Fairfield Osborn, leading American palaeontologist and converted critic, wrote:

If there is a Providence hanging over the affairs of prehistoric men, it certainly manifested itself in this case, because the three fragments of the second Piltdown man found by Dawson are exactly those which we would have selected to confirm the comparison with the original type. . . . Placed side by side with the corresponding fossils of the first Piltdown man they agree precisely; there is not a shadow of a difference.

Providence, unbeknownst to Osborn, walked in human form at Piltdown.

For the next 30 years, Piltdown occupied an uncomfortable but acknowledged place in human prehistory. Then, in 1949, Kenneth P. Oakley applied his fluorine test to the Piltdown remains. Bones pick up fluorine in proportion to their time of residence in a deposit and the fluorine content of the deposit. Both the skull and jaw of Piltdown contained barely detectable amounts of fluorine; they could not have lain long in the gravels. Oakley still did not suspect fakery. He proposed that Piltdown, after all, had been a recent interment into ancient gravels.

But a few years later, in collaboration with J. S. Weiner and W. E. le Gros Clark, Oakley finally considered the obvious alternative – that the 'interment' had been made in this century with intent to defraud. He found that the skull and jaw had been artificially stained, the flints and bone worked with modern blades, and the associated mammals, although genuine fossils, imported from elsewhere. Moreover, the teeth had been filed down to simulate human wear. The old anomaly – an apish jaw with a human cranium – was resolved in the most parsimonious way of all. The skull *did* belong to a modern human; the jaw *was* an orang-utan's.

But who had foisted such a monstrous hoax on scientists so anxious for such a find that they remained blind to an obvious resolution of its anomalies? Of the original trio, Teilhard was dismissed as a young and unwitting dupe. No one has ever (and rightly, in my opinion) suspected Smith Woodward, the superstraight arrow who devoted his life to the reality of Piltdown and who, past 80 and blind, dictated in retirement his last book, with its chauvinistic title, *The Earliest Englishman*.

Suspicion instead has focused on Dawson. Opportunity he certainly had, although no one has ever established a satisfactory motive. Dawson was a highly respected amateur with several important finds to his credit. He was overenthusiastic and uncritical, perhaps even a bit unscrupulous in his dealings with other amateurs, but no direct evidence of his complicity has ever come to light. Nevertheless, the circumstantial case is strong and well summarised by J. S. Weiner in *The Piltdown Forgery*.

Supporters of Dawson have maintained that a more professional scientist must have been involved, at least as a co-conspirator, because the finds were so cleverly faked. I have always regarded this as a poor argument, advanced by scientists largely to assuage their embarrassment that such an indifferently designed hoax was not detected sooner. The staining, to be sure, had been done consummately. But the 'tools' had been poorly carved and the teeth rather crudely filed – scratch marks were noted as soon as scientists looked with the right hypothesis in mind. Le Gros Clark wrote: 'The evidences of artificial abrasion immediately sprang to the eye. Indeed so obvious did they seem it may well be asked – how was it that they had escaped notice before?' The forger's main skill consisted in knowing what to leave out – discarding the chin and articulation.

Piltdown reappeared prominently in the news last November because yet another scientist has been implicated as a possible co-conspirator. Shortly before he died last year at age 93, J. A. Douglas, emeritus professor of geology at Oxford, made a tape recording suggesting that his own predecessor in the chair, W. J. Sollas, was the culprit. In support of this assertion, Douglas offered only three items scarcely ranking as evidence in my book: First, Sollas and Smith Woodward were bitter enemies. (So what. Academia is a den of vipers, but verbal sparring and elaborate hoaxing are responses of differing magnitude.) Secondly, in 1910, Douglas gave Sollas some mastodon bones that could have been used as part of the imported fauna. (But such bones and teeth are not rare.) Thirdly, Sollas once received a package of potassium bichromate and neither Douglas nor Sollas's photographer could figure out why he had wanted it. Potassium bichromate was used in staining the Piltdown bones. (It was also an important chemical in photography, and I do not regard the supposed confusion of Sollas's photographer as a strong sign that the professor had some nefarious usages in mind.) In short, I find the evidence against Sollas so weak that I wonder why the leading scientific journals of England and the United States gave it so much space. I would exclude Sollas completely, were it not for the paradox that his famous work on *Ancient Hunters* supports Smith Woodward's views on Piltdown in terms so obsequiously glowing that it could be read as subtle sarcasm.

So who did do it?
Only three hypotheses make much sense to me. First, Dawson was widely suspected and disliked by some amateur archaeologists (and equally acclaimed by others). Some compatriots regarded him as a fraud. Others were bitterly jealous of his standing among professionals. Perhaps one of his colleagues devised this complex and peculiar form of revenge. The second hypothesis, and the most probable in my view, holds that Dawson acted alone, whether for fame or to show up the world of professionals we do not know.

The third hypothesis is much more interesting. It would render Piltdown as a joke that went too far, rather than a malicious forgery. It represents the 'pet theory' of many prominent vertebrate palaeontologists who knew the man well. I have sifted all the evidence, trying hard to knock it down. Instead, I find it consistent and plausible, although not the leading contender. A. S. Romer, late head of the museum I inhabit at Harvard and America's finest vertebrate palaeontologist, often stated his suspicions to me. Louis Leakey also believed it. His autobiography refers anonymously to a 'second man', but internal evidence clearly implicates a certain individual to anyone in the know.

It is often hard to remember a man in his youth after old age imposes a different persona.

Teilhard de Chardin became an austere and almost Godlike figure to many in his later years; he was widely hailed as a leading prophet of our age. But he was once a fun-loving young student. He knew Dawson for three years before Smith Woodward entered the story. He may have had access, from a previous assignment in Egypt, to mammalian bones (probably from Tunisia and Malta) that formed part of the 'imported' fauna at Piltdown. I can easily imagine Dawson and Teilhard, over long hours in field and pub, hatching a plot for different reasons. Dawson to expose the gullibility of pompous professionals; Teilhard to rub English noses once again with the taunt that their nation had no legitimate human fossils, while France revelled in a superabundance that made her the queen of anthropology. Perhaps they worked together, never expecting that the leading lights of English science would fasten to Piltdown with such gusto. Perhaps they expected to come clean but could not.

Teilhard left England to become a stretcher bearer during the First World War. Dawson, on this view, persevered and completed the plot with a second Piltdown find in 1915. But then the joke ran away and became a nightmare. Dawson sickened unexpectedly and died in 1916. Teilhard could not return before the war's end. By that time, the three leading lights of British anthropology and palaeontology – Arthur Smith Woodward, Grafton Elliot Smith and Arthur Keith – had staked their careers on the reality of Piltdown. (Indeed they ended up as two Sir Arthurs and one Sir Grafton, largely for their part in putting England on the anthropological map.) Had Teilhard confessed in 1918, his promising career (which later included a major role in describing the legitimate Peking man) would have ended abruptly. So he followed the Psalmist and the motto of Sussex University, later established just a few miles from Piltdown – 'Be still, and know . . .' – to his dying day. Possible. Just possible.

All this speculation provides endless fun and controversy, but what about the prior and more interesting question: why had anyone believed Piltdown in the first place? It was an improbable creature from the start. Why had anyone admitted to our lineage an ancestor with a fully modern cranium and the unmodified jaw of an ape?

Indeed, Piltdown never lacked detractors. Its temporary victory was born in conflict and nurtured throughout by controversy. Many scientists continued to believe that Piltdown was an artefact composed of two animals accidentally commingled in the same deposit. In the early 1940s, for example, Franz Weidenreich, perhaps the world's greatest human anatomist, wrote (with devastating accuracy in hindsight): '*Eoanthropus* ("dawn man", the official designation of Piltdown) should be erased from the list of human fossils. It is the artificial combination of fragments of a modern human braincase with orang-utan-like mandible and teeth.' To this apostasy, Sir Arthur Keith responded with bitter irony: 'This is one way of getting rid of facts which do not fit into a preconceived theory; the usual way pursued by men of science is, not to get rid of facts, but frame theory to fit them.'

Moreover, had anyone been inclined to pursue the matter, there were published grounds for suspecting fraud from the start. A dental anatomist, C. W. Lyne, stated that the canine found by Teilhard was a juvenile tooth, just erupted before Piltdown's death, and that its intensity of wear could not be reconciled with its age. Others voiced strong doubts about the ancient manufacture of Piltdown's tools. In amateur circles of Sussex, some of Dawson's colleagues concluded that Piltdown must be a fake, but they did not publish their beliefs.

If we are to learn anything about the nature of scientific inquiry from

Piltdown – rather than just revelling in the joys of gossip – we will have to resolve the paradox of its easy acceptance. I think that I can identify at least four categories of reasons for the ready welcome accorded to such a misfit by all the greatest English palaeontologists. All four contravene the usual mythology about scientific practice – that facts are 'hard' and primary and that scientific understanding increases by patient collection and fitting together of these objective bits of pure information. Instead, they display science as a human activity, motivated by hope, cultural prejudice, and the pursuit of glory yet stumbling in its erratic path toward a better understanding of nature.

The imposition of strong hope upon dubious evidence.   Before Piltdown, English palaeonthropology was mired in a limbo now occupied by students of extraterrestrial life: endless fields for speculation and no direct evidence. Beyond some flint 'cultures' of doubtful human workmanship and some bones strongly suspected as products of recent interments into ancient gravels, England knew nothing of its most ancient ancestors. France, on the other hand, had been blessed with a superabundance of Neanderthals and Cro-Magnons and their associated art and tools. And French anthropologists delighted in rubbing English noses with this marked disparity of evidence. Piltdown could not have been better designed to turn the tables. It seemed to predate Neanderthal by a considerable stretch of time. If human fossils had a fully modern cranium hundreds of thousands of years before beetle-browed Neanderthal appeared, then Piltdown must be our ancestor and the French Neanderthals a side branch. Smith Woodward proclaimed: 'The Neanderthal race was a degenerate offshoot of early man while surviving modern man may have arisen directly from the primitive source of which the Piltdown skull provides the first discovered evidence.' This international rivalry has often been mentioned by Piltdown's commentators, but a variety of equally important factors have usually escaped notice.

Reduction of anomaly by fit with cultural biases.   A human cranium with an ape's jaw strikes us today as sufficiently incongruous to merit strong suspicion. Not so in 1913. Biases, largely cultural in origin, exert a strong influence on behalf of 'brain primacy' in human evolution. The argument rested on a false inference from contemporary importance to historical priority: We rule today by virtue of our intelligence. Therefore, in our evolution, an enlarged brain must have preceded and inspired all other alterations of our body. We should expect to find human ancestors with enlarged, perhaps nearly modern, brains and a distinctly simian body. (Ironically, nature followed an opposite path. Our earliest ancestors, the Australopithecines, were fully erect but still small brained.) Thus, Piltdown neatly matched a widely anticipated result. Grafton Elliot Smith wrote in 1924:

The outstanding interest of the Piltdown skull is in the confirmation it affords of the view that in the evolution of Man the brain led the way. It is the veriest truism that Man has emerged from the simian state in virtue of the enrichment of the structure of his mind. . . . . The brain attained what may be termed the human rank at a time when the jaws and face, and no doubt the body also, still retained much of the uncouthness of Man's simian ancestors. In other words, Man at first . . . was merely an Ape with an overgrown brain. The importance of the Piltdown skull lies in the fact that it affords tangible confirmation of these inferences.

Piltdown also buttressed some all too familiar racial views among white Europeans. In the 1930s and 1940s, following the discovery of Peking man in strata approximately equal in age with the Piltdown gravels, phyletic trees based on Piltdown and affirming the antiquity of white supremacy began to appear in the literature (although they were never adopted by Piltdown's chief champions, Smith Woodward, Smith, and Keith). Peking man (originally called *Sinanthropus*, but now placed in *Homo erectus*) lived in China with a brain two-thirds modern size, while Piltdown man, with its fully developed brain, inhabited England. If Piltdown, as the earliest Englishman, was the progenitor of white races, while other hues must trace their ancestry to *Homo erectus*, then whites crossed the threshold to full humanity long before other people. As longer residents in this exalted state, whites much excel in the arts of civilisation.

Reduction of anomaly by matching fact to expectation. We know, in retrospect, that Piltdown had a human cranium and an ape's jaw. As such, it provides an ideal opportunity for testing what scientists do when faced with uncomfortable anomaly. G. E. Smith and others may have advocated an evolutionary head start for the brain, but no one dreamed of an independence so complete that brains might become fully human before jaws changed at all! Piltdown was distressingly too good to be true.

If Keith was right in his taunt to Weidenreich, then Piltdown's champions should have modelled their theories to the uncomfortable fact of a human cranium and an ape's jaw. Instead, they modelled the 'facts' – another illustration that information always reaches us through the strong filters of culture, hope, and expectation. As a persistent theme in 'pure' description of the Piltdown remains, we learn from all its major supporters that the skull, although remarkably modern, contains a suite of definitely simian characters! Smith Woodward, in fact, originally estimated the cranial capacity at a mere 1070 cu. cm (compared with a modern average of 1400 to 1500), although Keith later convinced him to raise the figure nearer to the low end of our modern spectrum. Grafton Elliot Smith, describing the brain cast in the original paper of 1913, found unmistakable signs of incipient expansion in areas that mark the higher mental faculties in modern brains. He concluded:

We must regard this as being the most primitive and most simian human brain so far recorded; one, moreover, such as might reasonably have been expected to be associated in one and the same individual with the mandible which so definitely indicates the zoological rank of its original possessor.

Just a year before Oakley's revelation, Sir Arthur Keith wrote in his last major work (1948): 'His forehead was like that of the orang, devoid of a supraorbital torus; in its modelling his frontal bone presented many points of resemblance to that of the orang of Borneo and Sumatra.' Modern *Homo sapiens*, I hasten to add, also lacks a supraorbital torus, or brow ridge.

Careful examination of the jaw also revealed a set of remarkably human features for such an apish jaw (beyond the forged wear of the teeth). Sir Arthur Keith repeatedly emphasised, for example, that the teeth were inserted into the jaw in a human, rather than a simian, fashion.

Prevention of discovery by practice.   In former years, the British Museum did not occupy the vanguard in maintaining open and accessible collections – a happy trend of recent years, and one that has helped to lift the odour of mustiness (literally and figuratively) from major research museums. Like the stereotype of a librarian who protects books by guarding them from use, Piltdown's keepers severely restricted access to the original bones. Researchers were often permitted to look but not touch; only the set of plaster casts could be handled. Everyone praised the casts for their accuracy of proportion and detail, but the detection of fraud required access to the originals – artificial staining and wear of teeth cannot be discovered in plaster. Louis Leakey writes in his autobiography:

As I write this book in 1972 and ask myself how it was that the forgery remained unmasked for so many years, I have turned my mind back to 1933, when I first went to see Dr Bather, Smith Woodward's successor. . . . I told him that I wished to make a careful examination of the Piltdown fossils, since I was preparing a textbook on early man. I was taken into the basement to be shown the specimens, which were lifted out of a safe and laid on a table. Next to each fossil was an excellent cast. I was not allowed to handle the originals in any way, but merely to look at them and satisfy myself that the casts were really good replicas. Then, abruptly, the originals were removed and locked up again, and I was left for the rest of the morning with only the casts to study.

It is my belief now that it was under these conditions that all visiting scientists were permitted to examine the Piltdown specimens, and that the situation changed only when they came under the care of my friend and contemporary Kenneth Oakley. He did not see the necessity of treating the fragments as if they were the crown jewels but, rather, considered them simply as important fossils – to be looked after carefully, but from which the maximum scientific evidence should be obtained.

Henry Fairfield Osborn, although not known as a generous man, paid almost obsequious homage to Smith Woodward in his treatise on the historical path of human progress *Man Rises to Parnassus* (1927). He had been a sceptic before his visit to the British Museum in 1921. Then, on Sunday morning, 24 July, 'after attending a most memorable service in Westminster Abbey', Osborn 'repaired to the British Museum to see the fossil remains of the now thoroughly vindicated Dawn Man of Great Britain'. (He, at least, as head of the American Museum of Natural History, got to see the originals.) Osborn swiftly converted and proclaimed Piltdown 'a discovery of transcendant importance to the prehistory of man'. He then added: 'We have to be reminded over and over again that Nature is full of paradoxes and that the order of the Universe is not the human order.' Yet Osborn had seen little but the human order on two levels – the comedy of fraud and the subtler, yet ineluctable, imposition of theory upon nature. I am not distressed that the human order must veil all our interactions with the Universe for the veil, however strong its threads, is translucent.

Source: In *New Scientist*, **82**, No. 1149 (5 Apr. 1979): 42–4.

Chapter 5

5.0 Introduction

In their discussions of belief in relation to the physical sciences, the historians in this chapter take a variety of approaches. The first two papers show some of the epistemological positions apparently compatible with even the same physical hypothesis. Thus Edward Daub's study of the rise and fall of Maxwell's Demon draws attention to the similar device of Loschmidt – which the latter used, however, for a quite different epistemological purpose. It is the more anthropomorphic – or, as it turns out, Theomorphic – character of Maxwell's creation that Daub uses in support of his answer to the historiographical thought-experiment: why did Maxwell not foresee the energy requirements which make the demon inoperative?

Gerald Holton's paper demonstrates another striking example of a physicist's perception of his own theories, in showing that Mach reached a clearer understanding of Einstein's epistemological position long before Einstein himself did.

The last two papers constitute part of a debate which has become increasingly pressing in recent years, about whether the content of even the 'hardest' sciences (here, quantum theory) has been fundamentally determined by extra-scientific beliefs. Paul Forman brings forward a considerable amount of evidence to show that this was the case in Weimar Germany. John Hendry, in what is so far the fullest published analysis of the paper, does not accept Forman's argument, largely on historiographical grounds.

5.1 Edward E. Daub, Maxwell's Demon

In his presentation of the 'two cultures' issue, C. P. Snow relates that he occasionally became so provoked at literary colleagues who scorned the restricted reading habits of scientists that he would challenge them to explain the second law of thermodynamics. The response was invariably a cold negative silence.[1] The test was too hard. Even a scientist would be hard-pressed to explain Carnot engines and refrigerators, reversibility and irreversibility, energy dissipation and entropy increase, Gibbs free energy and the Gibbs rule of phase, all in the span of a cocktail party conversation. How much more difficult, then, for a non-scientist. Even Henry Adams, who sought to find an analogy for his theory of history in the second law of thermodynamics, had great difficulty in understanding the rule of phase.

When Adams sought help with his manuscript 'The Rule of Phase Applied to History', he, too, encountered a cold silence. After months of search he complained to his brother Brooks that he had yet to discover a physicist 'who can be trusted to tell me whether my technical terms are all wrong'.[2] James F. Jameson, editor of the *American Historical Review*, responding to Henry's plea to find him 'a critic . . . a scientific, physico-chemical proof-reader', also met several rebuffs before he found the right man, Professor Henry A. Bumstead of Yale, a former student of Gibbs.[3] Bumstead's twenty-seven pages of detailed commentary must have satisfied Adam's hunger for 'annihilation by a competent hand',[4] as his revised version appeared only posthumously.[5] In it the chastened historian wrote that 'Willard Gibbs helped to change the face of science, but his Phase was not the Phase of History'.[6] Attracted to Gibbs's terminology because of the purely verbal agreement between physical phases and the epochs of Comtean history,[7] Adams erroneously adopted the phase rule as a scientific analogy for the progressive mutations of history.[8] If Maxwell had read Adams's misinterpretation of Gibbs's thought, he might have repeated his quip that the value of metaphysics is inversely proportional to the author's 'confidence in reasoning from the names of things',[9] but he would doubtless have been amused at the antics Adams attributed to his demon in history.

Adams once wrote to his brother Brooks that 'an atom is a man' and that 'Clerk Maxwell's demon who runs the second law of Thermo-dynamics ought to be made President'.[10] On another occasion he found Maxwell's demon a useful illustration for the behaviour of the German nation. 'Do you know the kinetic theory of gases?' he asked a British friend. 'Of course you do, since Clerk Maxwell was an Oxford man, I suppose. Anyway, Germany is and always has been a remarkably apt illustration of Maxwell's conception of "sorting demons". By bumping against all its neighbours, and being bumped in turn, it gets and gives at last a common motion.'[11] But such an aggressive mobile demon as the German nation was very different from the one Maxwell had conceived, a being who did not jostle atoms but arranged to separate them, not for the sake of generating some common motion but rather to illustrate Maxwell's contention that the second law of thermodynamics was statistical in character.

Maxwell's tiny intelligence and the statistical second law
The fundamental basis for the second law of thermodynamics was Clausius's axiom that it is impossible for heat to pass from a colder to a warmer body unless some other change accompanies the process. To

show that this law was only statistically true, Maxwell proposed a thought experiment in which a gas at uniform temperature and pressure was separated by a partition, equipped with a frictionless sliding door and operated by a tiny intelligence who could follow the movements of individual molecules. Although the temperature of the gas was uniform, the velocities of the gas molecules need not be, since temperature is the average kinetic energy of the molecules. The velocities should in fact vary, because the molecules would inevitably be exchanging energy in collisions. The demon might therefore circumvent the axiom regarding the behaviour of heat merely by separating the faster molecules from the slower. By permitting only fast molecules to enter one half and only slow molecules to leave it, Maxwell's tiny intelligence could create a temperature difference and a flow of heat from lower to higher temperatures. Maxwell concluded that the second law 'is undoubtedly true as long as we can deal with bodies only in mass, and have no power of perceiving or handling the separate molecules of which they are made up'. In the absence of such knowledge, we are limited to the statistical behaviour of molecules.[12]

Maxwell did not reach this insight immediately upon conceiving the idea of a tiny intelligence regulating the motions of molecules. His thought progressed (as did his characterizations) in the letters to Tait, Thomson and Rayleigh, where he first discussed this quaint creature. Upon introducing him to Tait in 1867 as a 'very observant and neat-fingered being', Maxwell prefaced his description by suggesting to Tait, who was deeply engrossed writing his *Sketch of Thermodynamics* at the time, that in his book Tait might 'pick a hole – say in the second law of $\theta\triangle^{cs}$., that if two things are in contact the hotter cannot take heat from the colder without external agency'. If only we were clever enough, Maxwell suggested, we too might mimic the neat-fingered one.[13] A month later, discussing his newly designated 'pointsman for flying molecules', he teased Thomson with the provocative thought, 'Hence energy need not be always dizzypated as in the present wasteful world.'[14]

Not, however, until three years later, when the 'intelligence' had grown to a mature 'doorkeeper ... exceedingly quick', did Maxwell reach what became his enduring verdict: 'Moral. The 2nd law of thermodynamics has the same degree of truth as the statement that if you throw a tumblerful of water into the sea, you cannot get the same tumblerful of water out again.'[15] Thus was born Maxwell's prophetic insight that the second law of thermodynamics could never be given a mechanical interpretation based on the laws of pure dynamics which follow the motion of every particle. The second law was true only for matter *en masse*, and that truth was only statistical, not universal.

If Maxwell's demon was thus limited to operating a door for the sole purpose of demonstrating the statistical nature of the second law, where did Henry Adams get the idea of a Germanic bouncing demon who could generate a common motion? In his only public reference to the demon, Adams introduced the idea while criticizing the mechanistic view for omitting mind from the universe. Mind would be the only possible source for direction in an otherwise chaotic universe, Adams maintained, noting that 'The sum of motion without direction is zero, as in the motion of a kinetic gas where only Clerk Maxwell's demon of thought could create a value.'[16] This image of a demon who operates amidst molecules to create value from chaos stemmed from Adams's reading of William Thomson's ideas in 'The Sorting Demon of Maxwell'.[17]

Thomson's demon and the dissipation of energy

It was Thomson who baptized the creature of Maxwell's imagination in an essay in 1874. Maxwell had introduced his brainchild in the span of a few pages in his *Theory of Heat*, describing him simply as a 'being whose faculties are so sharpened that he can follow every molecule in its course',[18] and Thomson went on to christen him the 'intelligent demon'.[19] Whereas Maxwell had stationed his lonely being at the single minute hole in a partitioning wall, there to separate the fleet from the slow, Thomson recruited a whole army of demons to wage war with cricket bats and drive back an onrushing horde of diffusing molecules.[20] In his second essay, Thomson's description became even more anthropomorphic:

He is a being with no preternatural qualities, and differs from real living animals only in extreme smallness and agility. He can at pleasure stop, or strike, or push, or pull away any atom of matter, and so moderate its natural course of motion. Endowed equally with arms and hands – two hands and ten fingers suffice – he can do as much for atoms as a pianoforte player can do for the keys of the piano – just a little more, he can push or pull each atom in any direction.[21]

Thomson's amazing creature could even subdue the forces of chemical affinity by absorbing kinetic energy from moving molecules and then apply that energy to sever molecular bonds. 'Let him take in a small store of energy by resisting the mutual approach of two compound molecules, letting them press as it were on his two hands and store up energy as in a bent spring; then let him apply the two hands between the oxygen and double hydrogen constituents of a compound molecule of vapour of water, and tear them asunder.'[22] The exploits of Thomson's demon give the impression that his main role was to restore dissipated energy. Is motion lost in viscous friction? Simply sort out the molecules moving in one direction and motion reappears. Is heat lost by conduction? Simply separate the faster and slower moving molecules to restore the temperature gradient. Is chemical energy dissipated as heat? Simply use the kinetic energy of the molecules to tear asunder the chemical bonds.

Such an interpretation of the demon's activity appealed to Thomson, for he had been the first to suggest the rather dire image of a universe ruled by the inexorable second law of thermodynamics. In his 1852 paper on the universal dissipation of mechanical energy, Thomson had observed that mechanical energy is continually being dissipated into heat by friction and heat is continually being dissipated by conduction. In this state of affairs, the Earth and its life are caught in a vicious cycle of energy dissipation and decline unless there is action by some non-mechanical agency.[23] Maxwell's thought-experiment showed, however, that energy dissipation need not be irrevocable from the point of view of an acute and designing mind. '"Dissipation of Energy"', Thomson wrote, 'follows in nature from the fortuitous concourse of atoms. The lost motivity is not restorable otherwise than by an agency dealing with atoms; and the mode of dealing with the atoms is essentially a process of assortment.'[24]

Such an understanding of the nature of dissipation was not original with Thomson. Maxwell had drawn the same analogy in far clearer terms, though without reference to any demonic activity. In discussing the difference between dissipated and available energy, Maxwell showed that these concepts were relative to the extent of our knowledge:

It follows . . . that the idea of dissipation of energy depends on the extent of our knowledge. Available energy is energy which we can direct into any desired channel. Dissipated energy is energy which we cannot lay hold of and direct at pleasure, such as the energy of the confused agitation of molecules which we call heat. Now, confusion, like the correlative term order, is not a property of material things in themselves, but only in relation to the mind which perceives them. A memorandum-book does not, provided it is neatly written, appear confused to an illiterate person, or to the owner who understands it thoroughly, but to any other person able to read it appears to be inextricably confused. Similarly the notion of dissipated energy would not occur to a being who could not turn any of the energies of nature to his own account, or to one who could trace the motion of every molecule and seize it at the right moment. It is only to a being in the intermediate stage, who can lay hold of some forms of energy while others elude his grasp, that energy appears to be passing inevitably from the available to the dissipated state.[25]

It was thus Maxwell, not Thomson, who assigned the demon the role of illuminating the nature of dissipated energy and of showing that all energy remains available for a mind able to 'trace the motion of every molecule and seize it at the right moment'. No doubt Maxwell first conceived his tiny intelligence because he was concerned about energy dissipation.

Why else would Maxwell have suggested to Tait in 1867 that he should pick a hole in the second law of thermodynamics, namely, 'that if two things are in contact the hotter cannot take heat from the colder without external agency'? Why should Maxwell choose this problem and why did he suggest that if man was clever enough he might mimic Maxwell's thought-child? Why did Maxwell tease Thomson with the suggestion that energy need not always be 'dizzypated as in the present wasteful world'? Since Maxwell did not finally conclude that the second law is statistical until three years later, he must have had some other reason for beginning this chain of thought. The reason is to be found in

Maxwell's concern about energy dissipation. It is significant in this connection that Josef Loschmidt, the scientist on the Continent who most abhorred the image of a decaying universe, also invented a 'demon' to thwart dissipation, even before Maxwell.

Loschmidt's non-demon and the dynamical interpretation of the second law

Ludwig Boltzmann, Loschmidt's colleague in Vienna, gives the following report of Loschmidt's invention in his account of Loschmidt's varied efforts to obviate the dire theoretical consequences of the second law:

On another occasion he imagined a tiny intelligent being who would be able to see the individual gas molecules and, by some sort of contrivance, to separate the slow ones from the fast and thereby, even if all activity [*Geschehen*] in the universe had ceased, to create new temperature differences. As we all know, this idea, which Loschmidt only hinted in a few lines of an article, was later proposed in Maxwell's *Theory of Heat* and was widely discussed.[26]

Boltzmann's memory, however, failed him on two counts. Loschmidt had devoted more than a few lines to the topic, but he had conceived no tiny intelligent creature. Boltzmann was recalling, not Loschmidt's original idea, but the later arguments he had had with Loschmidt concerning Maxwell's creation. In one of these discussions Boltzmann told Loschmidt that no intelligence could exist in a confined room at uniform temperature, at which point Josef Stefan, who had been listening quietly to the dispute, remarked to Loschmidt, 'Now I understand why your experiments in the basement with glass cylinders are such miserable failures'.[27] Loschmidt had been trying to observe gravitational concentration gradients in salt solutions as evidence for refuting the second law and its prediction of final uniformity.

Loschmidt's non-demon appeared in 1869.[28] It represented an attempt to do exactly the kind of thing which Maxwell had suggested to Tait, namely, 'to pick a hole' in the second law of thermodynamics. Loschmidt was also aiming at Clausius's statement that 'It is impossible for heat to pass from a colder to a warmer body without an equivalent compensation.'[29] Although the axiom was admittedly supported by ordinary experience, Loschmidt proposed to show that it was not true for all conceivable cases. Imagine, he said, a large space V with molecules moving about at various velocities, some above and some below the mean velocity c, plus a small adjoining space v that is initially empty. Consider now, he continued, a small surface element of the wall separating the two compartments and the succession of molecules striking it. Given the initial conditions of all the molecules, the order of their collisions with that surface element should be so fixed and determined that the element could be instructed to open and close in a pattern that would admit only the faster molecules into the empty space v.

Thus we can obviously conceive of these exchanges as so ordered that only those molecules whose velocities lie above the average value *c* may trespass into *v*, and it would further be possible to allow their number so to increase, that the density of the gas in *v* may become greater than that in *V*. It is therefore not theoretically impossible, without the expenditure of work or other compensation, to bring a gas from a lower to a higher temperature or even to increase its density.[30]

Thus, Loschmidt's conception was both earlier and far less anthropomorphic than Maxwell's doorkeeper.

In some of his later writings, Maxwell moved in Loschmidt's direction. When Maxwell wrote to Rayleigh in 1870, just before the only public appearance of his idea in the *Theory of Heat*, he noted, 'I do not see why even intelligence might not be dispensed with and the thing made self-acting'. In a final undated summary statement entitled 'Concerning Demons', he reduced his creature to a valve:

Is the production of an inequality their only occupation? No, for less intelligent demons can produce a difference of pressure as well as temperature by merely allowing all particles going in one direction while stopping all those going the other way. This reduces the demon to a valve. As such value him. Call him no more a demon but a valve. . . .[32]

But although Maxwell had revised his thinking and moved from a tiny intelligence to a more mechanical device, he still claimed the same important and distinctive role for his thought experiment. What, he asked, is the chief end of my creature? 'To show that the 2nd Law of Thermodynamics has only a statistical certainty.'[33]

Loschmidt drew a very different conclusion from the ability of his non-demon to create temperature differences. Since the absolute validity of Clausius's axiom had been rendered doubtful, he argued, the second law of thermodynamics must be established on other grounds, namely, on those very dynamical foundations[34] which Maxwell's demon had led Maxwell to reject. Thus, despite their common origin in the desire to pick a hole in the second law of thermodynamics, Maxwell's demon and Loschmidt's non-demon performed strikingly different roles. In Maxwell's view, the demon did not undermine Clausius's axiom as a basis for the second law of thermodynamics. Since the axiom was statistically true, the predictions based upon it, namely, the irreversible increase in entropy and the increasing dissipation of energy, were valid conclusions. Since these truths were only statistical, however, Maxwell scorned supposed proofs of the second law based upon dynamical studies that traced the motions of individual atoms.

In his *Theory of Heat* Maxwell had raised the following question: 'It would be interesting to enquire how far those ideas . . . derived from the dynamical method . . . are applicable to our actual knowledge of concrete things, which . . . is of an essentially statistical character'.[35] Although that query stood unchanged throughout the four editions of Maxwell's book, his own conviction became clear in candid comments to Tait with regard to various attempts, notably by Clausius and

Boltzmann, to explain the second law of thermodynamics by means of Hamilton's equations in dynamics. In 1873 he jested in lyrical fashion:

But it is rare sport to see those learned Germans contending for the priority in the discovery that the second law of $\theta\triangle^{cs}$ is the Hamiltonsche Princip. . . . The Hamiltonsche Princip the while soars along in a region unvexed by statistical considerations while the German Icari flap their waxen wings in nephelococcygin, amid those cloudy forms which the ignorance and finitude of human science have invested with the incommunicable attributes of the invisible Queen of Heaven.[36]

In 1876, he was pungently clear. No pure dynamical statement, he said, 'would submit to such an indignity'.[37] There in essence lay the true scientific role of Maxwell's doorkeeper, to show the folly of seeking to prove a statistical law as though it expressed the ordered behaviour of traditional dynamic models.

In Loschmidt's thought, a similar devotion to mechanics led to a very different orientation. Since his non-demon disproved the unconditional validity of Clausius's axiom, the appearances, he felt, must be deceptive. The true basis for the second law must lie in traditional dynamical principles. Loschmidt stated his fundamental position most clearly in 1876 when he said: 'Since the second law of the mechanical theory of heat, just like the first, should be a principle of analytical mechanics, then it must be valid, not only under the conditions which occur in nature, but also with complete generality; for systems with any molecular form and under any assumed forces, both intermolecular and external'.[38] He then conceived a variety of models in which columns of atoms in thermal equilibrium would exhibit a temperature gradient and thus contradict the usual thermodynamic axiom proposed by Clausius. Loschmidt concluded that Clausius's axiom was an inadequate basis for the second law, since these molecular models, which fulfilled all the requisite conditions for proving the second law from Hamilton's principle, did not entail the truth of that axiom.[39]

Loschmidt rejoiced that the threatening implications of the second law for the fate of the universe were finally disproved:

Thereby the terrifying *nimbus* of the second law, by which it was made to appear as a principle annihilating the total life of the universe, would also be destroyed; and mankind could take comfort in the disclosure that humanity was not solely dependent upon coal or the Sun in order to transform heat into work, but would have an inexhaustible supply of transformable heat at hand in all ages.[40]

When Loschmidt pressed his case against Clausius's axiom even further, however, by raising the so-called reversibility paradox, he forced Boltzmann to conclude that no dynamical interpretation of the second law is possible.

Loschmidt's reversibility paradox and the statistical second law
The dynamical interpretations of the second law which Loschmidt favoured were restricted to those cases where entropy is conserved in the

universe. Such processes are generally called reversible because they are so ideally contrived that the original conditions may always be completely recovered. No entropy increase or dissipation of energy occurs in reversible cycles. Interpretations of the second law based on Clausius's axiom, however, consider another type of process, the irreversible case in which entropy irrevocably increases and energy dissipates. According to Clausius, there were three possible cases corresponding to negative, zero and positive changes in entropy. The negative entropy change was impossible because that would be equivalent to a flow of heat from a cold to a hot body, contrary to Clausius's axiom. Entropy was a quantity, therefore, which could only remain constant or increase, depending on whether the process was reversible or irreversible.[41] Loschmidt's dynamical interpretation would only countenance the reversible case. Boltzmann, however, sought a dynamical interpretation for the irreversible increase in entropy as well, and it was to refute that possibility that Loschmidt created his reversibility paradox.

In order to have a dynamical interpretation of the irreversible case, there would have to be a mathematical function which showed a unilateral change to a maximum, after which it would remain constant, thereby reflecting the irreversible increase of entropy to a maximum at equilibrium. Boltzmann had derived such a function from an analysis of the collisions between molecules. To refute such an idea, only a single counter-example is necessary. One need only demonstrate that there exists at least one distribution of molecular velocities and positions from which the opposite behaviour would proceed. Loschmidt provided just such a thought-experiment. Imagine, he said, a system of particles where all are at rest at the bottom of the container except for one which is at rest some distance above. Let that particle fall and collide with the others, thus initiating motion among them. The ensuing process would lead to increasingly disordered motion until the system reached an apparently static equilibrium, just as the proponents of irreversible entropy change to an equilibrium would predict. But now imagine the instantaneous reversal of every single velocity and the very opposite process becomes inevitable. At first, little change would be evident, but the system would gradually move back towards the initially ordered situation in which all particles were at rest on the bottom and only one rested at a height above them.[42] Boltzmann labelled this Loschmidt's paradox, and a real paradox it became, since Boltzmann managed to reverse all of Loschmidt's conclusions.

The paradox revealed to Boltzmann that his attempt to find a dynamical function of molecular motion which would mirror the behaviour of entropy could only lead to a dead end, for whatever the mathematical function might be, the mere reversal of velocities would also reverse the supposedly unidirectional behaviour of that function. Boltzmann concluded that no purely dynamical proof of the second law would ever be possible and that the irreversible increase of entropy must reflect, not a mechanical law, but states of differing probabilities. Systems move towards equilibrium simply because the number of molecular states which correspond to equilibrium is vastly greater than the number of more ordered states of low entropy. Boltzmann offered an analogy to the Quinterns used in Lotto. Each Quintern has an equal probability of appearing, but a Quintern with a disordered arrangement of numbers is far more likely to appear than one with an ordered arrangement such as 12345. Boltzmann therefore provided the key for quantifying the statistical interpretation of the second law in terms of the relative numbers of molecular states that correspond to equilibrium and non-equilibrium.[43]

Thus, the chief end of Maxwell's creature, 'to show that the 2nd Law of Thermodynamics has only a statistical certainty', became established as a cardinal principle of classical physics by way of Loschmidt's non-demon and his reversibility paradox. Although Maxwell's demon has served to popularize the statistical interpretation of the second law through generations of thermodynamics textbooks, a new thought-experiment involving information theory has now challenged the demon's traditional role. The arguments made there suggest that Maxwell's brainchild must, alas, be laid to rest.

The price of information
Consider the case, often discussed in information theory and originally introduced by Szilard in 1929,[44] where but a single molecule is involved. If the second law is merely statistical, then it certainly should fail to meet the test of this simplest of all arrangements. Pierce has described a modified version of Szilard's innovation as follows. Consider a piston equipped with a large trapdoor and with arrangements to allow the piston to lift weights by moving either left or right. Initially the piston is not connected to any weights; it is moved to the centre of the cylinder while the trapdoor is kept open, thus assuring that no collisions with the lone molecule in the cylinder can occur and that, therefore, no work will be required. Then the trapdoor is closed and the piston clamped into its central position. The molecule must be entrapped on one of the two sides of the piston, and Maxwell's demon informs us whether it is on the left or the right. With that information in hand, the piston is released and the molecule is made to do work by driving the weightless piston and a suitably suspended weight towards the empty side. The maximum possible work may be readily calculated; it would amount to $W = 0.693 \, kT$.[45] In Maxwell's day, the demon would have chalked up another victory, but not now.

'Did we get this mechanical work free?' Pierce succinctly asks. 'Not quite!'

In order to know which pan to put the weight on, we need one bit of information, specifying which side the molecule is on. . . . What is the very least energy needed to transmit one bit of information at the temperature T? . . . exactly $0.693 \, kT$ joule, just equal to the most energy the machine can generate. . . . Thus, we use up all the output of the machine in transmitting enough information to make the machine run![46]

Thus, the reign of Maxwell's brainchild in physics, designed to demonstrate that the second law of thermodynamics has only statistical validity, has come to an end.

The mood of the scientific community had changed. The second law was no longer subject to reproach for the indignities Maxwell and Loschmidt supposed it would inflict on pure dynamics. It was natural, therefore, to extend the province of the law and challenge all imagined contradictions. Szilard was the first to stress that any manipulator of molecules would have to rely on measurement and memory. If one assumed that the demon could perform such operations without causing any changes in the system, one would by that very assumption deny the second law, which requires equivalent compensations for all decreases in entropy.[47] Szilard therefore proposed that whatever negative entropy Maxwell's demon might be able to create should be considered as compensated by an equal entropy increase due to the measurements the demon had to make. In essence, Szilard made Maxwell's doorkeeper mortal – no longer granting this tiny intelligence the ability to 'see' molecules without actually seeing them, i.e. without the sensory exchanges of energy that all other existences require. Szilard took this step for the sake of a grander vision, the dream that the adoption of his principle would lead to the discovery of a more general law of entropy in which there would be a completely universal relation for all measurements.[48] Information theory has brought that vision to reality.

One puzzling question, however, remains. Why did Maxwell not realize that his creature required energy in order to detect molecules? Brillouin has suggested that Maxwell did not have an adequate theory of radiation at his disposal. 'It is not surprising', he said, 'that Maxwell did not think of including radiation in the system in equilibrium at temperature *T*. Black body radiation was hardly known in 1871, and it was thirty years before the thermodynamics of radiation was clearly understood.'[49] It is certainly true that a quantitative expression for radiant energy and the entropy of information would require an adequate theory of black body radiation, but the absence of such a detailed theory does not explain why Maxwell failed to realize that some energy exchanges were required.

If we were able to ask Maxwell, 'Why did you not require your tiny intelligence to use energy in gathering his information?', Maxwell would no doubt reply, 'Of course! Why didn't I think of that?'[50] Why didn't Maxwell think of that? Because his demon was the creature of his theology.

The demon and theology

Maxwell's demon is the very image of the Newtonian God who has ultimate dominion over the world and senses the world in divine immediacy. Newton wrote in his General Scholium:

It is allowed by all that the Supreme God exists necessarily, and by the same necessity he exists *always* and *everywhere*. Whence also he is all similar, all eye, all ear, all brain, all arm, all power to perceive, to understand, and to act; but in a manner not at all human, in a manner not at all corporeal, in a manner utterly unknown to us. As a blind man has no idea of colour, so we have no idea of the manner by which the all-wise God perceives and understands all things.[51]

How natural it was for Maxwell, faced with the idea of a universe destined towards dissipation, to conceive a being on the model of God for whom the universe always remains ordered and under his rule. A memorandum-book, Maxwell said, does not appear confused to its owner though it does to any other reader. Nor would the notion of dissipated energy occur to 'a being who could trace the motion of every molecule and seize it at the right moment'. Maxwell's demon was not mortal because he was made in the image of God. And like God, he could see without seeing and hear without hearing. In short, he could acquire information without any expenditure of energy.

Upon being asked by a clergyman for a viable scientific idea to explain how, in the *Genesis* account, light could be created on the first day although the Sun did not appear until the third day, Maxwell replied that he did not favour reinterpreting the text in terms of prevailing scientific theory. To tie a religious idea to a changeable scientific view, Maxwell said, would only serve to keep that scientific idea in vogue long after it deserved to be dead and buried.[52] Thus Maxwell would certainly sever the demon's ties to theology if he were faced with Szilard's requirements for the cost of information. Witnessing the demise of his creature, he would not long mourn at the grave but rather be grateful for the exciting years his thought-child had enjoyed. Steeped in the knowledge and love of biblical imagery, Maxwell would doubtless take pleasure in the thought that by becoming mortal his doorkeeper had prepared the way for new life in science.[53]

Source: Studies in History and Philosophy of Science, **1** (1970): 213–27.

Notes
1. C. P. Snow, *The Two Cultures and the Scientific Revolution*. New York, 1961: 15–16.
2. H. D. Cater (ed.), *Henry Adams and his Friends*. Boston, 1947: 640.
3. Ibid.: 646–7, 650n.
4. Ibid.: 647.
5. E. Samuels, *Henry Adams: The Major Phase*. Cambridge, 1964: 450: 'Adams minutely revised the essay during the next year or two to meet Bumstead's more specific criticisms. . . . The firm outlines of the script would suggest that all of these changes were made before his stroke in 1912. No indication remains that he submitted the revised essay to the *North American Review*. Not until 1919, a year after his death, did it appear in Brooks Adams's edition of Henry's 'philosophical writings', *The Degradation of the Democratic Dogma*. . . .'

6. H. Adams, *The Degradation of the Democratic Dogma*, B. Adams (ed.). New York, 1919: 267.
7. W. H. Jordy, *Henry Adams: Scientific Historian*. New Haven, 1952: 166.
8. Ibid.: 169, 170.
9. Maxwell–Tait Correspondence, Cambridge University Library; letter to Tait, 23 Dec. 1867: 'I have read some metaphysics of various kinds and find it more or less ignorant discussion of mathematical and physical principles, jumbled with a little physiology of the senses. The value of metaphysics is equal to the mathematical and physical knowledge of the author divided by his confidence in reasoning from the names of things.'
10. Letter to Brooks Adams, 2 May 1903, op. cit., note 2: 545.
11. H. Adams, *Letters of Henry Adams (1892–1918)*, N. C. Ford (ed), Boston, 1938. Letter to Cecil Spring Rice, 11 Nov. 1897: 135–6.
12. J. C. Maxwell, *The Theory of Heat*, 2nd edn. London, 1872: 308–9.
13. Letter from Maxwell to Tait, 11 Dec. 1867, quoted in C. G. Knott, *Life and Scientific Work of Peter Guthrie Tait*. Cambridge, 1911: 213–14.
14. Letter from Maxwell to William Thomson, 16 Jan. 1868, Edinburgh University Library.
15. Letter from Maxwell to Strutt, 6 Dec. 1870, quoted in R. J. Strutt, *John William Strutt*. London, 1924: 47.
16. Adams, op. cit., note 6: 279.
17. W. Thomson, 'The sorting demon of Maxwell', reprinted in *Popular Lectures and Addresses*, 1. London, 1889: 137–41.
18. Maxwell, op. cit., note 12: 308.
19. W. Thomson, 'Kinetic theory of the dissipation of energy', *Nature*, **9** (1874): 442.
20. Ibid.
21. Thomson, op. cit., note 17: 137–8.
22. Ibid.: 140.
23. W. Thomson, 'On a universal tendency in nature to the dissipation of mechanical energy', *Philosophical Magazine*, **4** (1852): 256–60.
24. Thomson, op. cit., note 17: 139.
25. J. C. Maxwell, 'Diffusion', *Encyclopedia Britannica*, 9th edn. New York, 1878, 7: 220.
26. L. Boltzmann, 'Zur Errinerung an Josef Loschmidt', in *Populäre Schriften*. Leipzig, 1905: 231. (I am indebted to R. Dugas, *La théorie physique au sens Boltzmann*. Neuchatel, 1959: 171, note 2, for this reference.)
27. Ibid.
28. J. Loschmidt, 'Der zweite Satz der mechanischen Wärmetheorie', *Akademie der Wissenschaften, Wien. Mathematisch-Naturwissenschaftliche Klasse, Sitzungsberichte*, **59**, Abth. 2 (1869): 395–418.
29. Ibid. 399. The editor must have slipped, however, since the text reads 'Die Wärme geht niemals aus einem heisseren in einen kälteren über ohne eine äquivalente Compensation'.
30. Ibid.: 401.
31. Quoted in Strutt, op. cit., note 15: 47.
32. Quoted in Knott, op. cit., note 13: 215. Knott seems to have erred in suggesting that this undated letter was penned at about the same time that Maxwell originally proposed his idea to Tait in Dec. 1867.
33. Ibid.
34. Loschmidt, op. cit., note 28: 401–6.

35. Maxwell, op. cit., note 12: 309.
36. Letter from Maxwell to Tait, Tripos, Dec. 1873, Maxwell–Tait Correspondence, Cambridge University Library.
37. Letter from Maxwell to Tait, 13 Oct. 1876, ibid. Martin Klein's recent article, 'Maxwell, his demon, and the second law of thermodynamics', *American Scientist*, **58** (1970): 84–97, treats these attempts to reduce the second law to dynamics in considerable detail.
38. J. Loschmidt, 'Ueber den Zustand des Wärmegleichgewichtes eines System von Körpern' *Sitzungsberichte* (see note 28) **73**, Abth. 2 (1876): 128.
39. Ibid.: 128–35.
40. Ibid.: 135.
41. R. Clausius, 'Ueber eine veränderte Form des zweiten Hauptsatzes der mechanischen Wärmetheorie', *Annalen der Physik*, **93** (1854): 481–506. Clausius did not use the word 'entropy' until 1865. In the original paper he spoke of the sum of transformation values $\int dQ/T = N$ for a cycle, where $N < 0$ was impossible, $N = 0$ was the reversible case, and $N > 0$ was the irreversible cycle.
42. Loschmidt, op. cit., note **38**: 137–9.
43. L. Boltzmann, 'Bemerkungen über einige Probleme der mechanischen Wärmetheorie' (1877), reprinted in *Wissenschaftliche Abhandlungen von Ludwig Boltzmann*, F. Hasenohrl (ed.) Leipzig, 1909, **II**: 120.
44. L. Szilard, 'Ueber die Entropieverminderung in einem thermodynamischen System bei Eingriffen intelligenter Wesen', *Zeitschrift für Physik*, 53 (1929): 840–56.
45. J. R. Pierce, *Symbols, Signals, and Noise*. New York, 1961: 198–201.
46. Ibid.: 201.
47. Szilard, op. cit., note 44: 842.
48. Ibid.: 843.
49. L. Brillouin, *Science and Information Theory*. New York, 1956: 164.
50. I am indebted to my colleague Richard Cole of the Kansas University Philosophy Department for this thought-experiment.
51. I. Newton, *Newton's Philosophy of Nature* H. S. Thayer (ed.), New York, 1953: 44.
52. L. Campbell and W. Garnett, *The Life of James Clerk Maxwell*. London, 1882: 394.
53. The support of the National Science Foundation is gratefully acknowledged.

5.2 Gerald Holton, Mach, Einstein and the Search for Reality

In the history of ideas of our century, there is a chapter that might be entitled 'The Philosophical Pilgrimage of Albert Einstein', a pilgrimage from a philosophy of science in which sensationism and empiricism were at the center, to one in which the basis was a rational realism. This essay, a portion of a more extensive study,[1] is concerned with Einstein's gradual philosophical reorientation, particularly as it has become discernible during the work on his largely unpublished scientific correspondence.[2]

The earliest known letter by Einstein takes us right into the middle of the case. It is dated 19 March 1901 and addressed to Wilhelm Ostwald.[3] The immediate cause for Einstein's letter was his failure to receive an assistantship at the school where he had recently finished his formal studies, the Polytechnic Institute in Zürich; he now turned to Ostwald to ask for a position at his laboratory, partly in the hope of receiving 'the opportunity for further education'. Einstein included a copy of his first publication, *Folgerungen aus den Capillaritätserscheinungen*[4] (1901), which he said had been inspired (*angeregt*) by Ostwald's work; indeed, Ostwald's *Allgemeine Chemie* is the first book mentioned in all of Einstein's published work.

Not having received an answer, Einstein wrote again to Ostwald on 3 April 1901. On 13 April 1901 his father, Hermann Einstein, sent Ostwald a moving appeal, evidently without his son's knowledge. Hermann Einstein reported that his son esteems Ostwald 'most highly among all scholars currently active in physics'.[5]

The choice of Ostwald was significant. He was, of course, not only one of the foremost chemists, but also an active 'philosopher-scientist' during the 1890s and 1900s, a time of turmoil in the physical sciences as well as in the philosophy of science. The opponents of kinetic, mechanical, or materialistic views of natural phenomena were vociferous. They objected to atomic theory and gained great strength from the victories of thermodynamics, a field in which no knowledge or assumption was needed concerning the detailed nature of material substances (for example, for an understanding of heat engines).

Ostwald was a major critic of the mechanical interpretation of physical phenomena, as were Helm, Stallo, and Mach. Their form of positivism – as against the sophisticated logical positivism developed later in Carnap and Ayer's work – provided an epistemology for the new phenomenologically based science of correlated observations, linking energetics and sensationism. In the second (1893) edition of his influential textbook on chemistry, Ostwald had given up the mechanical treatment of his first edition for Helm's 'energetic' one. 'Hypothetical' quantities such as atomic entities were to be omitted; instead, these authors claimed they were satisfied, as Merz wrote around 1904, with 'measuring such quantities as are presented in observation, such as energy, mass, pressure, volume, temperature, heat, electrical potential, etc., without reducing them to imaginary mechanisms or kinetic quantities'. They condemned such conceptions as the ether, with properties not accessible to direct observation, and they issued a call 'to consider anew the ultimate principles of all physical reasoning, notably the scope and validity of the Newtonian laws of motion and of the conceptions of force and action, of absolute and relative motion'.[6]

All these iconoclastic demands – except anti-atomism – must have been congenial to the young Einstein who, according to his colleague Joseph Sauter, was fond of calling himself 'a heretic'.[7] Thus, we may well suspect that Einstein felt sympathetic to Ostwald who denied in the *Allgemeine Chemie* that 'the assumption of that medium, the ether, is unavoidable. To me it does not seem to be so. . . . There is no need to inquire for a carrier of it when we find it anywhere. This enables us to look upon radiant energy as independently existing in space.'[8] It is a position quite consistent with that shown later in Einstein's papers of 1905 on photon theory and relativity theory.

In addition, it is worth noting that Einstein, in applying to Ostwald's laboratory, seemed to conceive of himself as an experimentalist. We know from many sources that in his student years in Zürich, Einstein's earlier childhood interest in mathematics had slackened considerably. In the *Autobiographical Notes*, Einstein reported: 'I really could have gotten a sound mathematical education. However, I worked most of the time in the physical laboratory, fascinated by the direct contact with experience.'[9] To this, one of his few reliable biographers adds: 'No one could stir him to visit the mathematical seminars. . . . He did not yet see the possibility of seizing that formative power resident in mathematics, which later became the guide of his work. . . . He wanted to proceed quite empirically, to suit his scientific feeling of the time. . . . As a natural scientist, he was a pure empiricist.'[10]

Ostwald's main philosophical ally was the prolific and versatile Austrian physicist and philosopher Ernst Mach (1838–1916), whose main work Einstein had read avidly in his student years and with whom he was destined to have later the encounters that form a main concern of this paper. Mach's major book, *The Science of Mechanics*,[11] first published in 1883, is perhaps most widely known for its discussion of Newton's *Principia*, in particular for its devastating critique of what Mach called the 'conceptual monstrosity of absolute space' – a conceptual monstriosity because it is 'purely a thought-thing which cannot be pointed to in experience'.[12] Starting from his analysis of Newtonian presuppositions, Mach proceeded in his announced program of eliminating all metaphysical ideas from science. As Mach said quite bluntly in the preface to the first edition of *The Science of Mechanics*: 'This work is not a text to drill theorems of mechanics. Rather, its intention is an enlightening one – or to put it still more plainly, an anti-metaphysical one.'

It will be useful to review briefly the essential points of Mach's philosophy. Here we can benefit from a good, although virtually unknown, summary presented by his sympathetic follower, Moritz Schlick, in the essay *Ernst Mach, der Philosoph*.

Mach was a physicist, physiologist, and also psychologist, and his philosophy . . . arose from the wish to find a principal point of view to which he could hew in any research, one which he would not have to change when going from the field of physics to that of physiology or psychology. Such a firm point of view he reached by going back to that which is given before all scientific research: namely, the world of sensations. . . . Since all our testimony concerning the so-called external world relies only on sensations, Mach held that we can and must take these sensations and complexes of sensations to be the sole contents [*Gegenstände*] of those testimonies, and, therefore, that there is no need to assume in addition an unknown reality hidden behind the sensations. With that, the existence *der Dinge an sich* is removed as an unjustified and unnecessary assumption. A body, a physical object, is nothing else than a complex, a more or less firm [we would say, invariant] pattern of sensations, i.e. of colors, sounds, sensations of heat and pressure, etc.

There exists in this world nothing whatever other than sensations and their connections. In place of the word 'sensations', Mach liked to use rather the more neutral word 'elements'. . . . [As is particularly clear in Mach's book *Erkenntnis und Irrtum*] scientific knowledge of the world consists, according to Mach, in nothing else than the simplest possible description of the connections between the elements, and it has as its only aim the intellectual mastery of those facts by means of the least possible effort of thought. This aim is reached by means of a more and more complete 'accommodation of the thoughts to one another'. this is the formulation by Mach of his famous 'principle of the economy of thought'.[13]

The influence of Mach's point of view, particularly in the German-speaking countries, was enormous – on physics, on physiology, on psychology, and on the fields of the history and the philosophy of science)[14] not to mention Mach's profound effect on the young Lenin, Hofmannsthal, Musil, among many others outside the sciences). Strangely neglected by recent scholarship – there is not even a major biography – Mach has in the last two or three years again become the subject of a number of promising studies. To be sure, Mach himself always liked to insist that he was beleaguered and neglected, and that he did not have, or wish to have, a philosophical system; yet his philosophical ideas and attitudes had become so widely a part of the intellectual equipment of the period from the 1880s on that Einstein was quite right in saying later that 'even those who think of themselves as Mach's opponents hardly know how much of Mach's views they have, as it were, imbibed with their mother's milk'.[15]

The problems of physics themselves at that time helped to reinforce the appeal of the new philosophical attitude urged by Mach. The great program of nineteenth-century physics, the reconciliation of the notions of ether, matter, and electricity by means of mechanistic pictures and hypotheses, had led to enormities – for example, Larmor's proposal that the electron is a permanent but movable state of twist or strain in the ether, forming discontinuous particles of electricity and possibly of all ponderable matter. To many of the younger physicists of the time, attacking the problems of physics with conceptions inherited from classical nineteenth-century physics did not seem to lead anywhere. And here Mach's iconoclasm and incisive critical courage, if not the details of his philosophy, made a strong impression on his readers.

Mach's early influence on Einstein

As the correspondence at the Einstein Archives at Princeton reveals, one of the young scientists deeply caught up in Mach's point of view was Michelange (Michele) Besso – Einstein's oldest and closest friend, fellow student, and colleague at the Patent Office in Bern, the only person to whom Einstein gave public credit for help (*manche wertvolle Anregung*) when he published his basic paper on relativity in 1905. It was Besso who introduced Einstein to Mach's work. In a letter of 8 April 1952 to Carl Seelig, Einstein wrote: 'My attention was drawn to Ernst Mach's *Science of Mechanics* by my friend Besso while a student, around the year 1897. The book exerted a deep and persisting impression upon me . . ., owing to its physical orientation toward fundamental concepts and fundamental laws.' As Einstein noted in his *Autobiographical Notes* written in 1946, Ernst Mach's *The Science of Mechanics* 'shook this dogmatic faith' in 'mechanics as the final basis of all physical thinking. . . . This book exercised a profound influence upon me in this regard while I was a student. I see Mach's greatness in his incorruptible skepticism and independence; in my younger years, however, Mach's epistemological position also influenced me very greatly.'[16]

As the long correspondence between those old friends shows, Besso remained a loyal Machist to the end. Thus, writing to Einstein on 8 December 1947, he still said: 'As far as the history of science is concerned, it appears to me that Mach stands at the center of the development of the last 50 or 70 years.' Is it not true, Besso also asked, 'that this introduction [to Mach] fell into a phase of development of the young physicist [Einstein] when the Machist style of thinking pointed decisively at observables – perhaps even, indirectly, to clocks and meter sticks?'

Turning now to Einstein's crucial first paper on relativity in 1905, we can discern in it influences of many, partly contradictory, points of view – not surprising in a work of such originality by a young contributor. Elsewhere I have examined the effect – or lack of effect – on that paper of three contemporary physicists: H. A. Lorentz,[17] Henri Poincaré,[18] and August Föppl. Here we may ask in what sense and to what extent Einstein's initial relativity paper of 1905 was imbued with the style of thinking associated with Ernst Mach and his followers – apart from the characteristics of clarity and independence, the two traits in Mach which Einstein always praised most.

In brief, the answer is that the Machist component – a strong component, even if not the whole story – shows up prominently in two related respects: first, by Einstein's insistence from the beginning of his relativity paper that the fundamental problems of physics cannot be understood until an epistemological analysis is carried out, particularly so with respect to the meaning of the conceptions of space and time[19]; and second, by Einstein's identification of reality with what is given by sensations, the 'events', rather than putting reality on a plane beyond or behind sense experience.

From the outset, the instrumentalist, and hence sensationist, views of measurement and of the concepts of space and time are strikingly evident. The key concept in the early part of the 1905 paper is introduced at the top of the third page in a straightforward way. Indeed, Leopold Infeld in his biography of Einstein called them 'the simplest sentence[s] I have ever encountered in a scientific paper'. Einstein wrote: 'We have to take into account that all our judgments in which time plays a part are always judgments of *simultaneous events*. If for instance I say, "that train arrived here at seven o'clock", I mean something like this: "The pointing of the small hand of my watch to seven and the arrival of the train are simultaneous events."'[20]

The basic concept introduced here, one that overlaps almost entirely Mach's basic 'elements', is Einstein's concept of *events [Ereignisse]* – a word that recurs in Einstein's paper about a dozen times immediately following this citation. Transposed into Minkowski's later formulation of relativity, Einstein's 'events' are the intersections of particular 'world lines', say that of the train and that of the clock. The time (t coordinate) of an event by itself has no operational meaning. As Einstein says: 'The 'time' of an event is that which is given simultaneously with the event by a stationary clock located at the place of the event.'[21] We can say that just as the *time* of an event assumes meaning only when it connects with our consciousness through sense experience (that is, when it is subjected to measurement-in-principle by means of a clock present at the same place), so also is the *place*, or space coordinate, of an event meaningful only if it enters our sensory experience while being subjected to measurement-in-principle (that is, by means of meter sticks present on that occasion at the same time).[22]

This was the kind of operationalist message which, for most of his readers, overshadowed all other philosophical aspects in Einstein's paper. His work was enthusiastically embraced by the groups who saw themselves as philosophical heirs of Mach, the Vienna Circle of neopositivists and its predecessors and related followers,[23] providing a tremendous boost for the philosophy that had initially helped to nurture it. A typical response welcoming the relativity theory as 'the victory over the metaphysics of absolutes in the conceptions of space and time . . . a mighty impulse for the development of the philosophical point of view of our time', was extended by Joseph Petzoldt in the inaugural session of the *Gesellschaft für positivistische Philosophie* in Berlin, 11 November 1912.[24] Michele Besso, who had heard the message from Einstein before anyone else, had exclaimed: 'In the setting of Minkowski's space-time framework, it was now first possible to carry through the thought which the great mathematician, Bernhard Riemann, had grasped: "The space-time framework itself is formed by the events in it."'[25]

To be sure, re-reading Einstein's paper with the wisdom of hindsight, as we shall do presently, we can find in it also very different trends, warning of the possibility that 'reality' in the end is not going to be left identical with 'events'. There are premonitions that sensory experiences, in Einstein's later work, will not be regarded as the chief building blocks of the 'world', that the laws of physics themselves will be seen to be built into the event-world as the undergirding structure 'governing' the pattern of events.

Such precursors appear even earlier, in one of Einstein's first letters in the Archives. Addressed to his friend Marcel Grossmann, it is dated 14 April 1901, when Einstein believed he had found a connection between

Newtonian forces and the forces of attraction between molecules: 'It is a wonderful feeling to recognize the unity of a complex of appearances which, to direct sense experience, seem to be separate things.' Already there is a hint here of the high value that will be placed on intuited unity and the limited role seen for evident sense experience.

But all this was not yet ready to come into full view, even to the author. Taking the early papers as a whole, and in the context of the physics of the day, we find that Einstein's philosophical pilgrimage did start on the historic ground of positivism. Moreover, Einstein thought so himself, and confessed as much in letters to Ernst Mach.

The Einstein-Mach letters

In the history of recent science, the relation between Einstein and Mach is an important topic that has begun to interest a number of scholars. Indeed, it is a drama of which we can sketch here four stages: Einstein's early acceptance of the main features of Mach's doctrine; the Einstein–Mach correspondence and meeting; the revelation in 1921 of Mach's unexpected and vigorous attack on Einstein's relativity theory; and Einstein's own further development of a philosophy of knowledge in which he rejected many, if not all, of his earlier Machist beliefs.

Happily, the correspondence is preserved at least in part. A few letters have been found, all from Einstein to Mach. Those of concern here are part of an exchange between 1909 and 1913, and they testify to Einstein's deeply felt attraction to Mach's viewpoint, just at a time when the mighty Mach himself – forty years senior to the young Einstein whose work was just becoming widely known – had for his part embraced the relativity theory publicly by writing in the second (1909) edition of *Conservation of Energy*: 'I subscribe, then, to the principle of relativity, which is also firmly upheld in my *Mechanics* and *Wärmelehre*.[26] In the first letter, Einstein writes from Bern on 9 August 1909. Having thanked Mach for sending him the book on the Law of Conservation of Energy, he adds: 'I know, of course, your main publications very well, of which I most admire your book on Mechanics. You have had such a strong influence upon the epistemological conceptions of the younger generation of physicists that even your opponents today, such as Planck, undoubtedly would have been called Mach followers by physicists of the kind that was typical a few decades ago.'

It will be important for our analysis to remember that Planck was Einstein's earliest patron in scientific circles. It was Planck who, in 1905, as editor of the *Annalen der Physik*, received Einstein's first relativity paper and thereupon held a review seminar on the paper in Berlin. Planck defended Einstein's work on relativity in public meetings from the beginning, and by 1913 had succeeded in persuading his German colleagues to invite Einstein to the Kaiser-Wilhelm-Gesellschaft in Berlin. With a polemical essay *Against the New Energetics* in 1896, he had made clear his position, and by 1909 Planck was one of the few opponents of Mach, and scientifically the most prominent one. He had just written a famous attack, *Die Einheit des physikalischen Weltbildes*. Far from accepting Mach's view that, as he put it, 'Nothing is real except the perceptions, and all natural science is ultimately an economic adaptation of our ideas to our perceptions', Planck held to the entirely antithetical position that a basic aim of science is 'the finding of a *fixed* world picture independent of the variation of time and people', or, more generally, 'the complete liberation of the physical picture from the individuality of the separate intellects'.[27]

At least by implication, Einstein's remarks to Mach show that he dissociated himself from Planck's view. It may also not be irrelevant that just at that time Einstein, who since 1906 had been objecting to inconsistencies in Planck's quantum theory, was preparing his first major invited paper before a scientific congress, the eighty-first meeting of the Naturforscherversammlung, announced for September, 1909, in Salzburg. Einstein's paper called for a revision of Maxwell's theory to accommodate the probabilistic character of the emission of photons – none of which Planck could accept – and concluded: 'To accept Planck's theory means, in my view, to throw out the bases of our [1905] theory of radiation.'

Mach's reply to Einstein's first letter is now lost, but it must have come quickly, because eight days later Einstein sends an acknowledgement:

Bern, 17 August 1909. Your friendly letter gave me enormous pleasure. . . . I am very glad that you are pleased with the relativity theory. . . . Thanking you again for your friendly letter, I remain, your student [indeed: *Ihr Sie verehrender Schüler*], A. Einstein.

Einstein's next letter was written as physics professor in Prague, where Mach before him had been for twenty-eight years. The post had been offered to Einstein on the basis of recommendations of a faction (Lampa, Pick) who regarded themselves as faithful disciples of Mach. The letter was sent out about New Year's 1911–12, perhaps just before or after Einstein's sole (and, according to Philipp Frank's account in *Einstein, His Life and Times*, not very successful) visit to Mach, and after the first progress toward the general relativity theory:

. . . I can't quite understand how Planck has so little understanding for your efforts. His stand to my [general relativity] theory is also one of refusal. But I can't take it amiss; so far, that one single epistemological argument is the only thing which I can bring forward in favor of my theory.[28]

Here, Einstein is referring delicately to the Mach Principle, which he had been putting at the center of the developing theory.[29] Mach responded by sending Einstein a copy of one of his books, probably the *Analysis of Sensations*.

In the last of these letters to Mach (who was now seventy-five years old, and for some years had been paralyzed), Einstein writes from Zürich on 25 June 1913:

Recently you have probably received my new publication on relativity and gravitation which I have at last finished after unending labor and painful doubt. [This must have been the *Entwurf einer verallgemeinerten Relativitätstheorie und einer Theorie der Gravitation*, written with Marcel Grossmann.[30]] Next year at the solar eclipse it will turn out whether the light rays are bent by the sun, in other words whether the basic and fundamental assumption of the equivalence of the acceleration of the reference frame and of the gravitational field really holds. If so, then your inspired investigations into the foundations of mechanics – despite Planck's unjust criticism – will receive a splendid confirmation. For it is a necessary consequence that inertia has its origin in a kind of mutual interaction of bodies, fully in the sense of your critique of Newton's bucket experiment.[31]

The paths diverge

The significant correspondence stops here, but Einstein's public and private avowals of his adherence to Mach's ideas continue for several years more. For example, there is his well-known, moving eulogy of Mach, published in 1916.[15] In August, 1918, Einstein writes to Besso quite sternly about an apparent – and quite temporary – lapse in Besso's positivistic epistemology; it is an interesting letter, worth citing in full:

28 August 1918.
Dear Michele:
   In your last letter I find, on re-reading, something which makes me angry: That speculation has proved itself to be superior to empiricism. You are thinking here about the development of relativity theory. However, I find that this development teaches something else, that it is practically the opposite, namely that a theory which wishes to deserve trust must be built upon generalizable facts.
   Old examples: Chief postulates of thermodynamics [based] on impossibility of perpetuum mobile. Mechanics [based] on a grasped [*ertasteten*] law of inertia. Kinetic gas theory [based] on equivalence of heat and mechanical energy (also historically). Special Relativity on the constancy of light velocity and Maxwell's equation for the vacuum, which in turn rest on empirical foundations. Relativity with respect to uniform [?] translation is a *fact of experience*.
   General Reality: *Equivalence of inertial and gravitational mass.* Never has a truly useful and deep-going theory really been found purely speculatively. The nearest case is Maxwell's hypothesis concerning displacement current; there the problem was to do justice to the fact of light propagation. . . . With cordial greetings, your Albert. [Emphasis in the original.]

   Careful reading of this letter shows us that already here there is evidence of divergence between the conception of 'fact' as understood by Einstein and 'fact' as understood by a true Machist. The impossibility of the perpetuum mobile, the first law of Newton, the constancy of light velocity, the validity of Maxwell's equations, the equivalence of inertial and gravitational mass – none of these would have been called 'facts of experience' by Mach. Indeed, Mach might have insisted that – to use one of his favorite battle words – it is evidence of 'dogmatism' not to regard all these conceptual constructs as continually in need of probing re-examinations; thus, Mach had written:

. . . for me, matter, time and space are still *problems*, to which, incidentally, the physicists (Lorentz, Einstein, Minkowski) are also slowly approaching.[32]

Similar evidence of Einstein's gradual apostasy appears in a letter of 4 December 1919 to Paul Ehrenfest. Einstein writes:

I understand your difficulties with the development of relativity theory. They arise simply because you want to base the innovations of 1905 on epistemological grounds (nonexistence of the stagnant either) instead of empirical grounds (equivalence of all inertial systems with respect to light).

Mach would have applauded Einstein's life-long suspicion of formal epistemological systems, but how strange would he have found this use of the world *empirical* to characterize the hypothesis of the equivalence of all inertial systems with respect to light! What we see forming slowly here is Einstein's view that the fundamental role played by experience in the construction of fundamental physical theory is, after all, not through the 'atom' of experience, not through the individual sensation or the protocol sentence, but through some creative digest or synthesis of '*die gesammten Erfahrungstatsachen*', the *totality* of physical experience.[33] But all this was still hidden. Until Mach's death, and for several years after, Einstein considered and declared himself a disciple of Mach.

In the meantime, however, unknown to Einstein and everyone else, a time bomb had been ticking away. Set in 1913, it went off in 1921, five years after Mach's death, when Mach's *The Principles of Physical Optics* was published at last. Mach's preface was dated July, 1913 – perhaps a few days or, at most, a few weeks after Mach had received Einstein's last, enthusiastic letter and the article on general relativity theory. In a well-known passage in the preface (but one usually found in an inaccurate translation), Mach had written:

I am compelled, in what may be my last opportunity, to cancel my views [*Anschauungen*] of the relativity theory.

I gather from the publications which have reached me, and especially from my correspondence, that I am gradually becoming regarded as the forerunner of relativity. I am able even now to picture approximately what new expositions and interpretations many of the ideas expressed in my book on Mechanics will receive in the future from this point of view. It was to be expected that philosophers and physicists should carry on a crusade against me, for, as I have repeatedly observed, I was merely an unprejudiced rambler endowed with original ideas, in varied fields of knowledge. I must, however, as assuredly disclaim to be a forerunner of the relativists as I personally reject the atomistic doctrine of the present-day school, or church. The reason why, and the extent to which, I reject [*ablehne*] the present-day relativity theory, which I find to be growing more and more dogmatical, together with the particular reasons which have led me to such a view – considerations based on the physiology of the senses, epistemological doubts, and above all the insight resulting from my experiments – must remain to be treated in the sequel [a sequel which was never published].[34]

Certainly, Einstein was deeply disappointed by this belated disclosure of Mach's sudden dismissal of the relativity theory. Some months later, during a lecture on 6 April 1922 in Paris, in a discussion with the anti-Machist philosopher Emile Meyerson, Einstein allowed in a widely reported remark that Mach was '*un bon mécanicien*', but '*déplorable philosophe*'.[35]

We can well understand that Mach's rejection was at heart very painful, the more so as it was somehow Einstein's tragic fate to have the contribution he most cared about rejected by the very men whose approval and understanding he would have most gladly had – a situation not unknown in the history of science. In addition to Mach, the list includes these four: *Henri Poincaré*, who, to his death in 1912, only once deigned to mention Einstein's name in print, and then only to register an objection; *H. A. Lorentz*, who gave Einstein personally every possible encouragement – short of fully accepting the theory of relativity for himself; *Max Planck*, whose support of the special theory of relativity was unstinting, but who resisted Einstein's ideas on general relativity and the early quantum theory of radiation; and *A. A. Michelson*, who to the end of his days did not believe in relativity theory, and once said to Einstein that he was sorry that his own work may have helped to start this 'monster'.[36]

Soon Einstein's generosity again took the upper hand and resulted, from then to the end of his life, in many further personal testimonies to Mach's earlier influence.[37] A detailed analysis was provided in Einstein's letter of 8 January 1948 to Besso:

As far as Mach is concerned, I wish to differentiate between Mach's influence in general and his influence on me. . . . Particularly in the *Mechanics* and the *Wärmelehre* he tried to show how concepts arose out of experience. He took convincingly the position that these conceptions, even the most fundamental ones, obtained their warrant only out of empirical knowledge, that they are in no way logically necessary. . . .

I see his weakness in this, that he more or less believed science to consist in a mere ordering of empirical material; that is to say, he did not recognize the freely constructive element in formation of concepts. In a way he thought that theories arise through *discoveries* and not through inventions. He even went so far that he regarded 'sensations' not only as material which has to be investigated, but, as it were, as the building blocks of the real world; thereby, he believed, he could overcome the difference between psychology and physics. If he had drawn the full consequences, he would have had to reject not only atomism but also the idea of a physical reality.

Now, as far as Mach's influence on my own development is concerned, it certainly was great. I remember very well that you drew my attention to his *Mechanics* and *Wärmelehre* during my first years of study, and that both books made a great impression on me. The extent to which they influenced my own work is, to say the truth, not clear to me. As far as I am conscious of it, the immediate influence of Hume on me was greater. . . . But, as I said, I am not able to analyze that which lies anchored in unconscious thought. It is interesting, by the way, that Mach rejected the special relativity theory passionately (he did not live to see the general relativity theory [in the developed form]). The theory was, for him, inadmissibly speculative. He did not know that this speculative character belongs also to Newton's mechanics, and to every theory which thought is capable of. There exists only a gradual difference between theories, insofar as the chains of thought from fundamental concepts to empirically verifiable conclusions are of different lengths and complications.[38]

Antipositivistic component of Einstein's work

Ernst Mach's harsh words in his 1913 preface leave a tantalizing mystery. Ludwig Mach's destruction of his father's papers has so far made it impossible to find out more about the 'experiments' (possibly on the constancy of the velocity of light) at which Ernst Mach hinted. Since 1921, many speculations have been offered to explain Mach's remarks.[39] They all leave something to be desired. Yet, I believe, it is not so difficult to reconstruct the main reasons why Mach ended up rejecting the relativity theory. To put it very simply, Mach had recognized more and more clearly, years before Einstein did so himself, that Einstein had indeed fallen away from the faith, had left behind him the confines of Machist empiriocriticism.

The list of evidences is long. Here only a few examples can be given, the first from the 1905 relativity paper itself: what had made it really work was that it contained and combined elements based on two entirely different philosophies of science – not merely the empiricist–operationist component, but the courageous initial postulation, in the second paragraph, of two thematic hypotheses (one on the constancy of light velocity and the other on the extension of the principle of relativity to all branches of physics), two postulates for which there was and can be no direct empirical confirmation.

For a long time, Einstein did not draw attention to this feature. In a lecture at King's College, London, in 1921, just before the posthumous publication of Mach's attack, Einstein still was protesting that the origin of relativity theory lay in the facts of direct experience:

> . . . I am anxious to draw attention to the fact that this theory is not speculative in origin; it owes its invention entirely to the desire to make physical theory fit observed fact as well as possible. We have here no revolutionary act, but the natural continuation of a line that can be traced through centuries. The abandonment of certain notions connected with space, time, and motion, hitherto treated as fundamentals, must not be regarded as arbitrary, but only as conditioned by observed facts.[40]

By June, 1933, however, when Einstein returned to England to give the Herbert Spencer Lecture at Oxford entitled 'On the Method of Theoretical Physics', the more complex epistemology that was in fact inherent in his work from the beginning had begun to be expressed. He opened this lecture with the significant sentence: 'If you want to find out anything from the theoretical physicists about the methods they use, I advise you to stick closely to one principle: Don't listen to their words, fix your attention on their deeds.' He went on to divide the tasks of experience and reason in a very different way from that advocated in his earlier visit to England:

We are concerned with the eternal antithesis between the two inseparable components of our knowledge, the empirical and the rational. . . . The structure of the system is the work of reason; the empirical contents and their mutual relations must find their representation in the conclusions of the theory. In the possibility of such a representation lie the sole value and justification of the whole system, and especially the concepts and fundamental principles which underlie it. Apart from that, these latter are free inventions of the human intellect, which cannot be justified either by the nature of that intellect or in any other fashion *a priori*.

In the summary of this section, he draws attention to the 'purely fictitious character of the fundamentals of scientific theory'. It is this penetrating insight which Mach must have smelled out much earlier and dismissed as 'dogmatism'.

Indeed, Einstein, in his 1933 Spencer Lecture – widely read, as were and still are so many of his essays – castigates the old view that 'the fundamental concepts and postulates of physics were not in the logical sense inventions of the human mind but could be deduced from experience by 'abstraction' – that is to say, by logical means. A clear recognition of the erroneousness of this notion really only came with the general theory of relativity.'

Einstein ends this discussion with the enunciation of his current credo, so far from that he had expressed earlier:

Nature is the realization of the simplest conceivable mathematical ideas. I am convinced that we can discover, by means of purely mathematical constructions, those concepts and those lawful connections between them which furnish the key to the understanding of natural phenomena. Experience may suggest the appropriate mathematical concepts, but they most certainly cannot be deduced from it. Experience remains, of course, the sole criterion of physical utility of a mathematical construction. But the creative principle resides in mathematics. In a certain sense, therefore, I hold it true that pure thought can grasp reality, as the ancients dreamed.[41]

Technically, Einstein was now at – or rather just past – the midstage of his pilgrimage. He had long ago abandoned his youthful allegiance to a primitive phenomenalism that Mach would have commended. In the first of the two passages just cited and others like it, he had gone on to a more refined form of phenomenalism which many of the logical positivists could still accept. He has, however, gone beyond it in the second passage, turning toward interests that we shall see later to have matured into clearly metaphysical conceptions.

Later, Einstein himself stressed the key role of what we have called thematic rather than phenomenic elements[42] – and thereby he fixed the early date at which, in retrospect, he found this need to have arisen in his earliest work. Thus he wrote in his *Autobiographical Notes* of 1946 that 'shortly after 1900 . . . I despaired of the possibility of discovering the true laws by means of constructive efforts based on known facts. The longer and the more despairingly I tried, the more I came to the conviction that only the discovery of a *universal formal principle* could lead us to assured results.'[43]

Another example of evidence of the undercurrent of disengagement from a Machist position is an early one: it comes from Einstein's article on relativity in the 1907 *Jahrbuch der Radioaktivität und Elektronik*[44] where Einstein responds, after a year's silence, to Walter Kaufmann's 1906 paper in the *Annalen der Physik*.[45] That paper had been the first publication in the *Annalen* to mention Einstein's work on the relativity theory, published there the previous year. Coming from the eminent experimental physicist Kaufmann, it had been most significant that this very first discussion was announced as a categorical, experimental disproof of Einstein's theory. Kaufmann had begun his attack with the devastating summary:

I anticipate right here the general result of the measurements to be described in the following: *the measurement results are not compatible with the Lorentz-Einsteinian fundamental assumption.*[46]

Einstein could not have known that Kaufmann's equipment was inadequate. Indeed, it took ten years for this to be fully realized, through the work of Guye and Lavanchy in 1916. So in his discussion of 1907, Einstein had to acknowledge that there seemed to be small but significant differences between Kaufmann's results and Einstein's predictions. He agreed that Kaufmann's calculations seemed to be free of error, but 'whether there is an unsuspected systematic error or whether the foundations of relativity theory do not correspond with the facts one will be able to decide with certainty only if a great variety of observational material is at hand'.[47]

Despite this prophetic remark, Einstein does not rest his case on it. On the contrary, he has a very different, and what for his time and situation must have been a very daring, point to make: he acknowledges that the theories of electron motion given earlier by Abraham and by Bucherer do give predictions considerably closer to the experimental results of Kaufmann. But Einstein refuses to let the 'facts' decide the matter: 'In my opinion both [their] theories have a rather small probability, because their fundamental assumptions concerning the mass of moving electrons are not explainable in terms of theoretical systems which embrace a greater complex of phenomena.'[48]

This is the characteristic position – the crucial difference between Einstein and those who make the correspondence with experimental fact the chief deciding factor for or against a theory: even though the 'experimental facts' at that time very clearly seemed to favor the theory of his opponents rather than his own, he finds the ad hoc character of their theories more significant and objectionable than an apparent disagreement between his theory and their 'facts'.[49]

So already in this 1907 article – which, incidentally, Einstein mentions in his postcard of 17 August to Ernst Mach, with a remark regretting that he has no more reprints for distribution – we have explicit evidence of a hardening of Einstein against the epistemological priority of experiment, not to speak of sensory experience. In the years that followed, Einstein more and more openly put the consistency of a simple and convincing theory or of a thematic conception higher in importance than the latest news from the laboratory – and again and again he turned out to be right.

Thus, only a few months after Einstein had written in his fourth letter to Mach that the solar eclipse experiment will decide 'whether the basic and fundamental assumption of the equivalence of the acceleration of the reference frame and of the gravitational field really holds', Einstein writes to Besso in a very different vein (in March 1914), before the first, ill-fated eclipse expedition was scheduled to test the conclusions of the preliminary version of the general relativity theory: 'Now I am fully satisfied, and I do not doubt any more the correctness of the whole system, may the observation of the eclipse succeed or not. The sense of the thing [*die Vernunft der Sache*] is too evident.' And later, commenting on the fact that there remains up to 10 per cent discrepancy between the measured deviation of light owing to the sun's field and the calculated effect based on the general relativity theory: 'For the expert, this thing is not particularly important, because the main significance of the theory does not lie in the verification of little effects, but rather in the great simplification of the theoretical basis of physics as a whole.'[50] Or again, in Einstein's 'Notes on the Origin of the General Theory of Relativity',[51] he reports that he 'was in the highest degree amazed' by the existence of the equivalence between inertial and gravitational mass, but that he 'had no serious doubts about its strict validity, even without knowing the results of the admirable experiment of Eötvös'.

The same point is made again in a revealing account given by Einstein's student, Ilse Rosenthal-Schneider. In a manuscript, 'Remininscences of Conversation with Einstein', dated 23 July 1957, she reports:

Once when I was with Einstein in order to read with him a work that contained many objections against his theory . . . he suddenly interrupted the discussion of the book, reached for a telegram that was lying on the windowsill, and handed it to me with the words, 'Here, this will perhaps interest you.' It was Eddington's cable with the results of measurement of the eclipse expedition [1919]. When I was giving expression to my joy that the results coincided with his calculations, he said quite unmoved, 'But I knew that the theory is correct'; and when I asked, what if there had been no confirmation of his prediction, he countered: 'Then I would have been sorry for the dear Lord – the theory *is* correct.'[52]

Minkowski's 'world' and the world of sensations

The third major point at which Mach, if not Einstein himself, must have seen that their paths were diverging is the development of relativity theory into the geometry of the four-dimensional space-time continuum, begun in 1907 by the mathematician Hermann Minkowski (who, incidentally, had had Einstein as a student in Zürich). Indeed, it was through Minkowski's semipopular lecture, 'Space and Time', on 21 September 1908 at the eightieth meeting of the Naturforscherversammlung,[53] that a number of scientists first became intrigued with relativity theory. We have several indications that Mach, too, was both interested in and concerned about the introduction of four-dimensional geometry into physics (in Mach's correspondence around 1910, for example, with August Föppl); according to Friedrich Herneck,[54] Ernst Mach specially invited the young Viennese physicist Philipp Frank to visit him 'in order to find out more about the relativity theory, above all about the use of four-dimensional geometry'. As a result, Frank, who had recently finished his studies under Ludwig Boltzmann and had begun to publish noteworthy contributions to relativity, published the 'presentation of Einstein's theory to which Mach gave his assent' under the title 'Das Relativitätsprinzip und die Darstellung der physikalischen Erscheinungen im vierdimensionalen Raum'.[55] It is an attempt, addressed to readers 'who do not master modern mathematical methods', to show that Minkowski's work brings out the 'empirical facts far more clearly by the use of four-dimensional world lines'. The essay ends with the reassuring conclusion: 'In this four-dimensional world the facts of experience can be presented more adequately than in three-dimensional space, where always only an arbitrary and one-sided projection is pictured.'

Following Minkowski's own papers on the whole, Frank's treatment can make it nevertheless still appear that in most respects the time dimension is equivalent to the space dimensions. Thereby one could think that Minkowski's treatment based itself not only on a functional and operational interconnection of space and time, but also – fully in accord with Mach's own views – on the primacy of ordinary, 'experienced' space and time in the relativistic description of phenomena.

Perhaps as a result of this presentation, Mach invoked the names of Lorentz, Einstein, and Minkowski in his reply of 1910 to Planck's first attack, citing them as physicists 'who are moving closer to the problems of matter, space, and time'. Already a year earlier, Mach seems to have been hospitable to Minkowski's presentation, although not without reservations. Mach wrote in the 1909 edition of *Conservation of Energy*[26]: 'Space and time are here conceived not as independent entities, but as forms of the dependence of the phenomena on one another'; he also added a reference to Minkowski's lecture of 1908.[53] But a few lines earlier, Mach had written: 'Spaces of many dimensions seem to me not so essential for physics. I would only uphold them if things of thought [*Gedankendinge*] like atoms are maintained to be indispensable, and if, then, also the freedom of working hypotheses is upheld.'

It was correctly pointed out by C. B. Weinberg[56] that Mach may eventually have had two sources of suspicion against the Minkowskian form of relativity theory. As was noted above, Mach regarded the fundamental notions of mechanics as problems to be continually discussed with maximum openness within the frame of empiricism, rather than as questions that can be solved and settled – as the relativists, seemingly dogmatic and sure of themselves, were in his opinion more and more inclined to do. In addition, Mach held that the questions of physics were to be studied in a broader setting, encompassing biology and psychophysiology. Thus Mach wrote: 'Physics is not the entire world; biology is there too, and belongs essentially to the world picture'.[57]

But I see also a third reason for Mach's eventual antagonism against such conceptions as Minkowski's (unless one restricted their application to 'mere things of thought like atoms and molecules, which by their very nature can never be made the objects of sensuous contemplations'[58]). If one takes Minkowski's essay seriously – for example, the abandonment of space and time separately, with identity granted only to 'a kind of union of the two' – one must recognize that it entails the abandonment of the conceptions of experiential space and experiential time; and that is an attack on the very roots of sensations-physics, on the meaning of actual measurements. If identity, meaning, or 'reality' lies in the four-dimensional space-time interval *ds*, one is dealing with a quantity which is hardly *denkökonomisch*, nor one that preserves the primacy of measurements in 'real' space and time. Mach may well have seen the warning flag; and worse was soon to come, as we shall see at once.

In his exuberant lecture of 1908,[53] Minkowski had announced that 'three-dimensional geometry becomes a chapter in four-dimensional physics. . . . Space and time are to fade away into the shadows, and only *eine Welt an sich* will subsist.' In this 'world' the crucial innovation is the conception of the '*zeitartige Vektorelement*', *ds*, defined as $(1/c)$ $\sqrt{c^2dt^2-dx^2-dy^2-dz^2}$ with imaginary components. To Mach, the word *Element* had a pivotal and very different meaning. As we saw in Schlick's summary, elements were nothing less than the sensations and complexes of sensations of which the world consists and which completely define the world. Minkowski's rendition of relativity theory was now revealing the need to move the ground of basic, elemental truths from the plane of direct experience in ordinary space and time to a mathematicized, formalistic model of the world in a union of space and time that is not directly accessible to sensation – and, in this respect, is reminiscent of absolute space and time concepts that Mach had called 'metaphysical monsters'.[59]

Here, then, is an issue which, more and more, had separated Einstein and Mach even before they realized it. To the latter, the fundamental task of science was economic and descriptive; to the former, it was speculative-constructive and intuitive. Mach had once written: 'If all the individual facts – all the individual phenomena, knowledge of which we desire – were immediately accessible to us, science would never have risen.'[60] To this, with the forthrightness caused perhaps by his recent discovery of Mach's opposition, Einstein countered during his lecture in Paris of 6 April 1922: 'Mach's system studies the existing relations between data of experience: for Mach, science is the totality of these relations. That point of view is wrong, and in fact what Mach has done is to make a catalog, not a system.'[61]

We are witnessing here an old conflict, one that has continued throughout the development of the sciences. Mach's phenomenalism brandished an undeniable and irresistible weapon for the critical reevaluation of classical physics, and in this it seems to hark back to an ancient position that looked upon sensuous appearances as the beginning and end of scientific achievement. One can read Galileo in this light, when he urges the primary need of *description* for the fall of bodies, leaving 'the causes' to be found out later. So one can understand (or rather, misunderstand) Newton, with his too-well-remembered remark: 'I feign no hypotheses.'[62] Kirchhoff is in this tradition. Boltzmann wrote of him in 1888:

The aim is not to produce bold hypotheses as to the essence of matter, or to explain the movement of a body from that of molecules, but to present equations which, free from hypotheses, are as far as possible true and quantitatively correct correspondents of the phenomenal world, careless of the essence of things and forces. In his book on mechanics, Kirchhoff will ban all metaphysical concepts, such as forces, the cause of a motion; he seeks only the equations which correspond so far as possible to observed motions.[63]

And so could, and did, Einstein himself understand the Machist component of his own early work.

Phenomenalistic positivism in science has always been victorious, but only up to a very definite limit. It is the necessary sword for destroying old error, but it makes an inadequate plowshare for cultivating a new harvest. I find it exceedingly significant that Einstein saw this during the transition phase of partial disengagement from the Machist philosophy. In the spring of 1917 Einstein wrote to Besso and mentioned a manuscript which Friedrich Adler had sent him. Einstein commented: 'He rides Mach's poor horse to exhaustion.' To this, Besso – the loyal Machist – responds on 5 May 1917: 'As to Mach's little horse, we should not insult it; did it not make possible the infernal journey through the relativities? And who knows – in the case of the nasty quanta, it may also carry Don Quixote de la Einsta through it all!'

Einstein's answer of 13 May 1917 is revealing: 'I do not inveigh against Mach's little horse; but you know what I think about it. It cannot give birth to anything living, it can only exterminate harmful vermin.'

Toward a rationalistic realism

The rest of the pilgrimage is easy to reconstruct, as Einstein more and more openly and consciously turned Mach's doctrine upside down – minimizing rather than maximizing the role of actual details of experience, both at the beginning and at the end of scientific theory, and opting for a rationalism that almost inevitably would lead him to the conception of an objective, 'real' world behind the phenomena to which our senses are exposed.

In the essay, 'Maxwell's Influence on the Evolution of the Idea of Physical Reality' (1931), Einstein began with a sentence that could have been taken almost verbatim from Max Planck's attack on Mach in 1909, cited above: 'The belief is an external world independent of the perceiving subject is the basis of all natural science.' Again and again, in the period beginning with his work on the general relativity theory, Einstein insisted that between experience and reason, as well as between the world of sensory perception and the objective world, there are logically unbridgeable chasms. He characterized the efficacy of reason to grasp reality by the word *miraculous*; the very terminology in these statements would have been an anathema to Mach.

We may well ask when and under what circumstances Einstein himself became aware of his change. Here again, we may turn for illumination to one of the hitherto unpublished letters, one written to his old friend, Cornelius Lanczos, on 24 January 1938:

Coming from sceptical empiricism of somewhat the kind of Mach's, I was made, by the problem of gravitation, into a believing rationalist, that is, one who seeks the only trustworthy source of truth in mathematical simplicity. The logically simple does not, of course, have to be physically true; but the physically true is logically simple, that is, it has unity at the foundation.

Indeed, all the evidence points to the conclusion that Einstein's work on general relativity theory was crucial in his epistemological development. As he wrote later in 'Physics and reality' (1936): 'the first aim of the general theory of relativity was the preliminary version which, while not meeting the requirements for constituting a closed system, could be connected in as simple a manner as possible with "directly observed facts"'.[64] But the aim, still apparent during the first years of correspondence with Mach, could not be achieved. In 'Notes on the origin of the general theory of relativity', Einstein reported:

I soon saw that the inclusion of non-linear transformation, as the principle of equivalence demanded, was inevitably fatal to the simple physical interpretation of the coordinate – i.e. that it could no longer be required that coordinate differences [ds] should signify direct results of measurement with ideal scales or clocks. I was much bothered by this piece of knowledge ... [just as Mach must have been].

The solution of the above mentioned dilemma [from 1912 on] was therefore as follows: A physical significance attaches not to the differentials of the coordinates, but only to the Riemannian metric corresponding to them.[65]

And this is precisely a chief result of the 1913 essay of Einstein and Grossmann,[66] the same paper which Einstein sent to Mach and discussed in his fourth letter. This result was the final consequence of the Minkowskian four-space representation – the sacrifice of the primacy of direct sense perception in constructing a physically significant system. It was the choice that Einstein had to make – against fidelity to a catalogue of individual operational experiences, and in favour of fidelity to the ancient hope for a unity at the base of physical theory.[67]

Enough has been written in other places to show the connections that existed between Einstein's scientific rationalism and his religious beliefs. Max Born summarized it in one sentence: 'He believed in the power of reason to guess the laws according to which God has built the world.'[68] Perhaps the best expression of this position by Einstein himself is to be found in his essay, *Über den gegenwärtigen Stand der Feld-Theorie*, in the *Festschrift* of 1929 for Aurel Stodola:

Physical Theory has two ardent desires, to gather up as far as possible all pertinent phenomena and their connections, and to help us not only to know *how* Nature is and *how* her transactions are carried through, but also to reach as far as possible the perhaps utopian and seemingly arrogant aim of knowing why Nature is *thus and not otherwise*. Here lies the highest satisfaction of a scientific person. . . . [On making deductions from a 'fundamental hypothesis' such as that of the kinetic-molecular theory,] one experiences, so to speak, that God Himself could not have arranged those connections [between, for example, pressure, volume, and temperature] in any other way than that which factually exists, any more than it would be in His power to make the number 4 into a prime number. This is the promethean element of the scientific experience. . . . Here has always been for me the particular magic of scientific considerations; that is, as it were, the religious basis of scientific effort.[69]

This fervor is indeed far from the kind of analysis which Einstein had made only a few years earlier. It is doubly far from the asceticism of his first philosophic mentor, Mach, who had written in his day book: 'Colors, space, tones, etc. These are the only realities. Others do not exist.'[70] It is, on the contrary, far closer to the rational realism of his first scientific mentor, Planck, who had written: 'The disjointed data of experience can never furnish a veritable science without the intelligent interference of a spirit actuated by faith. . . . We have a right to feel secure in surrendering to our belief in a philosophy of the world based upon a faith in the rational ordering of this world.'[71] Indeed, we note the philosophical kinship of Einstein's position with seventeenth-century natural philosophers – for example, with Johannes Kepler who, in the preface of the *Mysterium Cosmographicum*, announced that he wanted to find out concerning the number, positions, and motions of the planets, 'why they are as they are, and not otherwise', and who wrote to Herwart in April, 1599, that, with regard to numbers and quantity, 'our knowledge is of the same kind as God's, at least insofar as we can understand something of it in this mortal life'.

Not unexpectedly, we find that during this period (around 1930) Einstein's non-scientific writings began to refer to religious questions much more frequently than before. There is a close relation between his epistemology, in which reality does not need to be validated by the individual's sensorium, and what he called 'Cosmic Religion', defined as follows: 'The individual feels the vanity of human desires and aims, and the nobility and marvelous order which are revealed in nature and in the world of thought. He feels the individual destiny as an imprisonment and seeks to experience the totality of existence as a unity full of significance.'[72]

Needless to say, Einstein's friends from earlier days sometimes had to be informed of his change of outlook in a blunt way. For example, Einstein wrote to Moritz Schlick on 28 November 1930:

In general your presentation fails to correspond to my conceptual style insofar as I find your whole orientation so to speak too positivistic. . . . I tell you straight out: Physics is the attempt at the conceptual construction of a model of the *real world* and of its lawful structure. To be sure, it [physics] must present exactly the empirical relations between those sense experiences to which we are open; but only *in this way* is it chained to them. . . . In short, I suffer under the (unsharp) separation of Reality of Experience and Reality of Being. . . .

You will be astonished about the 'metaphysicist' Einstein. But every four-and-two-legged animal is de facto in this sense metaphysicist. [Emphasis in the original.]

Similarly, Philipp Frank, Einstein's early associate and later his biographer, reports that the realization of Einstein's true state of thought reached Frank in a most embarrassing way, at the Congress of German physicists in Prague in 1929, just as Frank was delivering 'an address in which I attacked the metaphysical position of the German physicists and defended the positivistic ideas of Mach'. The very next speaker disagreed and showed Frank that he had been mistaken still to associate Einstein's views with that of Mach and himself. 'He added that Einstein was entirely in accord with Planck's view that physical laws describe a reality in space and time that is independent of ourselves. At that time', Frank comments, 'this presentation of Einstein's views took me very much by surprise.'[73]

In retrospect it is, of course, much easier to see the evidences that this change was being prepared. Einstein himself realized more and more clearly how closely he had moved to Planck, from whom he earlier dissociated himself in three of the four letters to Mach. At the celebration of Planck's sixtieth birthday, two years after Mach's death, Einstein made a moving speech in which, perhaps for the first time, he referred publicly to the Planck–Mach dispute and affirmed his belief that 'there is no logical way to the discovery of these elementary laws. There is only the way of intuition' based on *Einfühlung* in experience.[74] The scientific dispute concerning the theory of radiation between Einstein and Planck, too, had been settled (in Einstein's favor) by a sequence of developments after 1911 – for example, by Bohr's theory of radiation from gas atoms. As colleagues, Planck and Einstein saw each other regularly from 1913 on. Among evidences of the coincidence of these outlooks there is in the Einstein Archives a handwritten draft, written on or just before 17 April 1931 and intended as Einstein's introduction to Planck's hard-hitting article 'Positivism and external reality'.[75] In lauding Planck's article, Einstein concludes: 'I presume I may add that both Planck's conception of the logical state of affairs as well as his subjective expectation concerning the later development of our science corresponds entirely with my own understanding.'[76]

This essay gave a clear exposition of Planck's (and one may assume, Einstein's) views, both in physics and in philosophy more generally. Thus Planck wrote there:

The essential point of the positivist theory is that there is no other source of knowledge except the straight and short way of perception through the senses. Positivism always holds strictly to that. Now, the two sentences: (1) *there is a real outer world which exists independently of our act of knowing* and (2) *the real outer world is not directly knowable* form together the cardinal hinge on which the whole structure of physical science turns. And yet there is a certain degree of contradiction between those two sentences. This fact discloses the presence of the irrational, or mystic, element which adheres to physical science as to every other branch of human knowledge. The effect of this is that a science is never in a position completely and exhaustively to solve the problem it has to face. We must accept that as a hard and fast, irrefutable fact, and this fact cannot be removed by a theory which restricts the scope of science at its very start. Therefore, we see the task of science arising before us as an incessant struggle toward a goal which will never be reached, because by its very nature it is unreachable. It is of a metaphysical character, and, as such, is always again and again beyond our achievement.[77]

From then on, Einstein's and Planck's writings on these matters are often almost indistinguishable from each other. Thus, in an essay in honor of Bertrand Russell, Einstein warns that the 'fateful "fear of metaphysics" . . . has come to be a malady of contemporary empiricistic philosophizing'.[78] On the other hand, in the numerous letters between the two old friends, Einstein and Besso, each to the very end touchingly and patiently tries to explain his position, and perhaps to change the other's. Thus, on 28 February 1952, Besso once more presents a way of making Mach's views again acceptable to Einstein. The latter, in answering on 20 March 1952, once more responds that the facts cannot lead to a deductive theory and, at most, can set the stage 'for intuiting a general principle' as the basis of a deductive theory. A little later, Besso is gently scolded (in Einstein's letter of 13 July 1952): 'It appears that you do not take the four-dimensionality of reality seriously, but that instead you take the present to be the only reality. What you call "world" is in physical terminology "spacelike sections" for which the relativity theory – already the special theory – denies objective reality.'

In the end, Einstein came to embrace the view which many, and perhaps he himself, thought earlier he had eliminated from physics in his basic 1905 paper on relativity theory: that there exists an external, objective, physical reality which we may hope to grasp – not directly, empirically, or logically, or with fullest certainty, but at least by an intuitive leap, one that is only guided by experience of the totality of sensible 'facts'. Events take place in a 'real world', of which the space–time world of sensory experience, and even the world of multidimensional continua, are useful conceptions, but no more than that.

For a scientist to change his philosophical beliefs so fundamentally is rare, but not unprecedented. Mach himself underwent a dramatic transformation quite early (from Kantian idealism, at about age seventeen or eighteen, according to Mach's autobiographical notes). We have noted that Ostwald changed twice, once to anti-atomism and then back to atomism. And strangely Planck himself confessed in his 1910 attack on Mach[27] that some twenty years earlier, near the beginning of his own career when Planck was in his late twenties (and Mach was in his late forties), he, too, had been counted 'one of the decided followers of the Machist philosophy', as indeed is evident in Planck's early essay on the conservation of energy (1887).

In an unpublished fragment apparently intended as an additional critical reply to one of the essays in the collection *Albert Einstein: Philosopher-Scientist* (1949), Einstein returned once more to deal – quite scathingly – with the opposition. The very words he used showed how complete was the change in his epistemology. Perhaps even without consciously remembering Planck's words in the attack on Mach of 1909 cited earlier – that a basic aim of science is 'the complete liberation of the physical world picture from the individuality of the separate intellects'[32] – Einstein refers to a 'basic axiom' in his own thinking:

It is the postulation of a 'real world' which so-to-speak liberates the 'world' from the thinking and experiencing subject. The extreme positivists think that they can do without it; this seems to me to be an illusion, if they are not willing to renounce thought itself.

Einstein's final epistemological message was that the world of mere experience must be subjugated by and based in fundamental thought so general that it may be called cosmological in character. To be sure, modern philosophy did not gain thereby a major, novel, and finished corpus. Physicists the world over generally feel that today one must steer more or less a middle course in the area between, on the one hand, the Machist attachment to empirical data or heuristic proposals as the sole source of theory and, on the other, the aesthetic-mathematical attachment to persuasive internal harmony as the warrant of truth. Moreover, the old dichotomy between rationalism and empiricism is slowly being dissolved in new approaches.[79] Yet by encompassing in his own philosophical development both ends of this range, and by always stating forthrightly and with eloquence his redefined position, Einstein not only helped us to define our own, but also gave us a virtually unique case study of the interaction of science and epistemology.

Source: Chapter 8 of Gerald Holton, *Thematic Origins of Scientific Thought*. Cambridge, Harvard University Press, 1973. Originally published in *Daedalus* (Spring 1968): 636–73.

Notes

1.   The main results of the study to this point are contained in Essays 5 through 10 in Gerald Holton, *Thematic Origins of Scientific Thought: Kepler to Einstein*. Cambridge, Harvard University Press, 1973.

2.   These documents are mostly on deposit at the Archives of the Estate of Albert Einstein at Princeton; where not otherwise indicated, citations made here are from those documents.

3.   This letter as well as the next two letters mentioned in the text (those of 3 April 1901 and 13 April 1901) have been published by Hans-Günther Körber, *Forchungen und Fortschritte*, **38** (1964): 75–6.

4.   Albert Einstein, 'Folgerungen aus den Capillaritätserscheinungen', *Annalen der Physik*, 4 (1901): 513–23.

5.   The only other known attempt on Einstein's part to obtain an assistantship at that time was a request to Kammerlingh-Onnes (12 April 1901), to which, incidentally, he also seems to have received no response.

6.   J. T. Merz, *A History of European Thought in the Nineteenth Century*, 4 vols. Edinburgh, William Blackwood & Sons, 1904–12; reprint edn, New York, Dover, 1965, II, pp: 184, 199.

7.   *Erinnerungen an Albert Einstein*, issued by the Patent Office in Bern, about 1965 (n.d., no pagination).

8.   Wilhelm Ostwald, 'Chemische Energie', *Lehrbuch der allgemeinen Chemie*, 2nd edn. Leipzig, Verlag von Wilhelm Engelmann, 1903, II, Part 1: 1014.

9.   Albert Einstein, 'Autobiographical notes', in P. A. Schilpp (ed.), *Albert Einstein: Philosopher-Scientist*. Evanston, Ill.' Library of Living Philosophers, 1949, p. 15.

10. Anton Reiser, *Albert Einstein*. New York; A. & C. Boni, 1930: 51–2.

11. Ernst Mach, *Die Mechanik in ihrer Entwicklung, historisch-kritisch dargestellt.* Leipzig, 1883.

12. Mach, op. cit., preface, 7th edn 1912.

13. Moritz Schlick, *Ernst Mach, der Philosoph*, in a special supplement on Ernst Mach in the *Neue Freie Presse* (Vienna), 12 June 1926. Einstein himself, in a brief and telling analysis, *Zur Enthüllung von Ernst Machs Denkmal*', published in the same issue (the day of the unveiling of a monument to Mach), wrote: 'Ernst Mach's strongest driving force was a philosophical one: the dignity of all scientific concepts and statements rests solely in isolated experiences [*Einzelerlebnisse*] to which the concepts refer. This fundamental proposition exerted mastery over him in all his research and gave him the strength to examine the traditional fundamental concepts òf physics (time, space, inertia) with an independence which at that time was unheard of.'

14. Among many evidences of Mach's effectiveness, not the least are his five hundred or more publications (counting all editions – for example, seven editions of his *The Science of Mechanics* in German alone during his lifetime), as well as his large exchange of letters, books, and reprints (of which many important ones 'carry the dedication of their authors', to cite the impressive catalogue of Mach's library by Theodor Ackermann, Munich, No. 634 [1959] and No. 636 [1960]). A glimpse of Mach's effect on those near him was furnished by William James, who in 1882 heard Mach give a 'beautiful' lecture in Prague. Mach received James 'with open arms. . . . Mach came to my hotel and I spent four hours walking and supping with him at his club, an unforgettable conversation. I don't think anyone ever gave me so strong an impression of pure intellectual genius. He apparently has read everything and thought about everything, and has an absolute simplicity of manner and winningness of smile when his face lights up, that are charming.' From James's letter, in Gay Wilson Allen, *William James, A Biography*. New York, Viking Press, 1967: 249.

The topicality of Mach's early speculations on what is now part of General Relativity Theory is attested to by the large number of continuing contributions on the Mach Principle. Beyond that, Mach's influence today is still strong in scientific thinking, though few are as explicit and forthright as the distinguished physicist R. H. Dicke of Princeton University, in his recent, technical book *The Theoretical Significance of Experimental Relativity*. London, Gordon and Breach, 1964: vii–viii: 'I was curious to know how many other reasonable theories [in addition to General Relativity] would be supported by the same facts. . . . The reason for limiting the class of theories in this way is to be found in matters of philosophy, not in the observations. Foremost among these considerations was the philosophy of Bishop Berkeley and E. Mach. . . . The philosophy of Berkeley and Mach always lurked in the background and influenced all of my thoughts.'

15. Albert Einstein, 'Ernst Mach', *Physikalische Zeitschrift,* **17** (1916): 101–4.
16. Einstein, 'Autobiographical notes': 21.
17. Gerald Holton, 'On the origins of the special theory of relativity', *American Journal of Physics,* **28** (1960): 627–36.
18. Gerald Holton, 'On the thematic analysis of science: the case of Poincaré and relativity', *Mélanges Alexandre Koyré*. Paris, Hermann, 1964: 257–68.
19. For evidences that this insistence on prior epistemological analysis of conceptions of space and time are Machist rather than primarily derived from Hume and Kant (who had, however, also been influential), see Einstein's detailed rendition of Mach's critique of Newtonian space and time, note 15; his discussion of Mach in note 9: 27–9; and in note 1.
20. Albert Einstein, 'Zur Electrodynamik bewegter Körper', *Annalen der Physik*, **17** (1905): 893.
21. Ibid.: 894.
22. Philipp Frank, 'Einstein, Mach, and logical positivism', in Schilpp, op. cit.: 272–3. 'The definition of simultaneity in the special theory of relativity is based on Mach's requirement that every statement in physics has to state relations between observable quantities. . . . There is no doubt that . . . Mach's requirement, the "positivistic" requirement, was of great heuristic value to Einstein.'
23. For example, see Philipp Frank, *Modern Science and Its Philosophy*. New York, George Braziller, 1955: 61–89; Viktor Kraft, *The Vienna Circle*, trans. Arthur Pap. New York, Philosophical Library, 1953; Richard von Mises, *Ernst Mach und die empiristische Wissenschaftsauffassung* (1938; printed as a fascicule of the series *Einheitswissenschaft*).
24. Joseph Petzoldt, 'Gesellschaft für positivistiche Philosophie', reprinted in *Zeitschrift für positivistische Philosophie*, **1** (1913): 4. In that same speech, Petzoldt sounded a theme that became widely favored in the positivistic interpretation of the genesis of relativity theory – namely, that the relativity theory was developed in direct response to the puzzle posed by the results of the Michelson experiment.

   In his interesting essay, 'Das Verhältnis der Machschen Gedankenwelt zur Relativitätstheorie', published as an appendix to the eighth German edition of Ernst Mach's *Die Mechanik in ihrer Entwicklung* (Leipzig, F. A. Brockhaus, 1921): 490–517. Petzoldt faithfully attempts to identify and discuss several Machist aspects of Einstein's relativity theory:

   (1) The theory 'in the end is based on the recognition of the coincidence

of sensations; and therefore it is fully in accord with Mach's world-view, which may be best characterized as a relativistic positivism' (p. 516).

(2) Mach's works 'produced the atmosphere without which Einstein's Relativity Theory would not have been possible' (p. 494), and in particular Mach's analysis of the equivalence of rotating reference objects in Newton's bucket experiment prepared for the next step, Einstein's 'equivalence of relativity moving coordinate systems' (p. 495).

(3) Mach's principle of economy is said to be marvelously exhibited in Einstein's succinct and simple statements of the two fundamental hypotheses. The postulate of the equivalence of inertial coordinate systems deals with 'the simplest case thinkable, which now also serves as a fundamental pillar for the General Theory. And Einstein chose also with relatively greatest simplicity the other basic postulate [constancy of light velocity]. ... These are the foundations. Everything else is logical consequence' (pp. 497–8).

25. Letter of Besso to Einstein, 16 February 1939. Among many testimonies to the effect of Einstein on positivistic philosophies of science, see P. W. Bridgman, 'Einstein's theory and the operational point of view', in Schilpp, op. cit.: 335–54.

26. Ernst Mach, *History and Root of the Principle of the Conservation of Energy.* Chicago, The Open Court Publishing Co., 1911: 95, translation by Philip E. B. Jourdain of the second edition (Ernst Mach, *Die Geschichte und die Wurzel des Satzes von der Erhaltung der Arbeit.* Leipzig, J. A. Barth, 1909). For a brief analysis of Mach's various expressions of adherence as well as reservations with respect to the principle of relativity, see Hugo Dingler, *Die Grundgedanken der Machschen Philosophie.* Leipzig, J. A. Barth, 1924: 73–86. Friedrich Herneck 'Nochmals über Einstein und Mach', *Physikalische Bläter,* **17** (1961): 275, reports that Frank wrote him he had the impression during a discussion with Ernst Mach around 1910 that Mach 'was fully in accord with Einstein's special relativity theory, and particularly with its philosophical basis'.

27. Republished in Max Planck, *A Survey of Physical Theory.* New York, Dover, 1960: 24. We shall read later a reaffirmation of this position, in almost exactly the same words, but from another pen.

After Mach's rejoinder 'Die Leitgedanken meiner naturwissenschaftlichen Erkenntnislehre und ihre Aufnahme durch die Zeitgenossen', *Scientia,* **7** (1910): 225, Plank wrote a second, much more angry essay, 'Zur Machschen Theorie der physikalischen Erkenntnis', *Vierteljahrschrift für wissenschaftliche Philosophie,* **34** (1910): 497. He ends as follows: 'If the physicist wishes to further his science, he must be a Realist, not an Economist [in the sense of Mach's principle of economy]; that is, in the flux of appearances he must above all search for and unveil that which persists, is not transient, and is independent of human senses.'

28. Philipp Frank, *Einstein, His Life and Times,* trans. George Rose, ed. and rev. Suichi Kusaka. New York, Alfred A. Knopf, 1947.

29. Later Einstein found that this procedure did not work; see Albert Einstein, 'Notes on the origin of the general theory of relativity', *Ideas and Opinions,* trans. Sonja Bargmann. London, Alvin Redman, 1954: 285–90; and other publications. In a letter of 2 February 1954 to Felix Pirani, Einstein writes: 'One shouldn't talk at all any longer of Mach's principle, in my opinion. It arose at a time when one thought that 'ponderable bodies' were the only physical reality and that in a theory all elements that are not fully

determined by them should be conscientiously avoided. I am quite aware of the fact that for a long time, I, too, was influenced by this fixed idea.'

30. Albert Einstein and Marcel Grossman, 'Entwurf einer verallgemeinerten Relativitätstheorie und einer Theorie der Gravitation', *Zeitschrift für Mathematik und Physik*, **62** (1913): 225–61.

31. For a further analysis and the full text of the four letters see Friedrich Herneck, 'Zum Briefwechsel Albert Einsteins mit Ernst Mach',*Forschungen und Fortschritte,* **37** (1963): 239–43 and 'Die Beziehungen zwischen Einstein und Mach, documentarisch Dargestellt', *Wissenschaftliche Zeitschrift der Friedrich-Schiller-Universität Jena, mathematisch-naturwissenschaftliche Reihe,* **15** (1966): 1–14; and Helmut Hönl, 'Ein Brief Albert Einsteins an Ernst Mach', *Physikalische Blätter,* **16** (1960): 571–80. Many other evidences, direct and indirect, have been published to show Mach's influence on Einstein prior to Mach's death in 1916. For example, recently a document has been found which shows that in 1911 Mach had participated in formulating and signing a manifesto calling for the founding of a society for positivistic philosophy. Among the signers, together with Mach, we find Joseph Petzoldt, David Hilbert, Felix Klein, Georg Helm, Sigmund Freud, and Einstein. (See Herneck, 'Nochmals über Einstein und Mach': 276.)

32. Ernst Mach, 'Die Leitgedanken meiner naturwissenschaftlichen Erkenntnislehre und ihre Aufnahme durch die Zeitgenossen', *Physikalische Zeitschrift,* **11** (1910): 605 (emphasis added). To be sure, Mach was not always a 'Machist' himself.

33. See Einstein, 'Time, space, and gravitation' (1948), *Out of My Later Years.* New York, Philosophical Library, 1950: 54–8. Einstein makes the distinction between constructive theories and 'theories of principle'. Einstein cites, as an example of the latter, the relativity theory, and the laws of thermodynamics. Such theories of principle, Einstein says, start with 'empirically observed general properties of phenomena'. See also 'Autobiographical notes': 53.

34. Ernst Mach, *Die Prinzipien der physikalischen Optik.* Leipzig, J. A. Barth, 1921. The English edition is *The Principles of Physical Optics*, trans. John S. Anderson and A. F. A. Young. London, Methuen, 1926; reprint edn. New York, Dover, 1953: vii–viii.

35. Einstein, in 'Séance du 6 avril, 1922: La Théorie de la relativité, *Bulletin de la Société Française de Philosophie,* **22** (1922): 91–113. In his 1913 preface rejecting relativity, Mach expressed himself perhaps more impetuously and irascibly than he may have meant. Some evidence for this possibility is in Mach's letters to Petzoldt. In early 1914, Mach wrote: 'I have received the copy of the positivistic *Zeitschrift* which contains your article on relativity; I liked it not only because you copiously acknowledge my humble contributions with respect to that theme, but also in general.' And within a month, Mach writes – rather more incoherently – to Petzoldt: 'The enclosed letter of Einstein [a copy of the last of Einstein's four letters, cited above] proves the penetration of positivistic philosophy into physics; you can be glad about it. A year ago, philosophy was altogether sheer nonsense. The details prove it. The paradox of the clock would not have been noticed by Einstein a year ago.'

I thank Dr John Blackmore for drawing my attention to the Mach–Petzoldt letters, and Dr H. Müller for providing copies from the Petzoldt Archive in Berlin.

36. R. S. Shankland, 'Conversations with Einstein', *American Journal of Physics*, **31** (1963): 56.
37. A typical example is his letter of 18 September 1930 to Armin Weiner: '. . . I did not have a particularly important exchange of letters with Mach. However, Mach did have a considerable influence upon my development through his writings. Whether or to what extent my life's work was influenced thereby is impossible for me to find out. Mach occupied himself in his last years with the relativity theory, and in a preface to a late edition of one of his works even spoke out in rather sharp refusal against the relativity theory. However, there can be no doubt that this was a consequence of a lessening ability to take up [new ideas] owing to his age, for the whole direction of thought of this theory conforms with Mach's, so that Mach quite rightly is considered as a forerunner of general relativity theory. . . .'
    I thank Colonel Bern Dibner for making a copy of the letter available to me from the Archives of the Burndy Library in Norwalk, Connecticut. Among other hitherto unpublished letters in which Einstein indicated his indebtedness to Mach, we may cite one to Anton Lampa, 9 December 1935: '. . . You speak about Mach as about a man who has gone into oblivion. I cannot believe that this corresponds to the fact since the philosophical orientation of the physicists today is rather close to that of Mach, a circumstance which rests not a little on the influence of Mach's writings.'
    Moreover, practically everyone else shared Einstein's explicitly expressed opinion of the debt of relativity theory to Mach; thus Hans Reichenbach wrote in 1921, 'Einstein's theory signified the accomplishment of Mach's program', ('Der gegenwärtige Stand der Relativitätsdiskussion', *Logos*, **10** (1922): 311). Even Hugo Dingler agreed: '[Mach's] criticism of the Newtonian conceptions of time and space served as a starting point for the relativity theory. . . . Not only Einstein's work, but even more recent developments, such as Heisenberg's quantum mechanics, have been inspired by the Machian philosophy.' Hugo Dingler, 'Ernst Mach', *Encyclopedia of the Social Sciences*, Edwin R. A. Seligman and Alvin Johnson (eds). New York, Macmillan, 1933, 9: 653. And H. E. Hering wrote an essay whose title is typical of many others: 'Mach als Vorläufer des physikalischen Relativitätsprinzips', *Kölner Universitätszeitung*, **1** (17 January 1920): 3–4. I thank Dr John Blackmore for a copy of the article.
38. In his article, 'Zur Enthüllung von Ernst Machs Denkmal', in the special supplement of *Neue freie Presse* cited in note 13, Einstein – then already disenchanted for some time with the Machist program – wrote immediately after the portion quoted in note 13: 'Philosophers and scientists have often criticized Mach, and correctly so, because he erased the logical independence of the concepts vis-à-vis the 'sensations,' [and] because he wanted to dissolve the Reality of Being, without whose postulation no physics is possible, in the Reality of Experience. . . .'
    There were additional sources, both published and unpublished, on the detailed aspects of the relation between Einstein and Mach, which, for lack of space, cannot be summarized here.
39. For example, by Einstein himself, by Joseph Petzoldt, and by Hugo Dingler. I assign relatively little weight to the possibility that the rift grew out of the difference between Einstein and Mach on atomism. Herneck provides the significant report that according to a letter from Philipp Frank, Mach was personally influenced by Dingler, whom Mach had

praised in the 1912 edition of the *Mechanik* and who was from the beginning an opponent of relativity theory, becoming one of the most 'embittered enemies' of Einstein (Herneck, 'Die Beziehungen zwischen Einstein und Mach': 14; see note 31). The copies of letters from Dingler to Mach in the Ernst-Mach-Institute in Freiburg indicate Dingler's intentions; nevertheless, there remains a puzzle about Dingler's role which is worth investigating. It is significant that in his 1921 essay (see note 24) Petzoldt devotes much space to a defense of Einstein's work against Dingler's attacks. See also Joachim Thiele, 'Analysis of Mach's Preface', in *NTM*, *Schriftenreihe für Geschichte der Naturwissenschaften, Technik und Medizin* (Leipzig), **2** (1965): 10–19.

40. Albert Einstein, 'Uber Relativitätstheorie', *Mein Weltbild*. Amsterdam, Querido Verlag, 1934: 214–20; republished as 'On the theory of relativity', *Ideas and Opinions*: 246–9. Herneck has given the texts of similar discussions on phonographic records by Einstein in 1921 and even in 1924; cf. Herneck's transcriptions of Einstein, 'Zwei Tondokumente Einsteins zur Relativitätstheorie', *Forschungen und Fortschritte*, **40** (1966): 133–4.

41. Quotations from Einstein, 'Zur Methode der theoretischen Physik, *Mein Weltbild*: 176–87, as reprinted in translation in 'On the method of theoretical physics', *Ideas and Opinions*: 270–6, except for correction of mistranslation of one line. There are a number of later lectures and essays in which the same point is made. See, for example, the lecture, 'Physics and reality', (1936, reprinted in *Ideas and Opinions*: 290–323), which states that Mach's theory of knowledge is insufficient on account of the relative closeness between experience and the concepts which it uses; Einstein advocates going beyond this 'phenomenological physics' to achieve a theory whose basis may be further removed from direct experience, but which in return has more 'unity in the foundations'. or see 'Autobiographical notes': 27: 'In the choice of theories in the future', he indicates that the basic concepts and axioms will continue to 'distance themselves from what is directly observable.'

   Even as Einstein's views developed to encompass the '*erlebbare, beobachtbare*' facts as well as the '*wild-spekulative*' nature of theory, so did those of many of the philosophers of science who also had earlier started from a strict Machist position. This growing modification of the original position, partly owing to 'the growing understanding of the general theory of relativity', has been chronicled by Frank, for example, in 'Einstein, Mach, and logical positivism', in Schilpp, op. cit.: 269–86.

42. For a discussion of thematic and phenomenic elements in theory construction, see my article, 'The thematic imagination in science', in *Science and Culture*, ed. Gerald Holton. Boston, Houghton Mifflin, 1965: 88–108.

43. Einstein, 'Autobiographical notes': 53. Emphasis added. On pp. 9–11, Einstein describes what may be a possible precursor of this attitude in his study of geometry in his childhood.

44. Einstein, 'Über das Relativitätsprinzip und die aus demselben gezogenen Folgerungen, *Jahrbuch der Radioaktivität und Elektronik*, **4** (1907): 411–62.

45. Walter Kaufmann. 'Über die Konstitution des Elektrons, *Annalen der Physik*, **19** (1906): 487–533. Emphasis in original.

46. Ibid.: 495.

47. See note 44: 439.

48. Ibid. Shortly after Kaufmann's article appeared, Max Planck ('Die Kaufmannschen Messungen der Ablenkbarkeit der β-Strahlen in ihrer Bedeutung für die Dynamik der Elektronen', *Physikalische Zeitschrift*, **7** (1906): 753–61), took it on himself publicly to defend Einstein's work in an analysis of Kaufmann's claim. He concluded that Kaufmann's data did not have sufficient precision for his claim. Incidentally, Planck tried to coin the term for the new theory that had not yet been named: '*Relativtheorie*'.

49. It should be remembered that Poincaré, with a much longer investment in attempts to fashion a theory of relativity, was quite ready to give in to the experimental 'evidence'. See note 18.

50. Carl Seelig, *Albert Einstein*. Zurich, Europa Verlag, 1954: 195.

51. Einstein, 'Einiges über die Entstehung der allgemeinen Relativitätstheorie', *Mein Weltbild*: 248–56, reprinted in translation as 'Notes on the origin of the general theory of relativity', *Ideas and Opinions*: 285–90.

52. 'Da könnt' mir halt der liebe Gott leid tun, die Theorie stimmt doch.' This semi-serious remark of a person who was anything but sacrilegious indeed illuminates the whole style of a significant group of new physicists. P. A. M. Dirac, 'The evolution of the physicist's picture of nature', *Scientific American*, **208** (1963): 46–7, speaks about this, with special attention to the work of Schrödinger, a spirit close to that of his friend, Einstein, despite the ambivalence of the latter to the advances in quantum physics. We can do no better than quote *in extenso* from Dirac's account:

'Schrödinger worked from a more mathematical point of view, trying to find a beautiful theory for describing atomic events, and was helped by deBroglie's ideas of waves associated with particles. He was able to extend deBroglie's ideas and to get a very beautiful equation, known as Schrödinger's wave equation, for describing atomic processes. Schrödinger got this equation by pure thought, looking for some beautiful generalization of deBroglie's ideas and not by keeping close to the experimental development of the subject in the way Heisenberg did.

'I might tell you the story I heard from Schrödinger of how, when he first got the ideas for this equation, he immediately applied it to the behavior of the electron in the hydrogen atom, and then he got results that did not agree with experiment. The disagreement arose because at that time it was not known that the electron has a spin. That, of course, was a great disappointment to Schrödinger, and it caused him to abandon the work for some months. Then he noticed that if he applied the theory in a more approximate way, not taking into account the refinements required by relativity, to this rough approximation, his work was in agreement with observation. He published his first paper with only this rough approximation, and in this way Schrödinger's wave equation was presented to the world. Afterward, of course, when people found out how to take into account correctly the spin of the electron, the discrepancy between the results of applying Schrödinger's relativistic equation and the experiment was completely cleared up.

'I think there is a moral to this story, namely, that it is more important to have beauty in one's equations than to have them fit experiment. If Schrödinger had been more confident of his work, he could have published it some months earlier, and he could have published a more accurate equation. The equation is now known as the Klein–Gordon equation, although it was really discovered by Schrödinger before he discovered his nonrelativistic treatment of the hydrogen atom. It seems that if one is working from the point of view of getting beauty in one's equations, and if one has really a sound insight, one is on a sure line of progress. If there is not complete agreement between the results of one's work and experiment, one should not allow oneself to be too discouraged, because the discrepancy may well be due to minor features that are not properly taken into account and that will get cleared up with further developments of the theory. That is how quantum mechanics was discovered . . .' (pp. 46–7).

53. Published several times – for example, by B. G. Teubner, Leipzig, 1909.

54. Friedrich Herneck, 'Zu einem Brief Albert Einsteins an Ernst Mach', *Physikalische Blätter*, **15** (1959): 564. Frank's remark is reported by Herneck in 'Ernst Mach und Albert Einstein', *Symposium aus Anlass des 50. Todestages von Ernst Mach*, ed. Frank Kerkhof. Freiburg, Ernst Mach Institut, 1966: 45–61.

55. Philipp Frank, 'Das Relativitätsprinzip und die Darstellung der physikal-ischen Erscheinungen im vierdimensionalen Raum', *Zeitschrift für physikalische Chemie*, **74** (1910): 466–95.

56. C. B. Weinberg, 'Mach's Empirio-Pragmatism in Physical Science'. Thesis, Columbia University, 1937.

57. Mach, 'Die Leitgedanken meiner naturwissenschaftlichen Erkenntnislehre und ihre Aufnahme durch die Zeitgenossen', *Scientia*, **7**: 225.

58. Ernst Mach, *Space and Geometry*. Chicago, Open Court Publishing Co., 1906: 138. Mach's attempts to speculate on the use of n-dimensional spaces for representing the configuration of such 'mere things of thought' – the derogatory phrase also applied to absolute space and absolute motion in Newton – are found in his first major book, *Conservation of Energy* (1st edn, 1872).

59. Cf. Joseph Petzoldt, 'Verbietet die Relativitätstheorie Raum und Zeit als etwas Wirkliches zu denken? *Verhandlungen der deutschen physikalischen Gesellschaft*, No. 21–4 (1918): 189–201. Here, and again in his 1921 essay (note 24), Petzoldt tries to protect Einstein from the charge – for example, made by Sommerfeld – that space and time no longer 'are to be thought of as real'.

60. Mach, *Conservation of Energy*: 54.

61. See note 35; also reported in 'Einstein and the philosophies of Kant and Mach', *Nature*, **112** (1923): 253.

62. That Einstein did not so misunderstand Newton can be illustrated, for example, in a comment reported by C. B. Weinberg: 'Dr Einstein further maintained that Mach, as well as Newton, tacitly employs hypotheses – not recognizing their non-empirical foundations.' (Weinberg, op. cit.: 55.) Dingler analyzed some of the nonempirical foundations of relativity theory in *Kritische Bemerkungen zu den Grundlagen der Relativitätstheorie*. Leipzig, S. Hirzel, 1921.

63. Cited by Robert S. Cohen in his very useful essay 'Dialectical materialism and Carnap's logical empiricism', *The Philosophy of Rudolf Carnap*, P. A. Schilpp (ed.). La Salle, Ill., Open Court Publishing Co., 1963: 109. I am also grateful to Professor Cohen for a critique of parts of this paper in earlier form.

64. Albert Einstein, 'Physics and reality', *Journal of the Franklin Institute*, **221** (1936): 313–47.

65. Einstein, 'Notes on the origin of the general theory of relativity': 288, 289.

66. See note 30: 230–1.

67. I am not touching in this essay on the effect of quantum mechanics on Einstein's epistemological development; the chief reason is that while from his 'heuristic' announcement of the value of a quantum theory in 1905, Einstein remained consistently skeptical about the 'reality' of the quantum theory of radiation, this opinion only added to the growing realism stemming from his work on general relativity theory. In the end, he reached the same position in quantum physics as in relativity; cf. his letter of 7 September 1944 to Max Born: 'In our scientific expectations we have become antipodes. You believe in the dice-playing God, and I in the perfect rule of law in an objectively existing world which I try to capture in a wildly speculative way.' (Reported by Max Born, 'Erinnerungen an Einstein', *Universitas, Zeitschrift für Wissenschaft, Kunst und Literatur,* **20** (1965): 795–807.)

68. Max Born, 'Physics and relativity', *Physics in My Generation.* London, Pergamon Press, 1956: 205.

69. Albert Einstein, Über den gegenwärtigen Stand der Feldtheorie', *Festschrift Prof. Dr. A. Stodola zum 70. Geburtstag,* E. Honegger (ed.) Zurich and Leipzig, Orell Füssli Verlag, 1929: 126–32. I am grateful to Professor Cornelius Lanczos and Professor John Wheeler for pointing out this reference to me.

70. Dingler, *Die Grundgedanken der Machschen Philosophie*: 98.

71. Max Planck, *The Philosophy of Physics,* trans. W. H. Johnson, New York, W. W. Norton, 1936: 122, 125.

72. Albert Einstein, 'Religion and science', *The New York Times Magazine,* 9 Nov. 1930; cf. *Mein Weltbild*: 39, and *Cosmic Religion.* New York, Covici-Friede, 1931: 48.

Possible reasons for Einstein's growing interest in these matters, partly related to the worsening political situation at the time, are discussed in Frank, *Einstein, His Life and Times* (see note 28). It is noteworthy that while Einstein was quite unconcerned with religious matters during the period of his early scientific publications, he gradually returned later to a position closer to that at a very early age, when he reported he had felt a 'deep religiosity. . . . It is quite clear to me that the religious paradise of youth . . . was a first attempt to free myself from the chains of the "merely personal"'. 'Autobiographical notes', in Schilpp, op. cit.: 3, 5. For a discussion, see Gerald Holton, 'Science and new styles of thought', *The Graduate Journal,* 7 (1967): 417–20.

73. Frank, *Einstein, His Life and Times*: 215. Einstein's change of mind was, of course, not acceptable to a considerable circle of previously sympathetic scientists and philosophers. See, for example, P. W. Bridgman, 'Einstein's theory and the operational point of view', in Schilpp, op. cit.: 335–354. An interesting further confirmation of Einstein's changed epistemological position became available after this paper was first published. Werner Heisenberg, in *Physics and Beyond.* New York, Harper & Row, 1971: 62–6, writes about his conversation with Einstein concerning physics and philosophy. See, for example p. 63 – a portion of a conversation set in 1925–26:

'But you don't seriously believe', Einstein protested, 'that none but observable magnitudes must go into physical theory?'

'Isn't that precisely what you have done with relativity?' I asked in some surprise. 'After all, you did stress the fact that it is impermissible to speak of absolute time, simply because absolute time cannot be observed; that only clock readings, be it in the moving reference system or the system at rest, are relevant to the determination of time.'

'Possibly I did use this kind of reasoning', Einstein admitted, 'but it is nonsense all the same. Perhaps I could put it more diplomatically by saying that it may be heuristically useful to keep in mind what one has actually observed. But in principle, it is quite wrong to try founding a theory on observable magnitudes alone. In reality the very opposite happens. It is the theory which decides what we can observe.'

See also Heisenberg's account of Einstein's critique of Mach, ibid.: 63–6.

74. Originally entitled 'Motiv des Forschens' (in *Zu Max Plancks 60 Geburtstag* [Karlsruhe, Müller, 1918]) reprinted in translation by James Murphy, as a preface to Max Planck, *Where Is Science Going?* London, Allen & Unwin, 1933: 7–12. In an earlier appreciation of Planck in 1913, Einstein had written only very briefly about his epistemology, merely lauding Planck's essay of 1896 against energetics, and not mentioning Mach.

75. Planck, 'Positivism and external reality', *International Forum*, **1**, No. 1 (1931): 12–16; **1**, No. 2 (1931): 14–19.

76. Einstein sent his introduction to the editor of the journal on 17 April 1931, but it appears to have come too late for inclusion.

77. Planck, 'Positivism and external reality': 15–17. Emphasis in original.

78. Albert Einstein, 'Bermerkungen zu Bertrand Russells Erkenntnis-Theorie', in P. A. Schilpp (ed.), *The Philosophy of Bertrand Russell*. Evanston, Ill., Library of Living Philosophers, 1944: 289.

79. Toward the end, Einstein himself acknowledged a similar point in his 'Remarks concerning the essays brought together in this co-operative volume', in Schilpp, *Albert Einstein*; 679–80: ' "Einstein's position . . . contains features of rationalism and extreme empiricism . . . " This remark is entirely correct. . . . A wavering between these extremes appears to me unavoidable.'

5.3 Paul Forman, Weimar Culture, Causality, and Quantum Theory, 1918–1927: Adaptation by German Physicists and Mathematicians to a Hostile Intellectual Environment

In perhaps the most original and suggestive section of his book on *The Conceptual Development of Quantum Mechanics* Max Jammer contended 'that certain philosophical ideas of the late nineteenth century not only prepared the intellectual climate for, but contributed decisively to, the formation of the new conceptions of the modern quantum theory'[1]; specifically, ' contingentism, existentialism, pragmatism, and logical empiricism, rose in reaction to traditional rationalism and conventional metaphysics. . . . Their affirmation of a concrete conception of life and their rejection of an abstract intellectualism culminated in their doctrine of free will, their denial of mechanical determinism or of metaphysical causality. United in rejecting causality though on different grounds, these currents of thought prepared, so to speak, the philosophical background for modern quantum mechanics. They contributed with suggestions to the formative stage of the new conceptual scheme and subsequently promoted its acceptance.'[2]

These are far-reaching propositions. Properly construed they are, I think, essentially correct. But it must be said that Jammer did not go very far toward demonstrating them. He displayed such anticausal sentiments among a variety of late nineteenth-century philosophers – French, Danish, and American – but adduced scarcely any evidence to bridge the wide gaps of a quarter century of time, a cultural tradition, and the disciplines of philosophy and physics, which separated their philosophical theses from the development of quantum mechanics by German-speaking Central-European physicists circa 1925. It is not my aim to fill in these gaps, but rather to examine closely the lay of the land on the far side of them. The result is, on the one hand, overwhelming evidence that in the years after the end of the First World War but before the development of an acausal quantum mechanics, under the influence of 'currents of thought', large numbers of German physicists, for reasons only incidentally related to developments in their own discipline, distanced themselves from, or explicitly repudiated, causality in physics.

Thus the most important of Jammer's theses – that extrinsic influences led physicists to ardently hope for, actively search for, and willingly embrace an acausal quantum mechanics – is here demonstrated for, but only for, the German cultural sphere. This cultural qualification is essential; it forms the basis of my attempt to provide, on the other hand, an answer to the question – in its general form crucial to all intellectual history – why and how these 'currents of thought', evidently of negligible effect upon physicists at the turn of the century, came to exert so strong an influence upon German physicists after 1918. For it seems to me that the historian cannot rest content with vague and equivocal expressions like 'prepared the intellectual climate for', or 'prepared, so to speak, the philosophical background for', but must insist upon a causal analysis, showing the circumstances under which, and the interactions through which, scientific men are swept up by intellectual currents.

Such an analysis may be either 'psychological' or 'sociological'. That is, it may either consider the mental makeup of the individual scientists concerned, stressing previous intellectual environments and conditioning experiences as determinative of present attitudes, or, on the contrary, it may ignore these factors, treating present mental posture as socially determined response to the immediate intellectual environment and current experiences. I have chosen the latter course, and sought a model in which certain 'field variables' and their derivatives at a given place and time are regarded as evoking corresponding attitudes. Though it may seem harsh to stress the social pressure and ignore the emotional pain, though it may seem unsatisfactory to break off our explanatory endeavors at the level of the individual decision, nonetheless I do think the 'sociological' the more general and fruitful approach.

The inquiry must begin, then, by characterizing the intellectual milieu in which the German physicists were working and quantum mechanics was developed. This is a formidable problem, above all on account of methodologic difficulties. And the task is especially unattractive to the historian of science, for it obliges him to deal with the 'expressions' of nonscientists as well as those of scientists, thus forcing the abandonment of the demarcation criterion by which he seeks to identify and delimit his subject. Nevertheless, with aid and guidance from previous studies by general intellectual historians, especially the work of Fritz K. Ringer, I have addressed this problem in Part I. I show that in the aftermath of Germany's defeat the dominant intellectual tendency in the Weimar academic world was a neo-romantic, existentialist 'philosophy of life', reveling in crises and characterized by antagonism toward analytical rationality generally and toward the exact sciences and their technical applications particularly. Implicitly or explicitly, the scientist was the whipping boy of the incessant exhortations to spiritual renewal, while the concept – or the mere word – 'causality' symbolized all that was odious in the scientific enterprise.

Now if, as is largely the case even at this late date, the interest of the historian of science is held exclusively by the substantive scientific achievements, he will immediately be struck by a remarkable paradox: this place and period of deep hostility to physics and mathematics was also one of the most creative in the entire history of these enterprises. Faced with this paradox many of us would be tempted to rub our hands with satisfaction, to regard it a welcome refutation of any attempt to impugn the autonomy of these sciences and the sufficiency of intellectualist-internalist history of them. But such an inference would be too hasty. Presupposing the hostility of the intellectual environment, the crucial question is the *nature* of the response of the exact scientists to this circumstance. I had myself previously assumed that in the face of antiscientific currents the predominant response in these highly professionalized sciences would be retrenchment, withdrawal into the science and the community of its practitioners, reaffirmation of the discipline's traditional ideology – i.e. its notion of the value, function, motive, goal, and future of scientific activity.[3] *Were* that the case, then, a fortiori, any attempt to attribute a strong and direct influence of that same intellectual environment upon the scientific discourse and dispositions of these same men would appear implausible.

Yet the historian who takes even the most casual notice of the valuations of physical science in contemporary American society, on the one hand, and the present ideological tendencies in these sciences, on the other hand, could scarcely maintain that the predominant response to a hostile intellectual environment is retrenchment. On the contrary, as sentiments of resentment and antagonism toward the scientific enterprise – coupled with a revival of existentialist *Lebensphilosophie* –

have become prominent in the last few years, so also have the expressions of and concessions to these same sentiments within the sciences themselves. We are indeed witnessing in America today a widespread and far-reaching accommodation of scientific ideology to a hostile intellectual milieu. As the distinguished physical chemist Franklin A. Long recently stated in both explanation and advocacy of this development: 'Faculty, and especially students, are sensitive to social problems, are eager to work on them, and are often prepared to change their previous ways of life to do so. The pressures of discipline orientation and the tradition of individual scholarship are strong among faculty members, but not strong enough to counter the pressures of social concern.' And in all of this 'responsiveness' there is an astonishing sincerity, a striking absence of cynical, calculated image projection, testifying to a surprising participation of the physical scientists themselves in those fundamentally, often manifestly, antiscientific sentiments.[4]

But our contemporary experience does not merely lead us to anticipate an ideological accommodation by the Weimar physicists and mathematicians; it also suggests a simple model for the circumstances under which such accommodation is likely to occur. We may suppose that when scientists and their enterprise are enjoying high prestige in their immediate (or otherwise most important) social environment, they are also relatively free to ignore the specific doctrines, sympathies, and antipathies which constitute the corresponding intellectual milieu. With approbation assured, they are free of external pressure, free to follow the internal pressure of the discipline – which usually means free to hold fast to traditional ideology and conceptual predispositions. When, however, scientists and their enterprise are experiencing a loss of prestige, they are impelled to take measures to counter that decline. Drawing upon Karl Hufbauer's factorization of prestige into image and values, one sees that such countermeasures will in general be attempts to alter the public image of science so as to bring that image back into consonance with the public's altered values. But if this is not mere image projection, then such alterations of the image of the scientist and his activity will also involve an alteration of the values and ideology of the science, and may even affect the doctrinal foundations of the discipline – as Theodore Brown has shown of the beleaguered College of Physicians in the latter seventeenth century.[5]

In Parts II and III, I apply this model to the German-speaking exact scientists working in academic environments in the Weimar period. Bearing in mind the radically rearranged scale of values ascendant in the aftermath of Germany's defeat, I explore in Part II the response of these scientific men at the ideological level. This response I have sought primarily in addresses by exact scientists to academically educated general audiences, and especially in their addresses to their assembled universities. The historian is fortunate that the institutions of German academic life provided frequent occasions for addresses before university convocations, and doubly fortunate that it was customary to publish such *Reden*. Conversely, the existence of these institutions is both an index and an instrument of the extraordinarily heavy social pressure which the German academic environment could and did exert upon the individual scholar or scientist placed within it. As I illustrate in Part II, there was in fact a strong tendency among German physicists and mathematicians to reshape their own ideology toward congruence with the values and mood of that environment – a repudiation of positivist conceptions of the nature of science, of utilitarian justifications of the pursuit of science, and, in some cases, of the very possibility and value of the scientific enterprise.

Was the tendency toward accommodation, which predominated in the response of this highly professionalized scientific community to its hostile intellectual environment, confined to the ideological level, or did it extend beyond it into the substantive doctrinal content of the science itself? Specifically, are there indications that German physicists and mathematicians were anxious to, and deliberately tried to, alter the character of their disciplines as cognitive enterprises and to alter specific concepts employed within them in order to bring their sciences in closer conformity with the values of the Weimar intellectual milieu? I strongly suspect that the intuitionist movement in mathematics, which won so many adherents and created so much furor in Germany in this period, was primarily an expression of just such inclinations and aims. I am convinced, and in Part III endeavor to demonstrate, that the movement to dispense with causality in physics, which sprang up so suddenly and blossomed so luxuriantly in Germany after 1918, was primarily an effort by German physicists to adapt the content of their science to the values of their intellectual environment.

The explanation of the creativity of this place and period must therefore be sought, in part at least, in the very hostility of the Weimar intellectual milieu. The readiness, the anxiousness of the German physicists to reconstruct the foundations of their science is thus to be construed as a reaction to their negative prestige. Moreover the nature of that reconstruction was itself virtually dictated by the general intellectual environment: if the physicist were to improve his public image he had first and foremost to dispense with causality, with rigorous determinism, that most universally abhorred feature of the physical world picture. And this, of course, turned out to be precisely what was required for the solution of those problems in atomic physics which were then at the focus of the physicists' interest.

I. Weimar culture as a hostile intellectual environment

I.1. As perceived by the physicists and mathematicians Through the summer of 1918 the German physical scientists, like the rest of the German public, continued to look forward with confidence and satisfaction to a victorious conclusion of the war in which they had been engaged four years. They, perhaps more than any other segment of the German academic world, also felt *self*-confidence and *self*-satisfaction due to their contributions to Germany's military success and to their anticipation of a postwar political and intellectual environment highly favourable to the prosperity and progress of their disciplines. The botanist looking about his institute, bleak and vacant, had to conclude that 'probably it will also remain so after the war, for youth will turn to technology and leave so 'unpractical' a discipline as botany lying by the wayside'.[6] The chemist, the physicist, the mathematician, however, emphasizing the great practical importance of their subjects during the war and the desirability and inevitability of still closer collaboration with technology in the future, looked forward to yet more, larger, and better stocked institutes and to substantially increased public esteem and academic prestige. 'The closer we appear to approach the victorious conclusion of the war', Felix Klein observed in June 1918 before an audience including leaders of German industry and the Prussian government, 'the more our thoughts are dominated by the question what, after peace is successfully won, ought then to come.' Klein's desiderata ranged from a mathematical institute for himself and his university, through a general reorientation of academic research in the exact sciences to achieve a 'preestablished harmony' with the requirements of industry and the military, to a corresponding re-orientation of German education at all levels.[7] And at least the first of these desiderata seemed assured as the Prussian Minister of Education, Friedrich Schmidt, came forward to announce a grant of 300,000 Mark. Who, participating in these festivities, could have foreseen that the Göttingen mathematical institute would not be built for another ten years, and then only with American money?[8]

When that 'victorious end' which seemed imminent in the summer of 1918 turned suddenly to utter defeat in the fall, the exact scientists found themselves confronting a dramatically transformed scale of public values and thus a drastically altered valuation of their field. That, certainly, was their perception of the situation. Had we no explicit testimony to this effect, we could nonetheless infer it from the defensive tone of the talks given by exact scientists before the assembled faculties and students at academic convocations. While during the latter years of the war such speeches convey self-assurance, confidence in the esteem and good will of the audience, in the Weimar period that is seldom the case. And while it is difficult to display this *tone*, one can at least point to passages alluding more or less explicitly to reproaches against exact science which the speaker clearly supposes to be in his audience's mind. Thus in November 1925 Wilhelm Wien described the great scientific discoveries of the early modern period, especially Newton's derivation of the motion of the planets from the laws of mechanics, as 'the first convincing demonstration of the causality [n.b.] of natural processes which revealed to man for the first time the possibility of comprehending nature by the logical force of his intellect'. But he then immediately conceded that this program, which the natural scientist finds so grand, has its limitations, and he proceeded to quote Schiller: 'Without feeling even for its creator's honour/ Like the dead stroke of the pendulum clock/ Nature devoid of God follows knavishly the law of gravity.'[9] The quotation is clearly in response to popular demand, as the astrophysicist, Hans Rosenberg, makes still clearer in his academic address on 18 January 1930: '"Your subject is, to be sure, the most sublime in space/ But, friend, the sublime does not reside in space", I hear Schiller-Goethe call out to us.'[10]

It is, of course, their audience which Wien, Rosenburg, *et al.* hear calling out these sentiments, and they seek to escape half the reproach by showing that they are themselves at least familiar with the classical literary expressions of German idealism. When, however, the physicist or mathematician was in the audience he had to listen to far sharper reproaches. In March 1921, Friedrich Poske came away from the funeral of the poet Carl Hauptmann smarting at the accusations against the exact natural sciences which he encountered there,[11] accusations apparently much like those which poor Max Born had to listen to daily from his wife, a would-be poet and playwright. Hedwig Born derived a masochistic pleasure from 'the feeling of being cast upon an icy lunar landscape' which the company of 'objective' natural scientists aroused in her.[12] Nor did she hesitate to let her husband's colleagues know that 'it is always like a revelation to me whenever behind the *physicist* I suddenly discover the human being; there are, I mean, also inhuman physicists'.[13] Certainly there is no reason to think that Einstein's explanation – 'what you call "Max's materialism" is simply the causal [n.b.] mode of considering things' – alleviated Mrs Born's disquiet.[14]

Painful as it may have been for the theoretical physicist to have to live with such attitudes, the accusation of *Entseelung*, of destruction of the soul, of the world was not the worst he encountered. As Max von Laue saw it in the summer of 1922, the school of Rudolf Steiner 'raises the most serious charges against today's natural science. It is represented as bearing the guilt for the world crisis [n.b.] in which we stand at present, and the whole of the intellectual and material misery bound up with that crisis is charged to natural science's account.'[15] The counterattack which Laue published was read 'with much pleasure' by his mentor and colleague Max Planck, who thought it 'will certainly achieve good effects in wider circles'.[16]

Clearly Planck saw Rudolf Steiner as merely providing the occasion and the ostensible target for rebutting a set of attitudes which he and Laue felt to be widespread among the German educated public. Planck himself adverted to these attitudes and to their danger for science in an address in the Prussian Academy of Sciences a few weeks later.[17] Early in the following year he complained bitterly in a public lecture that 'precisely in our age, which plumes itself so highly on its progressiveness, the belief in miracles in the most various forms – occultism, spiritualism, theosophy, and all the numerous shadings, however they may be called – penetrates wide circles of the public, educated and uneducated, more mischievously than ever, despite the stubborn defensive efforts directed against it from the scientific side'. Compared to this movement, the agitation of Planck's former *bête noir*, the Monist League, has had, he now allows, 'only very meagre success'.[18]

It is thus not surprising that the remnants of this largely defunct positivist–monist movement thoroughly agreed with Planck that the Weimar intellectual environment was fundamentally and explicitly antagonistic to science. Drawing upon the universally accepted analogy between contemporary Germany and the period following its defeat by Napoleon, Wilhelm Ostwald thought it evident that 'In Germany today we suffer again from a rampant mysticism, which, as at that time, turns against science and reason as its most dangerous enemies'.[19] And even where, as with the theory of relativity, there was great public interest in particular results of physical research, that interest was never, to my knowledge, construed by the physicists as evidencing appreciation and approbation of their enterprise. Rather, it struck Einstein as 'peculiarly ironical that many people believe that in the theory of relativity one may find support for the anti-rationalistic tendency of our days'.[20]

Arnold Sommerfeld was thus clearly speaking for most of his colleagues when, responding to a request from the most prestigious of the South German monthlies for a contribution to a special number on astrology, he asked:

Doesn't it strike one as a monstrous anachronism that in the twentieth century a respected periodical sees itself compelled to solicit a discussion about astrology? That wide circles of the educated or half-educated public are attracted more by astrology than by astronomy? That in Munich probably more people get their living from astrology than are active in astronomy? Certainly in Germany this anachronism is based in part upon the misery of the present. The belief in a rational [vernünftig] world order was shaken by the way the war ended and the peace dictated; consequently one seeks salvation in an irrational [unvernünftig] world order. But the reason must lie deeper, for astrology, spiritualism, and Christian Science are flourishing among our enemies also. We are thus evidently confronted once again with a wave of irrationality and romanticism like that which a hundred years ago spread over Europe as a reaction against the rationalism of the eighteenth century and its tendency to make the solution of the riddle of the universe a little too easy. Even though I [wir] have no illusions about being able to hold back this wave by means of arguments based upon reason, nonetheless I [wir] want to throw myself decisively against it.[21]

Although the German physical scientists, regardless of their special discipline, agreed that irrationalism and mysticism were characteristic of the postwar mood, altogether it was the mathematicians and the theoretical physicists who, more than the experimental physicists or the chemists, felt themselves to be the particular objects of odium, both public and private. One cannot withhold a certain sympathy for the Nazi Theodor Vahlen as he confesses in 1923 before the assembled members of his university how 'a friendly attitude toward mathematics is so rare that, if we run across it, it really strikes us as especially remarkable'.[22] This feeling of facing an antagonistic environment, inside and outside the university, was so generally shared among mathematicians that Gerhard Hessenberg could appeal to it in trying to persuade the theoretical physicist Arnold Sommerfeld to take a course of action which would antagonize an experimental physicist (Friedrich Paschen) to whom Sommerfeld looked for much of his raw material: 'But we poor scapegoats of mathematicians have gotten to hear so much evil about ourselves these days – behind our backs as well as to our faces – what difference does a little bit more or less make. . . .'[23] Indeed, these 'antimathematical currents', 'this onslaught against mathematics' which sprang forth after the war seemed so strong and threatening that in 1920 the German mathematicians joined together in a defense organization, the Mathematischer Reichsverband, whose special task was to protect the position of mathematics in the schools.[24]

The result, then, of this first approach to the problem of establishing the tenor of the intellectual environment within which the Weimar physical scientists worked so productively is unambiguous: the environment was *perceived* by the physical scientists to be markedly hostile. Is it therefore necessary to carry our inquiry any further? One might, after all, argue that it is vain to ask whether these perceptions corresponded to 'reality' and that moreover the answer would be of no consequence for the behaviour of the physical scientists. Nonetheless the accuracy of inaccuracy of these perceptions is certainly an important datum about these men, a datum which is essential for any attempt to infer their perceptions, and the effects upon their science, of a given intellectual environment. For the purposes of this paper, moreover, it is important to go farther afield in exploring the attitudes toward physical science in Weimar Germany; we need a more detailed specification of those attitudes if we are to determine how far and in what sense the ideology and ideas of the physical scientists may be regarded as responses to their intellectual environment.

[. . .]

II. Adaptation of ideology to the intellectual environment

II.1. Introduction Spengler epitomizes for us a set of attitudes, widely diffused among educated Germans, explicitly hostile to the ideology of the exact sciences and to particular concepts employed within them. In the remainder of this paper I explore some aspects of the response of the representatives of these sciences in German-speaking Central Europe – in the first instance of the response at the level of ideology; i.e. I explore the effect of this intellectual environment upon the professed justifications of scientific activity, upon the epistemological stance of the exact scientists, and upon their elan, their esprit, their confidence in the future of their discipline.

[. . .]

II.3. Capitulation to Spenglerism

[. . .]

A most interesting and most suggestive case of a clearly discernible Spenglerian influence is to be found in Richard von Mises' inaugural (and farewell) lecture as Professor of Mechanics at the *Technische Hochschule* Dresden – delivered in February 1920, after von Mises had accepted a chair of applied mathematics at the University of Berlin. Considerable stir was created by von Mises' contention – or rather concession to the intellectual milieu – that the 'age of technology', to which the *Technische Hochschulen* owed their rise, was on its way out. His advice to these institutions was that they do their best to get onto the wave of the future by entering the field which was designed to replace technology in the 'culture consciousness', namely speculative natural science, particularly relativity and atomic physics. In these subjects, he asserted, we have had for the past two decades a period like that of Copernicus, Galileo, and Kepler. 'It is not a question of new facts of any sort, nor of new theoretical propositions, nor even of new methods of research, but if I may say it – taking this word in its philosophical sense – of new intuitions [Anschauungen] of the world.' Atomic physics has taken up again 'the question of the old alchemists'; 'numerical harmonies, even numerical mysteries play a role, reminding one no less of the ideas of the pythagoreans than of some of the cabbalists'.[25]

Astonishing as these remarks are from the mouth of a convinced positivist, impossible as it may be to find their like two years earlier, much as they remind us of Spengler's prediction that a new mysticism was the fate and salvation of natural science, still *prima facie* evidence of a connection with the *Decline of the West* is wanting. In fact, the immediate precedent and probable inspiration for von Mises' reference to numerical harmonies and mysteries is an article on 'A number mystery in the theory of the Zeeman effect' by Arnold Sommerfeld which appeared in *Die Naturwissenschaften* a few weeks earlier, as well as the preface to Sommerfeld's *Atomic Structure and Spectral Lines*, which had appeared late in 1919.[26] There Sommerfeld had spoken of 'the mysterious organ upon which nature plays the spectral music' of the atomic spheres. In the future Sommerfeld was to go considerably further in this direction. A ceremonial address at a public session of the Bavarian Academy of Sciences in July 1925 offered Sommerfeld the opportunity to stress that 'hand in hand with this turn toward the arithmetical goes a certain inclination of modern physics toward pythagorean number mysticism. Precisely the most successful researchers in the field of theoretical spectral analysis – Balmer, Rydberg, Ritz – were pronounced number mystics. . . . If only Kepler could have experienced today's quantum theory! He would have seen the most daring dreams of his youth realized. . . .'[27]

It is true that, having indulged himself in such rhetoric at some length, Sommerfeld concluded with the hope that he will 'not be suspected of speaking in favor of mysticism in the ordinary sense, as it comes out in the astrological, metaphysical, and spiritualistic impulses of our time'. Nothing is farther from his intent, he insisted; he was not speaking of human things, but only of laws of nature, and he meant rather to be attacking 'conventionalism', 'positivism', and 'Machian philosophy'.[28] Yet it is perfectly clear that, despite the disclaimers, Sommerfeld was indeed catering to the antirational as well as the antipositivist inclinations of his audience, that he was trying to project an image of physics that would find favor with his auditors and raise the prestige of the discipline in their eyes. And one cannot help but be struck at the close correspondence between this image and that which Spengler sketched in the final pages of the *Decline of the West*.

But let us return to von Mises – upon whom a direct influence of Spengler can in fact be established by September 1921. When at this time von Mises added an appendix to the republication of his lecture of February 1920, his tone had changed entirely, his optimism and enthusiasm had disappeared. Von Mises had largely, and *explicitly*, adopted Spengler's perspective and assumptions. It is 'at least highly probable that the towering structure, under construction for the past five centuries, of a Western culture oriented entirely toward cognition and performance will collapse in the following centuries. From this standpoint one must count the theory of relativity and modern atomic physics as among the last building stones destined to crown the structure.' Accepting Spengler's doctrine that cultures, as 'living organisms', are fundamentally incommensurable, von Mises declared it 'entirely out of the question' that the culture which succeeds ours will 'continue the exact sciences in our sense'. Nor can such views be dismissed as pessimism – 'as if the man, who conscious of his old age and the inevitability of his death, is a pessimist because he faces the fact and acts accordingly'.[29] What, one wonders, would it mean for a physicist or mathematician to 'act accordingly'? Could it possibly mean that he strives to alter the content of his science and the very nature of the scientific enterprise, in order to fulfill Spengler's prophesies?

What is perhaps most striking and appalling about von Mises of September 1921 is the failure of nerve, the complete loss – just as Spengler predicted – of the esprit, the self-confidence which we expect from the mathematical physicist. And in this von Mises was by no means unique. One can find, on the contrary, many examples – most often in addresses before general academic audiences – of theoretical physicists and applied mathematicians denigrating the capacity of their discipline to attain true, or even valuable, knowledge. The earliest such is, perhaps, the passage which in the spring of 1918 Hermann Weyl placed as a conclusion to the first edition of *Space-Time-Matter*.[30] Theoretical physics is, Weyl maintained, entirely analogous to formal logic. 'True' propositions must conform to logic, but logic is incapable of judging the 'truth' of the propositions it manipulates; so also reality conforms to the laws of physics, but physics is incapable of informing us about the reality which its laws govern. Is, perhaps, this reality – these 'darker depths' than the mathematician can grasp with his methods[31] – Spengler's 'immediate becoming, life itself'? Indeed, as we shall see in Part III, it is.

A still more striking example of these same 'annihilating doubts' is offered us by Gustav Doetsch in his inaugural lecture as Privatdozent for applied mathematics at the University of Halle, 27 January 1922. There, in conclusion, pointing back to his exposition of the 'Meaning of Applied Mathematics', Doetsch burst forth:

Such *rationalistic dogmatism* is the characteristic expression of *that* intellectual epoch which is at this moment perishing [im Untergehen]. It is the spirit, one could say, of the *age of natural science*, which, essentially, coincided with the 19th century, and which in our days is sinking with violent convulsions into its grave in order to make room for a new spirit, a new life-feeling . . . this epoch, at whose beginning we unquestionably find ourselves today, is fed up with this rationalistic attitude. Whether we direct our attention toward expressionism in art, or to more recent philosophical tendencies, which in many ways have not yet emerged entirely distinctly, or to any other area of life and thought whatsoever, we find everywhere an ever stronger *aversion* for *that* spirit which believed that it had to express, and that it could express, everything whatsoever in dry words, in one formula – an aversion deriving from the unconscious feeling: *this* path has never and will never lead us to the *essence* of things, we must try to get 'nearer' to the object, to transfer ourselves inside of it itself. Whether the new path leads to the goal, or whether it can only get us closer, may be left undecided. Here my intent was only to point out in the *domain of natural science itself*, which has served as a model for so many others, that the *mathematical* treatment of the material of experience does not begin to impart information about the essence of the world, that is, to yield true cognition.[32]

And still Doetsch was not quite finished. After this tirade against his discipline he quoted Hegel's dictum that mathematics is 'kein Begreifen', and clinched his case by observing that if Hegel 'should not be regarded as the proper person to bring applied mathematics to a correct estimation of itself, then I refer to the words of our most brilliant contemporary mathematician, Hermann Weyl, in whose famous work, *Raum-Zeit-Materie*. . . .' It was, of course, rather daring of Doetsch to speak his mind *so* freely – although his general academic audience must have been very happy indeed to hear their views confirmed by a mathematician. It was, however, foolhardy of Privatdozent Doetsch to publish such sentiments in the journal of the German Society of Mathematicians, where it was read by, and had necessarily to offend, senior and influential colleagues. Indeed it may well have cost him a chair.[33]

And finally, to place in evidence an example from the latter part of our period, consider the picture of 'The Peculiar Nature of the Mathematician's Mind' which Max Dehn, Professor of Pure and Applied Mathematics, held up before his assembled university in January 1928. Painted in pure Spenglerian style, the characteristic mental tone of the contemporary [German] mathematician is skepsis, mistrust of reason, self-inculpation, pessimism, and resignation:

This somewhat sceptical attitude of many a contemporary mathematician is reinforced by what is going on in the neighboring field of physics. Here it appears to be the case that the physical phenomena no longer admit of being construed consistently [widerspruchslos] in a mathematical four-dimensional space-time manifold. Up to now we were able to provide physics with sufficiently freely built scaffolding for its ever bolder constructions. Now, however, in certain reflections arising from important investigations of the finest structure of matter, physics is perhaps in the process of cutting itself loose from mathematics. [Dehn is, *inter alia*, two years behind the times – see sections III.4–6]

All this has impelled many of us to be somewhat sceptical in more general questions as well. The fundamental conviction of every philosopher that the world can be comprehended consistently [widerspruchslos] by the human reason is, for the mathematician, no longer certain. . . . This attitude is, to be sure, not entirely original; it is reminiscent of the thought of the later Eleatics at the time of the foundation crisis in ancient Greece.

Out of this skepsis there develops a certain resignation, a kind of mistrust for the power of the human mind in general.

. . . because of the boundedness of human intellectual power a limit is set to abstraction, to the departure from the intuition. Beyond this limit no further development is possible. But contemporary mathematics is by no means dead, and naturally, even in topology, for example, a man can and hopefully will come who simplifies the processes so much . . . that a new development sets in. . . . Such achievements will, however, scarcely arise in the course of an organized routine. But if the mathematician is already complaining for this very reason – that in consequence of the modern development finally even the pursuit of his science has become organized – then he must properly say to himself: *mea maxima culpa*. For through mathematics the constructive power of the human being first unfolded, and thus brought forth the age of technology. And if, confronted by this disaster which he has brought about, the mathematician is seized with despair, then, for the third time, resignation saves him.[34]

The foregoing examples – especially the cases of von Mises and Doetsch – demonstrate most clearly that there were mathematical physicists who went so far in assimilating the values and mood of their intellectual milieu as to effectively repudiate their own discipline. They show, moreover, that this process of ideological adaptation to the intellectual environment was, either explicitly or implicitly, in large measure a capitulation to Spenglerism. These cases are extreme, of course, and as such atypical. Yet the stages by which von Mises advanced to this extreme, and the readiness of even a Sommerfeld to flirt with the very antiscientific tendencies he deplored, makes it difficult to avoid the conclusion that most German mathematicians and physicists largely participated in, or accommodated their persona to, a generally Spenglerian point of view.

This conclusion is supported by the combination of ample evidence that Spengler's book was read by many, if not most, German physicists and mathematicians and the remarkable paucity of public criticism by representatives of these disciplines. In reviewing the *Untergang* in 1919 Troeltsch had emphasized the desirability of such criticism, but reported that 'to be sure, when I asked one of our most eminent mathematicians and physicists [Planck?] to give his opinion of the book, and briefly described Spengler's principal theses, he refused to read any part of it'.[35] But that reaction was either untypical or changed very quickly, for I have seen explicit references to Spengler, either suggesting or demonstrating acquaintance with his book, by Max Born, Albert Einstein, Franz Exner, Philipp Frank, Gerhard Hessenberg, Pascual Jordan, Konrad Knopp, Richard von Mises, Friedrich Poske, Hermann Weyl, and Wilhelm Wien.[36] This list, which I expect could be substantially lengthened, is already of such extraordinary length as to make it virtually certain that Spengler's theses, and not merely the public enthusiasm for them, were generally known to the Weimar physicists and mathematicians. And yet they did or said remarkably little to oppose them. Reviewing the literature of the 'controversy over Spengler' in 1922 Manfred Schroeter found that 'both the cornerposts of the book, the first and sixth chapters, mathematics and physics, have remained almost unanswered'. Indeed, Schroeter was able to find very few criticisms by mathematicians, and only one by a physicist – Wilhelm Wien.[37]

Where and when a physicist or mathematician came forward to attack Spengler it was almost invariably in defense of that most basic tenet of the scientific ideology, the autonomy, objectivity, and universality of scientific knowledge. This notion Spengler claimed to have exploded by demonstrating that there are no immanent, invariant criteria of knowledge, that the science of a period is dependent in toto upon its *Lebensgefühl*. Yet for every opponent of Spengler's thesis one can cite another exact scientist who, more or less explicitly and more or less fully, identified himself with this doctrinal touchstone of Spenglerism. And once again von Mises provides evidence that the repudiation of this tenet of the scientific ideology was, in some cases and to some extent, a capitulation to the *Untergang, per se*. Thus in February 1920 von Mises had still been a good enough positivist to deny the influence of political and social conditions or the associated *Lebensgefühl* on the quantity, vitality, direction, or content of the higher intellectual productions. By September 1921, however, he had, as we have seen, gone over to Spengler in this respect as well.[38]

II.4. A craving for crises   The exploration of the forms and extent of the ideological adaptation by the physical scientist to his environment must not stop at that unmarked and undefinable frontier where motivation and metaphysics end and the scientific activity itself begins. For to the ideology belongs not merely the general conceptions of the nature and goals of scientific activity, not merely the morale and esprit of the scientist, but also the scientist's perception of the state of his discipline, his hopes, fears, and expectations for its future development. Here, then, we return to the notion and mood of crisis, the conviction of a crisis of culture and of science, which was an essential component of the persona of the Weimar academics.

But before inquiring how far the German mathematical-physical community was likewise infected by this mood, how far a craving for crises affected the exact scientist's perception of the significance and bearing of specific scientific problems, it is worthwhile to emphasize how ready the mathematicians and physicists were to serve themselves with the crisis rhetoric when addressing a general academic audience. For as the notion of crisis became a cliché, it also became an entrée, a ploy to achieve instant 'relevance', to establish rapport between the scientist and his auditors. By applying the word 'crisis' to his own discipline the scientist has not only made contact with his audience, but has *ipso facto* shown that his field – and he himself – is 'with it', sharing the spirit of the times. A presumption is thus insinuated, and often explicitly stated, that in the course of this crisis his science will shed all those characteristics which the academic audience finds most objectionable.

But now, unless we are willing to charge duplicity and suppose that the physicists and mathematicians were engaged in a cynical manipulation of their image, I think we must allow that their accommodation to the intellectual environment penetrated deeper than the rhetoric. Indeed, the rhetoric itself reacts back upon the persona of the scientist, upon his view of the conceptual situation in his science, of the extent and character of the reconstruction necessary or desirable. In fact, in this period, both mathematics and physics – but above all *German* mathematicians and physicists – went through deep and far-reaching crises, whose very definitions showed the most intimate relation with the principal currents of the Weimar intellectual milieu.

'The New Crisis in the Foundations of Mathematics' proclaimed by Hermann Weyl was precipitated virtually out of thin air in the two or three years following Germany's defeat. With extraordinary suddenness the German mathematical community began to feel how insecure were the foundations upon which the entire structure of mathematical analysis rested, how dubious the methods by which that edifice had been erected. Now, with quasi-religious enthusiasm, considerable numbers of German mathematicians rallied to L. E. J. Brouwer's standard calling for a complete reconstruction of mathematics, a redefinition of the enterprise, which, appropriately enough, went under the name 'intuitionism'.[39] The seriousness of this movement and its consequences may be judged by the vehemence of David Hilbert's counterattack in the spring of 1922. 'If Weyl notices an "inner untenability of the foundations upon which the construction of the empire [Reich] rests", and worries himself over "the threatening dissolution of the polity [Staatswesen] of analysis", then he is seeing ghosts.' Weyl and Brouwer are trying 'to erect a repressive dictatorship [Verbotsdiktatur]'; to follow 'such reformers' is to risk losing the most valuable treasures of mathematics. 'No, Brouwer is not, as Weyl believes, the revolution, but only the repetition, with old means, of an attempted putsch . . . and now especially, with the government [Staatsmacht] so well armed and secured by Frege, Dedekind, and Cantor, condemned to failure from the start.'[40]

Can one read this rhetoric and not suppose that both Weyl and Hilbert at the very least saw close parallels between the crisis in mathematics and the political crises then wracking Germany, that their sense of the significance of the mathematical issues was colored by their perceptions of the political issues, that perhaps this crisis in mathematics depended for its very existence upon the social–intellectual atmosphere in the aftermath of Germany's defeat? Looking back thirty years afterward, Weyl almost conceded as much, and in fact the 'crisis' itself, never resolved, eventually simply ceased to be felt.

Turning to physics one finds once again a notable internal crisis. This is the 'crisis of the old quantum theory' which gripped atomic physicists – first and foremost the Germans – in the years before the introduction of the quantum mechanics in 1925/26.[41] I have myself devoted some effort to the intriguing problem of isolating the particular difficulties and frustrations which led at a particular moment to a conviction that 'the whole system of concepts of physics must be reconstructed from the ground up', as Max Born asserted in the summer of 1923.[42] And while it is undoubtedly true that the internal developments in atomic physics were important in precipitating this widespread sense of crisis among German-speaking Central European physicists, and that these internal developments were necessary to give the crisis a sharp focus, nonetheless it now seems evident to me that these internal developments were not in themselves sufficient conditions. The *possibility* of the crisis of the old quantum theory was, I think, dependent upon the physicists' own craving for crises, arising from participation in, and adaptation to, the Weimar intellectual milieu.

Of this predisposition to perceive the state of physics as critical, we have many examples between the summer of 1921 and the summer of 1922, which is to say in the year immediately preceding that in which the crisis of the old quantum theory was precipitated. Taking only those cases in which the crisis is proclaimed in the title itself, there is Richard von Mises's lecture 'On the Present Crisis in Mechanics' of September 1921, Johannes Stark's pamphlet on *The Present Crisis in German Physics* of June 1922, Joseph Petzoldt's remarks 'Concerning the Crisis of the Causality Concept' of July 1922, and Albert Einstein's popular article 'On the Present Crisis in Theoretical Physics', dated August 1922.[43] Very roughly speaking each of these physicists is pointing in the same direction, viz. toward the quantum theory. There, of course, the agreement ends; each is putting his finger upon a largely, or completely, different 'problem'. But that very circumstance – the widespread but initially poorly focused application of the word and notion of a crisis –

suggests most strongly that the crisis of the old quantum theory, far from being forced upon the German physicists, was more than welcome to them.

And here again, as with 'intuitionism' in mathematics, one cannot help but be struck by the extraordinary convenience of the chief slogan of this crisis: the failure of mechanics. However appropriate this slogan may have been as a diagnosis of the internal difficulties in theoretical atomic physics, it certainly was *most* appropriate as a code word signaling the physicists' intent to rid their discipline of its most obnoxious elements. Conversely, the almost universal conviction among the German atomic physicists that this crisis was going to last a long, long time – although in fact it was 'resolved' within two or three years by the discovery of the quantum mechanics – can be understood in part as a reluctance to contemplate giving up their fashionable and praiseworthy plight, but also in part as an expression of a Spenglerian pessimism: 'in my heart I am once again convinced that this quantum mechanics' – which I, Werner Heisenberg, have just discovered – 'is the answer, for which reason Kramers accuses me of optimism'.[44]

III. "Dispensing with causality':[45] adaptation of knowledge to the intellectual environment

[. . .]

III.2. The first Intimations of an issue, 1919–1920
If one examines the annual indices to German books and periodicals in the first decades of the century, one finds a remarkable number of articles and tracts with the word 'causality' in their title. Most striking, however, is the spate of such tracts in the five years 1918–22.[46] Typically, these are short answers to the riddle of the universe, the revelations of enthusiasts rather than the ruminations of academics. (They show, *inter alia*, that Spengler was not alone in seeing causality as the key to that riddle.) Yet the German academics too were anxious not to be left out of this discussion; in 1915 the Prussian Academy of Sciences had offered a prize for the best history of the causal problem since Descartes, awarding it in 1919 to a devoted student of the noted determinist Benno Erdmann.[47]

It is also at just this time – I know of no example earlier than 1919 – that intimations of this issue appear in the private correspondence and public addresses of German physicists. In June 1919, replying to a lost letter from Max Born, Einstein asked ironically, 'Is a hardboiled x-brother and determinist allowed to say with tears in his eyes that he has lost faith in humanity. Precisely the instinctive behaviour of our contemporaries in political matters is suited to maintain a vivid belief in determinism.'[48] Here Einstein is, on the one hand, gently ridiculing Born for feeling sorry for himself and his country by reminding Born of the public image of the theoretical physicist – hardboiled determinist – and, on the other hand, Einstein is making a small joke which can only be to some point if it were a recognized fact that the law of causality was under attack in the social sphere and under discussion among physicists.

Einstein could still joke about the matter with Born in early December,[49] but at the end of January 1920 his tone had become most serious, for in the meantime Born, in a long letter also lost, had evidently confessed to Einstein that he was willing to entertain the idea of acausality, supporting himself upon arguments of his subordinate, Einstein's former student, Otto Stern.[50] 'That business of causality plagues me a great deal too', Einstein conceded, shaken by Born's defection but also anxious not to give offense by too categorical an assertion of his own 'very very great reluctance to forgo *complete* causality'. Yet even more interesting than these remarks themselves is the association of ideas which their precise location in Einstein's long letter – clearly a point by point reply to Born's – reveals. They occur toward the end, immediately following not unsympathetic remarks on Oswald Spengler, whose *Decline of the West*, published the year before, included barbs directed at both Einstein and Born. 'Sometimes in the evening', Einstein allowed, 'one likes to entertain one of his propositions, and in the morning smiles about it.'[51]

What proposition might Einstein have had in mind? Might he not have been thinking of Spengler's most fundamental proposition, the axis of the system, 'the opposition of the destiny-idea and the causality-principle'? Is the juxtaposition in Einstein's letter of Spengler's *Untergang* and the issue of causality in atomic physics pure chance? Is it not more likely that in Born's mind and/or in Einstein's mind there is an intimate, although perhaps not fully conscious, association between the physicists' new and sudden inclination to forgo causality and Spengler's enormously popular culture-criticism in which the physicist 'whose entire mental existence is founded upon the principle of causality' symbolizes the late and decadent fear of the irrational, the incomprehensible?

Such an association between Spengler and the issue of acausality is certainly explicit in Wilhelm Wien's public lecture on 'The Connections of Physics with Other Disciplines', which he delivered just one month later, at the end of February 1920, in the Prussian Academy of Sciences. Previously I used this lecture to illustrate a chameleonlike adaptation of the physicists' ideology to changes in the intellectual environment. Here, however, it stands as the first of a series of attempts to draw a clear line between that environment and physics as a cognitive enterprise. The apparent contradiction reflects rather the distinction I made in Part II between the peripheral and the central features of scientific ideology. Although Wien was ready to advance a new conception of the wellsprings and social-intellectual function of scientific activity designed to make the enterprise seem worthwhile in the public's eye, he was unwilling to compromise those ideological tenets which he regarded as essential to the scientific method and its cognitive goals. The true source and value of natural science lies, to be sure, in 'an inner need of the human mind', but that need is for a particular kind of knowledge; it is a 'longing to comprehend the causality of the course of phenomena [die Kausalität des Geschehens]'.

Wien's motivation for incorporating causality in the very definition of natural science becomes quite evident when, at the end of his lecture, he comes to the *Decline of the West*. While conceding that there is ample evidence of the accuracy of Spengler's characterization of our present cultural situation, Wien rejects in principle the notion of historical laws, of any necessary course of history. All such laws can be and are repeatedly violated 'by irrational expressions of the human spirit'. Turning Spengler against Spengler, Wien emphasizes that if we suppose a generally valid law of aging of cultures and use it to predict the future of our own, 'then we reintroduce a covert causality into history'. But to adapt Kant's well-known epigram, 'I would like to assert that there is so much the more true historical science in history the less it contains of physics. . . . Causality is the foundation of the physical world picture, but it is a category [Denkform] of our mind [Geist] and cannot be employed again for the analysis of the same spirit [Geist], whose effects it is the task of history to portray.'[52]

Embracing causality in order to effect a separation between his discipline and his milieu, Wien has then the task of repulsing Spengler's attempt to make physics culture-bound, and especially Spengler's contention that contemporary physics, as its nerve failed, was renouncing causality. It is nature which has compelled the physicists to resort increasingly to the use of statistics; it is neither a sign of 'decadence' nor of any renunciation of causality. On the contrary, every utilization of statistics 'postulates causality', but because of the great complexity the causal interconnections cannot be traced in detail. Where in 1914 Wien stressed that of all the natural sciences it is theoretical physics in which the personality has the greatest scope and importance, both in constructing theories and influencing the course of scientific development, now, contra Spengler, he is at pains to emphasize that 'however strongly the shaping of physical modes of thought depends upon the constitution of the physicist, it is nonetheless decisively determined by the nature of the things themselves'. Archimedes' results accord entirely with our own, and our results will in all probability be utilizable by physicists of a later culture.[53]

The attention which Wien gives to Spengler, his focus upon the issue of causality, and his consistent effort to isolate physics – as a cognitive enterprise for which causality is the defining characteristic – from its acausal, irrational historical milieu, all suggest that he sensed an intimate connection between the treasonable murmurings against causality among his colleagues and Spengler's brilliant expression of certain powerful currents in the contemporary milieu.

III.3 Conversions to acausality, 1919–1925

a. The Earliest Converts: Exner and Weyl

[. . .]

Apart from Exner, the earliest to speak out against causality was Hermann Weyl. Weyl was a phenomenologist of quite a different sort from his Machian ex-brothers. As Privatdozent at Göttingen shortly before the war Weyl had fallen under the influence of Edmund Husserl's program of 'pure phenomenology'. This Platonizing phenomenology of the mind, based upon intense introspection, had originated in epistemological concerns but in this period was degenerating into existentialism. Dating from 1917 is the first avowed intrusion of Weyl's philosophical outlook into his scientific work – his own attempt to place the continuum on an intuitionist foundation. But as I indicated in Section II.4, Weyl soon became the principal champion of Brouwerian intuitionism in Germany. That Weyl saw an intimate connection between intuitionism in mathematics and acausality in physics emerges quite clearly from his initial manifesto against causality. 'The Relation of the Causal to the Statistical Approach in Physics', printed in August 1920.[54] 'Are statistics merely a shortcut to certain consequences of causal laws', Weyl asks, 'or do they imply that no rigorous causal interconnection governs the world and that, instead, "chance" is to be recognized alongside law as an independent power restricting the validity of the law? The physicists are today entirely of the first opinion.' And yesterday, in the spring of 1918, Weyl had been too, having in his proposed extension of general relativity made an 'attempt', as he admits, 'at carrying through the idea of a pure physics of law for the entirety of the world'.[55] But now Weyl has changed his mind and is placing himself in opposition to the prevailing opinion. Why? He has certain dissatisfactions with classical statistical mechanics and the treatment of fluctuation phenomena, but the real issue, he admits is that

finally and above all, it is the essence of the continuum that it cannot be grasped as a rigid [starr] existing thing, but only as *something which is in the act of an inwardly directed unending process of becoming*. . . . In a *given* continuum, of course, this process of becoming can have reached only a certain point, i.e. the quantitative relations in an intuitively given piece *S* of the world [regarded as a four-dimensional continuum of events] are merely approximate, determinable only with a certain latitude, not merely in consequence of the limited precision of my sense organs and measuring instruments, but because *they are in themselves afflicted with a sort of vagueness*. . . . And only 'at the end of all time', so to speak, . . . would the unending process of becoming *S* be completed, and *S* sustain in itself that degree of definiteness which mathematical physics postulates as its ideal. . . . Thus the rigid [starr] pressure of natural causality relaxes, and there remains, without prejudice to the validity of natural laws, *room for autonomous decisions [Entscheidungen], causally absolutely independent of one another,* whose locus I consider to be the elementary quanta of matter. These 'decisions' are what is *actually real* in the world.[56]

I have quoted Weyl at some length, both because he goes on at some length and because a mere ascription of such radically existentialist views and motives would very likely be dismissed as incredible. Yet, clearly, these motives are primary. Weyl has resolved to abandon the ideal of a pure field physics – for which he had labored so hard and achieved such striking success – and adopted matter, or rather its free will, as the ultimate reality. The field and its laws, like geometry before Einstein, were now a mere backdrop. Why? Because it seemed necessary in order to escape the determinism which the field conception involved. Here, in the fall of 1919 and the summer of 1920 Weyl says not a word about Planck's quantum of action. It has evidently not yet occurred to him that the quantum theory could be dragged in to provide an ostensible physical basis for his existentialist repudiation of causality. It was only in the fall of 1920, when preparing the fourth edition of *Space-Time-Matter*, that Weyl seized upon the quantum theory as compelling him to say 'clearly and distinctly that physics in its present state is simply no longer capable of supporting the belief in a closed causality of material nature resting upon rigorously exact laws'. There Weyl also added that crucial existentialist consideration which had been with him for some time – the repudiation of determinism restores the uni-directionality of time, 'the most fundamental fact of our experience of time', which field physics denied us a priori.[57] Thus, 'not only is matter restored to its old claim to reality, but also the genuine idea of causality, of *Verursachung*, as we experience it most immediately in our will, awakes to new life. Branded as fetishism by Mach . . .' etc., etc.[58]

It seems pretty clear – and indeed it is characteristic of the acausalists – that the sort of primary reality which Weyl would have matter enjoy is simply not a sort of reality which is accessible to physical cognition. Thus by the summer of 1924, in carrying his 'Leibnizian agent-theory of matter' to its logical conclusion, Weyl was led back to the field as the primary *physical* reality:

the material particle itself is not even a point in space, but is something entirely outside the category of extension. . . . It is analogous to the Ego, whose actions, despite the fact that it is itself nonextensional, always have their origin, through its body, at a definite place in the world continuum. Yet whatever this field exciting agent may be in its inner essence – perhaps life and will – in physics we consider it only in terms of the field actions which are excited by it and we are able to characterize it numerically (charge, mass) only by virtue of these field actions.[59]

Weyl was now able to reconcile himself to this resurrection of the field because he thought he had finally found an escape from the proposition that the classical field theories embody and impose the Laplacian conception of causality. In a semipopular article in the form of a dialogue, Weyl argued that 'according to the general theory of relativity the concept of the relative notion of several bodies with respect to one another is just as little tenable as that of the absolute motion of a single body'. Consequently, the principle of causality cannot involve these untenable states of motion, and so reduces to the assertion that 'the world of events only depends upon, and must be unambiguously determined by, the charge and mass of all material particles. Since this is obviously absurd . . . that principle of causality must be abandoned.'[60]

b. 1921, Summer and fall: von Mises, Schottky, Nernst, *et al.*

The quasi-religious conversions to acausality, of which Weyl's is the earliest example, became a common phenomenon in the German physical community during the summer and fall of 1921. As if swept up in a great awakening, one physicist after the other strode before a general academic audience to renounce the satanic doctrine of causality and to proclaim the glad tidings that the physicists are about to release the world from bondage to it. The cases known to me are: Walter Schottky in June, Richard von Mises in September, Walther Nernst in October.[61]

The conversion of von Mises to acausality is particularly interesting not only because it shows the suddenness with which this regeneration could take place and its essential independence of the difficulties encountered in atomic physics, but also because it provides *prima facie* evidence of a direct connection between the repudiation of causality by a loyal scion of Austrian positivism and his capitulation to the *Weltschmerz* of Spengler's *Decline of the West*. In von Mises' inaugural (and farewell) address as delivered in February 1920 at the Technische Hochschule Dresden, and as printed in August 1920, causality was still handled unself-consciously and unpejoratively as equivalent to physical explanation. 'We see now in our time, how a new and simply enormous field of phenomena, the multiplicity of the chemical elements, is drawn into the realm of causal explanation.' And von Mises takes it for granted that the goal of atomic physics, as of all natural science, is and must be 'to explain all these phenomena on the basis of a very few principles, to reveal their causality.'[62] But when one turns to the thoroughly Spenglerian appendix which von Mises added in September 1921 to the republication of this lecture, one finds his attitude towards causality – as toward so much else – entirely transformed. Every electrical, every thermal, every optical process is a statistical phenomenon and as such fundamentally incompatible with the concept of causality. So long as we base ourselves upon that concept 'the quantum theory and everything connected therewith must appear as an insoluble riddle. Whoever traces back the history of physical cognition cannot help but recognize that here an essential *alteration of our mode of thinking*, of the entire scheme of "physical explanation", is inexorably demanded and is gradually being prepared.'[63]

Admittedly, von Mises has invoked the quantum theory as the occasion for the repudiation of causality. But he was not willing that it be *more* than the occasion, that, in particular, his own discipline of applied classical mechanics remain saddled with the stigma of causality. In this same month, September 1921, at the first of the annual German physics–mathematics congresses, von Mises read his colleagues a lecture

– or better, made a public confession before an assembly of his peers –
regarding 'The Present Crisis in Mechanics.'

Stated in the briefest form, this question – in whose negative answer I discern the
crisis in the present state of mechanics – runs thus: can we still assume that all
phenomena of motion and equilibrium which we observe in visible bodies are
explicable within the framework of the Newtonian axioms and their
extensions. In other words, can the temporal course of every motion of an
arbitrarily delimited portion of mass be unambiguously determined by
specifying the initial state and assuming some appropriate force law to be acting?
. . . All that I want to try to show here is that the accumulated facts which we
possess today make it evident that it is highly improbable that this goal of
classical mechanics could ever be attained, and that other, perfectly definite and
no longer unfamiliar, considerations are destined to relieve or to supplement the
rigid causal structure [den starren Kausalaufbau] of the classical theory . . .
whether the sacrifice be great or small, whether we find it difficult or easy, it
seemed to me unavoidable for once clearly and frankly to state that within the
purely empirical mechanics there are phenomena of motion and equilibrium
which will forever escape an explanation on the basis of the differential
equations of mechanics. . . .[64]

One cannot help but be struck by the 'me too' tone of von Mises'
repudiation of 'the stiff causal structure' of classical mechanics and his
representation of that renunciation as an act of moral virtue. Yet it is
also precisely this tone which suggests that a conversion to acausality
carried with it significant social approbation, social rewards so
substantial that von Mises could not bear to let the atomic physicists
monopolize them.

[. . .]

III.6 Causality's last stand, 1925–1926   We are now approaching the
end of the development which I have been trying to trace, that is of the
rise of a will to believe that causality does not obtain at the atomic level
before the invention of an acausal quantum mechanics. With the
introduction of Heisenberg's matrix mechanics in the fall of 1925 and of
Schrödinger's wave mechanics in the spring of 1926, physicists realized
relatively quickly that that belief no longer had to rest primarily upon
ethical considerations or to involve a purely gratuitous renunciation of
the possibility of exact knowledge of atomic processes. The grounds of
argument and belief were thereby substantially altered. I will not
attempt here to treat the growing realization of this new situation in any
detail, but only emphasize once again how conscious the physicists were
of the fact that they were playing before an audience hostile to
causality=mechanism=rationalism, and how anxious many were to play
up to that audience.

Not all, however, did so. During this period it was Wilhelm Wien who
assumed again the role of champion of causality. In January 1925 he had
taken his case to the general public through the pages of the Leipzig
Illustrierte Zeitung where his denial that the quantum theory has, will,

or could lead to an abandonment of the law of causality threaded its way among pictures of cabinet meetings and catastrophes, opera balls and carnival costumes. 'The notion that nature is comprehensible . . . is identical with the conviction that all natural processes can be reduced to causality, to invariably valid natural laws.' Of all purely philosophical notions the concept of causality has had the greatest impact on the development of humanity. It is responsible for the suppression of superstition, for modern natural science, and for the revolutions in technology and industry (n.b., the audience was nonacademic). Although the problem of the interaction of atoms and radiation 'has brought all of theoretical physics into a crisis which will occupy it for a long time', the present form of the quantum theory can only be transitional, for 'a statistics without a causal foundation will never be recognized by physics as something final'.[65]

During the academic year 1925–26 Wien fully exploited the platform available to him as rector of the University of Munich, speaking out in defense of causality in both his official addresses.[66] Although his inaugural lecture of November 1925 contained no reference to the current situation in physics, Wien nonetheless took the opportunity, as we saw in Section 1.1, to stress the historical importance of causality, equating it once again with the conviction that nature can be comprehended by the logical force of the human intellect, and then went on to criticize Langbehn, Chamberlain, and Spengler for their antirationalism and pessimism. The slightly equivocal tone of this lecture had, however, disappeared entirely in June 1926 when, towards the end of his term as rector, Wien spoke at the annual founder's day ceremonies on 'The Past, Present, and Future of Physics', or, more accurately, on causality in the past, present, and future of physics. The theme first appears on page 4 of the printed text as the capacity of the human intellect to grasp the causality of natural processes, continues on pages 6–8 where it is emphasized that, even when the laws are statistical, causality must reign at the level of the elementary processes, and reaches a climax on pages 10 and 11 where Bohr is attacked directly and by name.

Here one must recall that, supporting himself in part upon Heisenberg's discovery of a way to do atomic physics while renouncing the goal of a detailed picture of intra-atomic motions and mechanisms, Bohr had recently been expressing far more openly and categorically his hope and belief that such pictures were impossible in principle, that physics was faced 'with an essential failure of the pictures in space and time on which the description of natural phenomena has hitherto been based'.[67] Quoting these words, Wien then sought to reprimand and silence Bohr and all others of like convictions with that same demand for self-censorship which Planck had advanced so successfully in 1922: 'The physicists have always openly displayed before all the world the difficulties with which they have to contend. . . . But we must be very careful with pronouncements whose significance extends far beyond the limits of the field of physics.' And Wien then went on to assert in the strongest terms that there is no physical field which is closed to our understanding, and that physicists will not rest until they have subjected atomic processes to the law of causality.[68]

At this point, having dealt with Bohr and causality, Wien turned upon his colleague, the Professor of Theoretical Physics, Arnold Sommerfeld – without, of course, naming *him*. Although Wien had readily adapted his justifications for doing physics to the changing public values, he had nonetheless been concerned to shield the enterprise itself from the influence of the Weimar cultural milieu. Sommerfeld's 'Atomystik', on the contrary, dressed up for the public with pythagorean numerical harmonies and number mysteries, was not merely an attempt to use the quantum theory to play up to the ambient antirationalism, but represented an actual research program. 'The number mysticism', Wien hoped and expected, 'would be supplanted by the cool logic of physical thought; not perhaps to everyone's joy. For mysticism often exerts upon many minds a greater force of attraction than the cold and sober physical mode of thought. It is far from my intent to attack mysticism as such. There are many areas of the life of the soul from which mysticism cannot be excluded; but in physics it does not belong. A physics in which mysticism governs, or even collaborates, relinquishes the ground from which it draws its strength, and ceases to deserve its name.' Wien then concluded his lecture by reaffirming once again his confidence that 'insight into the causal interconnections of natural processes will continue to be possible', suggesting that those who express doubts on this score are just suffering from mental exhaustion, and perhaps also on that account are inclined to harken to pessimistic words about the *Untergang des Abendlandes* or the *Zusammenbruch der Naturwissenschaft*.[69]

The confidence and corresponding aggressiveness which Wien manifested on the issue of causality in the spring of 1926 derived chiefly from Erwin Schrödinger's papers on wave mechanics which Wien was then publishing in his journal, the *Annalen der Physik*. Having repudiated causality for social–ethical reasons in 1922–24, by the fall of 1925 Schrödinger had converted back to causality for what were most probably personal-political reasons.[70] He now conceived and developed the wave mechanics as a causal space–time description of atomic processes in opposition to the Copenhagen–Göttingen matrix mechanics. To accept their contention that such a description is not possible 'would be equivalent to a complete surrender'. For, Schrödinger argued in February 1926 in his second paper, 'we really cannot change the forms of thought, and what cannot be understood within them cannot be understood at all. There are such things – but I do not believe that the structure of the atom is one of them'.[71]

Yet at just that moment in June 1926 when Wien, armed with Schrödinger's theory, was striking out so vigorously, the anticipated victory was being transformed into defeat by Max Born's statistical interpretation of the wave function, building an abandonment of causality right into the foundations of the wave mechanics.[72] 'The true state of affairs', Heisenberg declared in the spring of 1927, 'can be characterized thus: Because all experiments are subject to the laws of quantum mechanics, . . . quantum mechanics establishes definitively the fact that the law of causality is not valid.'[73] And once again, when one sees how rapidly this failure of causality was accepted by physicists not merely as a definitive feature of the theory, but equally of reality, one can scarcely escape the conclusion that such a result, far from being regretted, was greeted with relief and satisfaction. The atomic physicists had fulfilled the obligation which Nernst – and their social–intellectual milieu – had laid upon them.

That conclusion is surely also suggested by the physicists' general anxiousness to carry the good news to the educated public – Heisenberg published a popular article retailing his conclusions even before his 'technical' paper was printed[74] – but also from the terms in which they presented these glad tidings. In a public lecture at the University of Hamburg early in 1927 Arnold Sommerfeld raised 'the question which is discussed so much these days, whether the rigid pattern [starre Form] of causality which we have inherited from the 18th century' – read enlightenment, utilitarianism, materialism, etc. – 'and from the rationalistic science of mechanics, is appropriate to our contemporary body of experience'.[75] And when the question is posed in this form there is no doubt either about the answer which his audience wished to hear. Or again, consider the terms in which Max Born discussed the same question in the *Vossische Zeitung*, Berlin's highbrow liberal newspaper, in the spring of 1928. After defining causality as determinism, and adding that all previous laws of physics had that characteristic, Born observed that 'such a conception of nature is deterministic and mechanistic. There is no place in it for freedom of any sort, whether of the will or of a higher power. And it is that which makes this view so highly valued by all "good rationalists"'. But happily physics has now discovered new laws which give it an entirely different character.[76] That character, Bohr had stressed repeatedly in his lectures at Como and at the Solvay Congress the previous fall, is an 'inherent "irrationality"'; indeed 'the inevitability of the feature of irrationality characterizing the quantum postulate' was accepted most willingly by Bohr, who showed no sympathy for Schrödinger's attempt 'to remove the irrational element expressed in the quantum postulate'.[77]

It is true that Sommerfeld himself, even as he raised the question of 'the rigid pattern of causality', stressed that it was not his intent to call into question 'the lawlike definiteness of the physical processes', and elsewhere, as we saw in Section I.1, was at this time actually writing against the less academic forms of the contemporary romantic reaction. But it seems to me that this circumstance only strengthens the inference that an acausal quantum mechanics was particularly welcome to the German physicists because of the irresistible opportunity it offered of improving their public image. Now they too could polemicize against the rigid, rationalistic concept of causality and hope to recover lost prestige thereby.

III.7 Conclusion   In an interview with Einstein in 1932, James Gardner Murphy, an Irish literary man with wide acquaintance among the German theoretical physicists, remarked that 'it is now the fashion in physical science to attribute something like free will even to the routine processes of inorganic nature'. 'That nonsense', Einstein replied, 'is not merely nonsense. It is objectionable nonsense. . . . Quantum physics has presented us with very complex processes and to meet them we must further enlarge and refine our concept of causality.' Murphy: 'You'll have a hard job of it, because you'll be going out of fashion . . . scientists live in the world just like other people. Some of them go to political meetings and the theater and mostly all that I know, at least here in Germany, are readers of current literature. They cannot escape the influence of the *milieu* in which they live. And that *milieu* at the present time is characterized largely by a struggle to get rid of the causal chain in which the world has entangled itself.'[78]

Murphy's assertion of the inescapability of the influence of the milieu is the more worthy of our attention as it is but a paraphrase of a passage from a lecture by Schrödinger, 'Is Natural Science Conditioned by the Milieu?', published earlier that year. Murphy's own contribution is the specific identification of hostility toward causality as the dominant characteristic of the contemporary milieu, and the implication that the scientist's attitude toward this particular concept had virtually been determined thereby.

Schrödinger's and Murphy's analysis is, as the foregoing investigation has shown, remarkably accurate, at least for the German-speaking Central European physicists. Their craving for crises, their readiness to adapt their ideology to the values of their social–intellectual environment argue a substantial and largely indiscriminate participation in the attitudes of their academic milieu, a readiness to swim along in the intellectual currents of the day. This circumstance is the more surprising if one bears in mind that the values characteristic of these intellectual currents which set in so strongly after Germany's defeat were fundamentally antithetical to the scientific enterprise. Indeed the mathematical physicist, the personification of analytical rationality, was often singled out as the prime exemplar of a despicable way of grasping the world. Above all, with astonishing unanimity, it was the physicist's attempt to subject the world to the rigid, dead hand of the law of causality – to use the rhetoric Spengler made so popular – which was taken to epitomize all that was most detestable in the scientific enterprise. These two circumstances – hostile environment and accommodation to its values – were then found to be linked by much direct and indirect evidence suggesting that the accommodation was in response to the hostility. Stated in terms of Karl Hufbauer's distinctions: suddenly deprived by a change in public values of the approbation and prestige which they had enjoyed before and during the First World War, the German physicists were impelled to alter their ideology and even the content of their science in order to recover a favorable public image. In particular, many resolved that one way or another, they must rid themselves of the albatross of causality.

In support of this general interpretation I illustrated and emphasized the fact that the program of dispensing with causality in physics was, on the one hand, advanced quite suddenly *after* 1918 and, on the other hand, that it achieved a very substantial following among German physicists *before* it was 'justified' by the advent of a fundamentally acausal quantum mechanics. I contended, moreover, that the scientific context and content, the form and level of exposition, the social occasions and the chosen vehicles for publication of manifestoes against causality, all point inescapably to the conclusion that substantive problems in atomic physics played only a secondary role in the genesis of this acausal persuasion, that the most important factor was the social-intellectual pressure exerted upon the physicists as members of the German academic community.

And here, saving perhaps the case of Hermann Weyl, it was not a question of 'philosophical' influences in any serious intellectual sense. By far the single most influential 'thinker' was Spengler, and that only because the *Untergang des Abendlandes*, the concentrated expression of the existentialist *Lebensphilosophie* that was diffused through the intellectual atmosphere, was read with attention by most German mathematicians and physicists on account of the prominent role Spengler had given their sciences. Thus, excepting Franz Exner, the philosophical theses of the latter nineteenth century to which Jammer has drawn attention, while they may perfectly well have some ultimate responsibility for the ideational content of the *Lebensphilosophie* of the Weimar period, played, *per se, an sich*, a negligible role in the sudden rise of anticausal sentiment among German physicists after the First World War. Rather, it was only as and when this romantic reaction against exact science had achieved sufficient popularity inside and outside the university to seriously undermine the social standing of the physicists and mathematicians that they were impelled to come to terms with it.

There are, moreover, many indications that this accommodationist strategy met with considerable success. The 'objectionable nonsense' about the free will of electrons which philosophers, aided and abetted by physicists, were talking in the late 1920s, constituted in fact a very favorable press. Although distasteful to Einstein, this image of modern physics was exactly suited to the taste of the educated public of the Weimar period. And I would emphasize that much of the nonsense announced with great fanfare by philosophers in the late 1920s owed nothing whatsoever to the quantum mechanics discovered in 1925–26, but was based wholly and solely upon the manifestoes against causality issued by physicists before that date. Such, for example, were the articles which Ludwig von Bertalanffy published in 1927 gloating over the fact that 'in physics itself views are coming to be accepted which in biology would be designated as vitalistic. . . . The causal world picture of the physicist is dissolving – into its place steps one which recognizes individuality, even for the molecular process. . . . Indeed that allusion of Nernst's to the freedom of will of the theologians can even be employed to support one of Spengler's most controversial ideas: that modern physics, renouncing rigorous causality and exact laws of nature, will give way to a new mysticism.'[79]

One must admit that Bertalanffy's equation of the renunciation of causality with mysticism is not wholly unjustified. For as we saw, the manifestoes by physicists against causality before 1925 were issued not in spite of, but much rather because of, the general belief that 'an abandonment of determinism would signify a renunciation of the comprehensibility of nature'. Far from engaging in any critical analysis of the concept of causality, directed toward the relaxation of determinism without renouncing *a priori* the comprehensibility of nature, these physicists actually reveled in that consequence, stressed the failure of analytical rationality, implicitly repudiated the cognitive enterprise in which physics had theretofore been engaged.

For this reason the acausality movement could not but arouse opposition within the German physics community. Indeed one has here the most characteristic difference between those physicists who hastened to renounce causality and those who clung to it even after the discovery of quantum mechanics. For Exner, Schottky, Nernst, and Bohr the failure of causality was essentially a failure of the human intellect; Weyl, von Mises, and Reichenbach went even further, expressing an existentialist revulsion against intellectuality. On the other hand those few physicists – strikingly and significantly few – who came forward to publicly oppose dispensing with causality all based their cases upon the value of rationality and their faith in the capacity of the human intellect to comprehend the natural world: so Einstein, Petzoldt, Planck, Schrödinger (after his reconversion), and W. Wien (vis-à-vis inorganic nature). And for this reason also I have not been able to, nor indeed wished to, maintain a perfectly neutral stance in my exposé. Although a readiness to view atomic processes as involving a 'failure of causality' proved to be, and remains, a most fruitful approach, before the introduction of a rational acausal quantum mechanics the movement to dispense with causality expressed less a research program than a proposal to sacrifice physics, indeed the scientific enterprise, to the *Zeitgeist*. My sympathies have consequently been with the conservatives in their defense of reason, rather than with the 'progressives' in their denigration of it.

But if this social-intellectual phenomenon is to be comprehended, in part at least, by means of a dichotomy between progressives and conservatives, then correlations might be anticipated between a physicist's position on the causality issue and his general intellectual-political orientation. And in fact, paralleling Ringer's observation that early in the Weimar period the 'modernist' academics tended to be 'methodologically adventurous', one finds that, by and large, those physicists who were readiest and earliest to repudiate causality had either distinctly 'progressive' political views by the standards of their social class and the German academic world, and/or had an unusually close interest in, or contact with, contemporary literature. Nernst, who in his youth had wished to become a poet and who retained his interest in literature throughout his life, was also one of the few German physicists who publicly associated themselves with the cause of parliamentary democracy. Von Mises, although politically conservative and national-istic, was on his way to becoming the foremost authority on the young Rilke. Born and Weyl were both well disposed toward the German republic, at least at its birth – in itself a sufficiently unusual sentiment in the German academic world – and both had literary wives. On the other hand, with the notable exception of Einstein, those who defended causality tended to be highly principled political conservatives and/or interested in classical literature. Such were Planck, Schrödinger, and Max von Laue – who kept their knowledge of Greek well polished. Standing to their right was W. Wien. And finally to the causalist camp one may add the outright reactionaries: Ernst Gehrcke, Erwin Lohr, Philipp Lenard, and Johannes Stark.

This very circumstance – that the alignment within the German physics community over the issue of causality correlates closely with the intellectual and political temper of the individual physicist – reminds us, however, that the 'sociological' model employed in this paper cannot be the whole truth. It provides a general framework, and seems to work especially well in certain extreme cases. But in order to account for its special applicability to some physicists and its special inapplicability to others one must invoke precisely those factors which are excluded from the model – individual personality and intellectual biography. The mechanisms advanced for the entrainment of the German physicists and mathematicians by the *Zeitgeist* is thus clearly not sufficient. And it may be that examination of other episodes of entrainment in the late nineteenth and early twentieth centuries will prove that it is also not necessary. But be that as it may, it seems difficult to deny that the shifts in scientific ideology and the anticipated shifts in scientific doctrine exposed in this paper were *in effect* adaptations to the Weimar intellectual environment. Moreover, whatever similarities one may find in the mental posture of non-German exact scientists in the same period, there is one feature which cannot, I think, be found outside the German cultural sphere: a repudiation of 'causality'.

Source: *Historical Studies in the Physical Sciences*, **3** (1971): 1–116. The extracts appearing here have been left, for uniformity of reference, under their original section numbering. Footnotes have been abbreviated and renumbered.

Notes

1. M. Jammer, *The Conceptual Development of Quantum Mechanics*. New York, McGraw-Hill, 1966, section 4.2. 'The philosophical background of non-classical interpretations'; on: 166-7.

2. Ibid.: 180. The search for philosophic precedents and influences has otherwise focused almost exclusively upon Bohr's doctrine of complementarity. This issue, which I am not directly concerned with here, has been recently examined once again and the literature reviewed by Gerald Holton, 'The roots of complementarity; *Daedalus,* **99** (Fall, 1970): 1015-55.

3. P. Forman, *The Environment and Practice of Atomic Physics in Weimar Germany*. Ph.D. dissertation, Berkeley, 1967; Ann Arbor, University Microfilms, 1968: 11-24.

4. F. A. Long, 'Interdisciplinary problem-oriented research in the university [editorial]', *Science*, **171** (12 Mar. 1971): 961. Marvin L. Goldberger, 'Physics and environment: how physicists can contribute', *Physics Today* (Dec. 1970): 26-30, and the reply by John Boardman, ibid. (Feb. 1971): 9. The new mood, especially the neo-Spenglerianism, in the scientific community is discussed by Bentley Glass in his presidential address to the AAAS. 28 Dec. 1970. 'Science: endless horizons or golden age?' *Science*, **171** (8 Jan. 1971): 23-9.

5. K. Hufbauer, 'Social support for chemistry in Germany during the eighteenth century: how and why did it change?' *Historical Studies in the Physical Sciences,* **3** (1971): 205-31: T. M. Brown. 'The College of Physicians and the acceptance of iatro-mechanism in England, 1665-95', *Bulletin of the History of Medicine,* **44** (1970): 12-30.

6. Karl v. Goebel to Th. Herzog, Munich 19 July 1917, in Goebel, *Ein deutsches Forscherleben in Briefen aus sechs Jahrzehnten, 1870-1932*, Ernst Bergdolt (ed.), 2nd edn. Berlin, 1940: 170.

7. F. Klein, 'Festrede zum 20. Stiftungstage [22 June 1918] der Göttinger Vereinigung zur Förderung der Angewandten Physik und Mathematik', *Jahresbericht der Deutschen Mathematiker-Vereinigung,* **27** (1918): Part I: 217-28; on: 217, 219.

8. *Jahresbericht der D. M.-V.,* 27 (1918): Part 2, 47. In 1926 the International Education Board of the Rockefeller Foundation appropriated $275,000 for a mathematical institute. (Geo. W. Gray, *Education on an International Scale. A History of the International Education Board, 1923-1938* [New York, 1941]: 30; Otto Neugebauer, 'Uber die Einrichtung des Mathematischen Institutes der Universität Göttingen', *Minerva-Zeitschrift,* **4** [1928]: 107-11.)

9. W. Wien, *Universalität und Einzelforschung. Rektorats-Antrittsrede, gehalten am 28. November 1925*. Münchener Universitätsreden, Heft 5. Munich, 1926: **14**.

10. H. Rosenberg. *Die Entwicklung des räumlichen Weltbildes der Astronomie. Rede zur Reichsgründungsfeier . . . am 18. Januar 1930*. Kiel, 1930: 26.

11. F. Poske at the Hauptversammlung of the Deutscher Verein zur Förderung des mathematischen und naturwissenschaftlichen Unterrichtes, 31 Mar. 1921. (Unterrichtsblätter für Mathematik und Naturwissenschaft, **27** [1921]: 34.)

12. H. Born, 'Albert Einstein ganz privat', *Helle Zeit–dunkle Zeit. In memoriam Albert Einstein*, C. Seelig (ed.). Zurich, 1956: 35-9, on 36.

13. H. Born to H. A. Kramers, 29 Sep. 1925. (Archive for History of Quantum Physics, Sources for History of Quantum Physics Microfilm 8, Section 2; for descriptions and locations of this archive, see Thomas S. Kuhn, *et al.*, *Sources for History of Quantum Physics. An Inventory and Report*, Memoirs of the American Philosophical Society, Vol. 68 [Philadelphia, 1967].)

14. Einstein to H. Born, 1 Sept. 1919, in Albert Einstein, Hedwig and Max Born, *Briefwechsel, 1916–1955*, edited and annotated by M. Born. Munich, 1969: 32.

15. M. v. Laue, 'Steiner und die Naturwissenschaft', *Deutsche Revue*, 47 (1922): 41–9; reprinted in Laue's *Aufsätze und Vortrage = Gesammelte Schriften und Vorträge*, Band III. Braunschweig, 1962: 48–56, on 48.

16. Planck to Laue, 8 July 1922: 'Ihren Aufsatz über R. Steiner habe ich mit vielem Vergnügen gelesen. Er . . . wird gewiss in weiteren Kreisen gute Wirkung erzielen.' Handschriftensammlung, Bibliothek, Deutsches Museum, Munich.

17. M. Planck, 'Ansprache des vorsitzenden Sekretärs, gehalten in der öffentlichen Sitzung zur Feier des Leibnizischen Jahrestages, 29. Juni 1922'. *Preuss. Akad. d. Wiss., Sitzungsber.* (1922): lxxv–lxxvii, reprinted in *Max Planck in seinen Akademie-Ansprachen: Erinnerungsschrift der Deutschen Akademie der Wissenschaften zu Berlin*. Berlin, 1948: 41–8.

18. M. Planck, *Kausalgesetz und Willensfreiheit. Offentlicher Vortrag gehalten in der Preuss. Akad. d. Wiss. am 17. Februar 1923*. Berlin, 1923; reprinted in Planck, *Vorträge und Evinnerungen*. Stuttgart, 1949: 139–68: on 162–3. And again, eight years later, 'It is astonishing how many people, particularly from educated circles . . . fall under the sway of these new religions, iridescing with every hue from the most confused mysticism on out to the crassest superstition.' ('Wissenschaft und Glaube. Weihnachtsartikel vom Jahre 1930', ibid.: 246–9; also quoted at length in Hans Hartmann, *Max Planck als Mensch und Denker* [1953: reprinted Frankfurt, 1964]: 52–5, on 52–3.)

19. W. Ostwald, *Lebenslinien. Eine Selbstbiographie*. Berlin, 1926–27:, 3: 442. And again, ibid., 2: 309. 'It is at present considered modern to speak all conceivable evil of the intellect.'

20. A. Einstein, *Vossische Zeitung*, 10 July 1921, as quoted by Siegfried Grundmann, 'Der Deutsche Imperialismus, Einstein und die Relativitätstheorie (1914–33)', *Relativitätstheorie und Weltanschauung*. Berlin, 1967: 155–286, on 194.

21. A. Sommerfeld, 'Über kosmische Strahlung', *Südd. Monatsheftc*, 24 (1927): 195–8; reprinted in Sommerfeld's *Gesammelte Schriften*. Braunschweig, 1968, 4, 580–3. Cf. Lewis M. Branscomb, Director of the US National Bureau of Standards, *Science*, 171 (12 Mar. 1971): 972: 'Astrology is booming; there are three professional astrologers in this country for every astronomer.'

22. Th. Vahlen. *Wert und Wesen der Mathematik. Festrede . . . am 15. V. 1923*, Greifswalder Universitätsreden 9. Greifswald, 1923: 1. And in this, if in nothing else, Konrad Knopp agreed with Vahlen: 'We mathematicians . . . have not been able to obtain, or even merely to retain, the position in public life which mathematics merits.' ('Mathematik und Kultur, Ein Vortrag', *Preussische Jahrbücher*, 211 [1928]: 283–300, on 283).

23. G. Hessenberg to A. Sommerfeld, 16 June 1922. (Sources for History of Quantum Physics Microfilm 33, Section 1.)

24. Georg Hamel, as president, at the first general assembly of the Mathematischer Reichsverband, Jena, 23 Sep. 1921, *Jahresbericht der Deutschen Mathematiker-Vereinigung*, 31 (1922): Part 2: 118. And again at the second general assembly, Leipzig, 22 Sept. 1922, the *Arbeitsausschuss* stressed in its report that 'With respect to its place and prestige [*Geltung*] in the schools, mathematics finds itself in a defensive position. The contemporary intellectual currents, directed against intellectualism and rationalism, are decidedly unfavorable to mathematics.' (Ibid., 32 [1923]: Part 2, 11–12.)

25. R. v. Mises, *Naturwissenschaft und Technik der Gegenwart. Eine akademische Rede mit Zusätzen*, Abhandlungen und Vorträge aus dem Gebiete der Mathematik, Naturwissenschaft und Technik, Heft 8. Leipzig, 1922: 2, 5, 16 respectively.

26. A. Sommerfeld, 'Ein Zahlenmysterium in der Theorie des Zeemaneffektes', *Naturwiss.*, 8 (23 Jan. 1920): 61–4.

27. A. Sommerfeld, *Die Bedeutung der Röntgenstrahlen für die heutige Physik. Festrede, gehalten in der öffentlichen Sitzung der B. Akademie der Wissenschaften . . . am 15. Juli 1925*. Munich, 1925, reprinted in Sommerfeld's *Ges. Schr.* (op. cit., note 21), 4: 564–79, on 573–4.

28. Ibid.: 575–6.

29. R. v. Mises, op. cit., (note 25): 32.

30. H. Weyl, *Raum-Zeit-Materie. Vorlesungen über allgemeine Relativitätstheorie*, 1st edn. Berlin, 1918: 226–7; again, somewhat more fully, in the 3rd edn. Berlin, 1919: 262–3. In the fall of 1920, when preparing the fourth edition (Berlin, 1921). Weyl struck this conclusion, replacing it by an attack on causality.

31. Ibid., 1st edn: 9; 3rd edn: 9; 4th edn: 9.

32. G. Doetsch, 'Der Sinn der angewandten Mathematik', *Jahresbericht der Deutschen Mathematiker-Vereinigung*, 31 (1922): 222–33, on 231–2. 'Antrittsvorlesung gelegentlich der Umhabilitierung von der T. H. Hannover an die Universität Halle a/S. am 27. Januar 1922.'

33. Ibid.: 233.

34. M. Dehn, *Über die geistige Eigenart des Mathematikers. Rede anlässlich der Grundungsfeier de Deutschen Reiches am 18. Januar 1928*, Frankfurter Universitätsreden 28. Frankfurt, 1928: 15, 18. This address by one of Hilbert's oldest students (Ph.D Göttingen, 1899) seems to have drawn considerable attention. Otto Neugebauer quoted it, without citing it, in concluding his exposition of the elaborate installations of the new Göttingen mathematical institute (op. cit., note 8).

35. E. Troeltsch, *Gesammelte Schriften*, Vol. 4: *Aufsätze zur Geistesgeschichte und Religionssoziologie*, Hans Baron (ed.), Tübingen, 1925 (reprinted 1961): 682.

36. Born–Einstein, *Briefwechsel* (op. cit., note 14): 44; the preface to the second edition of Franz Exner's *Vorlesungen über die physikalischen Grundlagen der Wissenschaften*. Leipzig-Vienna, 1922; P. Frank, *Das Kausalgesetz und seine Grenzen*, Schriften zur wissenschaftlichen Weltauffassung, Band 6. Vienna, 1932: 54; etc.

37. M. Schroeter, *Der Streit um Spengler. Kritik seiner Kritiker*. Munich, 1922: 56–7, 70.

38. R. v. Mises, op. cit. (note 25): 3, 32.

39. H. Weyl, *Das Kontinuum*. Berlin, 1918; reprinted New York, 1962; 'Der

circulus vitiosus in der heutigen Begründung der Analysis', *Jahresber. d. Dtsch. Mathematiker-Vereinigung,* **28** (1919): 85–92, reprinted in K. Chandrasekharan, (ed.), *Gesammelte Abhandlungen von Hermann Weyl,* **2** (Berlin, 1968): 43–50; 'Über die neue Grundlagenkrise der Mathematik,' *Math. Zeitschr.,* **10** (1921): 39–79, reprinted in *Ges. Abhl.,* **2**: 143–80.

40. D. Hilbert, 'Neubegründung der Mathematik. Erste Mitteilung,' *Abhandlungen aus dem Math. Seminar der Hamburgischen Universität,* 1 (1922): 157–77, reprinted in Hilbert's *Gesammelte Abhandlungen,* 3 (Berlin, 1935; reprinted New York, 1965): 157–77, on 159–60. Hilbert had delivered this tirade against his most brilliant pupil as a lecture at a number of universities before printing it.

41. Thomas S. Kuhn, 'The Crisis of the Old Quantum Theory, 1922–1925', address delivered at the American Philosophical Society, April 1966. Friedrich Hund, *Geschichte der Quantentheorie.* Mannheim, 1967: 103.

42. M. Born, 'Quantentheorie und Störungsrechnung', *Naturwiss.,* **11** (1923): 537–42, quoted in my essay on 'The doublet riddle and atomic physics *c.* 1924', *Isis,* **59** (1968): 156–74.

43. R. v. Mises, 'Über die gegenwärtige Krise der Mechanik', *Zeitschr. f. angewandte Math. u. Mech.,* 1 (1921): 425–31, reprinted in *Naturwiss.,* **10** (1922): 25–9, and reprinted once again in v. Mises' *Selected Papers,* **2** (Providence, R.I., 1964): 478–87, lecture, Math.-Phys. Congress, Jena, Sept. 1921; J. Stark, *Die gegenwärtige Krisis in der Deutschen Physik.* Leipzig, 1922, preface dated 'Anfang Juni 1922'; J. Petzoldt, 'Zur Krisis des Kausalitätsbegriffs', *Naturwiss.,* **10** (1922): 693–5, dated 2 July 1922, to which Walter Schottky replied 6 Oct. 1922 under the same title, ibid.: 982; A. Einstein, 'Über die gegenwärtige Krise der theoretischen Physik', *Kaizo* (Tokyo). 1 (Dec. 1922): 1–8, dated Aug. 1922.

44. '. . . mich des Optimismus anklagt. . . .' W. Heisenberg to Wolfgang Pauli, 29 June 1925, as quoted by B. L. van der Waerden, *Sources of Quantum Mechanics.* Amsterdam, 1967; reprinted New York, 1969: 27. Heisenberg's paper propounding this quantum mechanics, which was indeed 'schon richtig', appeared as 'Über quantentheoretische Umdeutung kinematischer und mechanischer Beziehungen', *Zeitschr. f. Phys.,* **33** (18 Sept. 1925): 879–93, received 29 July 1925, it is translated in v. d. Waerden, op. cit.: 261–76, and reprinted in M. Born, W. Heisenberg, P. Jordan, *Zur Begründung der Matrizenmechanik,* Dokumente der Naturwissenschaft, Abteilung Physik, Band 2, Armin Hermann (ed.). Stuttgart, 1962: 31–45.

45. This is the title of the opening chapter of Albrecht Mendelssohn-Bartholdy, *The War and German Society.* New Haven, 1937. The eminent emigré legal scholar there asserted that 'War canceled causality. It seemed to do so, at least, to the German people . . . the people as a whole, regardless of their interest in politics, their state of eduation, or their profession or walk in life, realize the change quite clearly, long before it could be measured by historians or sociologists' (p. 20).

46. *Deutsches Bücherverzeichnis,* **6** (1915–20): 770; **10** (1921–25): 1298, lists ten such.

47. Else Wentscher, *Geschichte des Kausalproblems in der neueren Philosophie.* Leipzig, 1921, preface.

48. Einstein–Born, *Briefwechsel* (op. cit., note 14): 29–30.

49. Ibid.: 38.

50. Einstein–Born, 27 Jan. 1920, *Briefwechsel:* 42–5.

51. Einstein, 27 Jan. 1920, loc cit.: 'Der Spengler hat auch mich nicht verschont. Man lasst sich gern manchmal am Abend von ihm etwas suggieren und lachelt am Morgen daruber. . . .'
    'Spengler didn't spare me either' is a puzzle, for, so far as I can see, in the original edition relativity is handled rather unpejoratively as 'die letzte Form der faustichen Natur': 599–601, while Born is not mentioned at all. It is then only in the revised edition (1923) that relativity is described as 'a ruthlessly cynical working hypothesis', and the space given it much reduced: 544–5 (Eng. edn.: 419), while Born, mistaken for a chemist ignorant of mathematics, receives his just deserts in a footnote on pp. 205–6 (Eng. edn, p. 156).
52. W. Wien, 'Über die Beziehungen der Physik zu andern Wissenschaften. Öffentlicher Vortrag, gehalten in der Preussischen Akademie der Wissenschaften in Berlin am 27 Februar 1920', *Aus der Welt der Wiss.*: 16–40, on : 20, 35, 38–9.
53. Ibid.: 37.
54. H. Weyl, 'Das Verhältnis der kausalen zur statistischen Betrachtungsweise in der Physik', *Schweizerische Medizinische Wochenschrift*, **50** (19 Aug. 1920): 737–41, reprinted in Weyl's *Ges. Abhl.*, **2**, 113–22.
55. H. Weyl, *Ges. Abhl.*, **2**: 116–17; 'Gravitation und Elektrizität', *Preuss. Akad. der Wiss.*, *Berlin, Sitzungsber.* (30 May 1918): 465–80, reprinted *Ges. Abhl.*, **2**: 29–42, and trans., with additional notes, in H. A. Lorentz, *et al.* *The Principle of Relativity*. London, 1923; reprinted New York, 1952: 201–16.
56. H. Weyl, *Ges. Abhl.*, **2**: 121–2.
57. H. Weyl, *Raum-Zeit-Materie*, 4th edn. Berlin, 1921: 283–4; *Space-Time-Matter*, trans. from 4th edn by H. L. Brose. London, 1922; reprinted New York, 1952: 310–12. The preface to this edition is dated Nov. 1920. Again, with a more precise statement of 'causality', in the 5th edn. Berlin, 1923: 286–7.
58. H. Weyl, 'Feld and Materie', *Annalen der Physik*, **65** (1921): 541–63, received 28 May 1921; reprinted in Weyl's *Ges. Abhl.*, **2**: 237–59, on 255.
59. H. Weyl, 'Was ist Materie?' *Naturwiss.*, **12** (11, 18, 25 July 1924): 561–9, 585–93, 604–11; *Ges. Abhl.*, **2**: 486–510, on 510.
60. H. Weyl, 'Massenträgheit und Kosmos. Ein Dialog', *Naturwiss.*, **12** (14 Mar. 1924): 197–204; *Ges. Abhl.*, **2**: 478–85.
61. W. Schottky, 'Das Kausalproblem der Quantentheorie als eine Grundfrage der modernen Naturforschung überhaupt. Versuch einer gemeinverständlichen Darstellung', *Naturwiss.*, **9** (24 and 30 June 1921): 492–6, 506–11; R. von Mises, 'Über die gegenwärtige Krise der Mechanik' (op. cit., note 143); W. Nernst, *Zum Gültigkeitsbereich der Naturgesetze*. Berlin, 1921, reprinted in *Naturwiss.*, **10** (26 May 1922): 489–95. This is Nernst's inaugural lecture as rector of the University of Berlin, 15 Oct. 1921.
62. R. von Mises, *Naturwissenschaft und Technik der Gegenwart* (op. cit., note 25): 19.
63. Ibid.: 30.
64. R. v. Mises, op. cit. (note 43), *Selected Papers*, **2**: 482, 487.
65. W. Wien, 'Kausalität und Statistik', *Illustrierte Zeitung* (Leipzig), Nr. 4169 (Feb. 1925): 192, 194, 196.
66. W. Wien, *Universalität und Einezelforschung. Rektorats-Antrittsrede gehalten am 28. November 1925*, Münchener Universitätsreden, Heft 5.

Munich, 1926; *Vergangenheit, Gegenwart und Zukunft der Physik. Rede gehalten beim Stiftungsfest der Universität München am 19. Juni 1926,* Münchener Universitätsreden, Heft 7. Munich, 1926. In his one other published academic address, *Goethe und die Physik. Vortrag gehalten in der Münchener Universität am 9. Mai 1923* (Leipzig, 1923), on p. 5, Wien had made a point of owning his allegiance to causality: 'Accustomed to seek the law of causality everywhere, the physicists ever and again give themselves great pains to uncover the reasons which led Goethe to his unfavorable attitude towards physics.'

67. W. Heisenberg, op. cit. (note 44). N. Bohr, 'Atomic theory and mechanics', lecture at the Sixth Congress of Scandinavian Mathematicians, 31 Aug. 1925, and revised before publication in *Nature*, **116** (5 Dec. 1925): 845-52; reprinted in Bohr, *Atomic Theory and the Description of Nature.* Cambridge, 1934: 25-51; quotation from: 34-5. The German text. 'Atomtheorie und Mechanik', appeared in *Naturwiss.*, **14** (Jan. 1926): 1-10.

68. W. Wien, *Vergangenheit, Gegenwart und Zukunft der Physik* (op. cit., note 66): 10.

69. Ibid.: 15, 18. We may perhaps read this as a veiled allusion to the breakdown which Bohr suffered in 1921 and which often threatened to recur.

70. V. V. Raman and Paul Forman, 'Why was it Schrödinger who developed de Broglie's ideas?' *Historical Studies in the Physical Sciences,* **1** (1969): 291-314.

71. E. Schrödinger, 'Quantisierung als Eigenwertproblem (Zweite Mitteilung)', *Ann. d. Phys.*, **79** (April 1926): 489-527.

72. Max Born, *Zur statistischen Deutung der Quantentheorie*, Dok. der Naturwiss., Abt. Physik, Bd 1, Armin Hermann (ed.). Stuttgart, 1962.

73. W. Heisenberg, 'Über der anschaulichen Inhalt der quantentheoretischen Kinematik und Mechanik', *Zeitschr. f. Phys.*, **43** (1927): 172-98, received 23 Mar. 1927: 197.

74. W. Heisenberg, 'Über die Grundprinzipien, der "Quantenmechanik"', *Forschungen und Fortschritte*, **3** (10 April 1927): 83.

75. A. Sommerfeld, 'Zum gegenwärtigen Stande der Atomphysik. Vortrag, gehalten auf Einladung der naturwissenschaftlichen Fakultät zu Hamburg', *Physikalische Zeitschr.*, **28** (1927): 231-9, received 18 Feb. 1927, reprinted in Sommerfeld's *Gesammelte Schriften*, **4** (Braunschweig, 1968): 584-92, on 588.

76. M. Born, *Vossische Zeitung*, 12 Apr. 1928.

77. N. Bohr, 'The quantum postulate and the recent development of atomic theory', *Nature*, **121** (14 Apr. 1928): 580-90, reprinted in Bohr, *Atomic Theory and the Description of Nature.* Cambridge, 1934): 52-91, on 580, 586, 590, and 54, 75, 91, respectively; German translation in *Naturwiss.*, **16** (1928): 245-57.

78. 'Epilogue: a Socratic dialogue. Planck-Einstein-Murphy', in Max Planck, *Where Is Science Going?* trans. James Murphy. New York, Norton, 1932: 201-21, on 201-5.

79. L. v. Bertalanffy, 'Über die Bedeutung der Umwälzung in der Physik für die Biologie', *Biologisches Zentralblatt*, **47** (Nov. 1927): 653-62, on 653-6.

5.4 John Hendry, Weimar Culture and Quantum Causality

The Forman theses

The result is . . . overwhelming evidence that in the years after the end of the First World War but before the development of an acausal quantum mechanics, under the influence of 'currents of thought', large numbers of German physicists, for reasons only incidentally related to developments in their own discipline, distanced themselves from, or explicitly repudiated, causality in physics.

Thus the most important of Jammer's theses – that extrinsic influences led physicists to ardently hope for, actively search for, and willingly embrace an acausal quantum mechanics – is here demonstrated for, but only for, the German cultural sphere.[1]

The quotation is from Paul Forman's 1971 paper on 'Weimar culture, causality, and quantum theory, 1918–27: adaptation by German physicists and mathematicians to a hostile intellectual environment'. This paper is widely recognized as being of the utmost importance, both as the first attempt to analyze the dramatic ideological changes that accompanied the development of quantum mechanics, and as a major milestone in science historiography. Despite widespread discussion in seminars, however, there is still no general agreement as to the validity or otherwise of its claims. The dominant feeling appears to be one of doubt, of 'case not proven', but since no-one has yet recovered the ground traversed by Forman opinions remain only vaguely grounded and, for the most part, private.[2] In view of this situation, and in the light, especially, of a recent follow-up paper in which Forman restates his contentions as a basis for further analysis,[3] there would appear to be some need for a thorough review of both the nature and the content of the arguments upon which these are based.

As the opening quotation suggests, Forman in fact has several quite different theses. His prime concern is with the introduction and reception of acausality in quantum theory, but he does not approach this problem directly. Adopting what he terms a 'sociological' approach, according to which 'present mental posture [is a] socially determined response to the immediate intellectual environment and current experiences', while individual psychological factors are ignored, he offers instead a 'causal analysis' from which, he claims, his main conclusion may be drawn. His first step is to demonstrate that the Weimar intellectual milieu was hostile to physics, and especially to causality:

I show that in the aftermath of Germany's defeat the dominant intellectual tendency in the Weimar academic world was a neo-romantic, existentialist 'philosophy of life', revelling in crises and characterized by antagonism toward analytic rationality generally, and toward the exact sciences and their technical implications particularly. Implicitly or explicitly, the scientist was the whipping boy of the incessant exhortations to spiritual renewal, while the concept – or the mere word – 'causality' symbolized all that was odious in the scientific enterprise.[4]

Symbolic of this milieu, and of particular influence, was Spengler's book, *The Decline of the West*, in which 'over and over again Spengler equates causality, conceptual analysis, and physics, and flays them across the stage of world history'.[5]

The second step in Forman's analysis is a demonstration of a general 'accommodation' of physicists and mathematicians to the milieu. He argues that 'most German mathematicians and physicists largely participated in, or accommodated their persona [*sic*] to, a generally Spenglerian point of view',[6] and that:

There was in fact a strong tendency among German physicists and mathematicians to reshape their own ideology toward a congruence with the values and mood of that environment – a repudiation of positivist conceptions of the nature of science, and, in some cases, of the very possibility and value of the scientific enterprise.[7]

Forman's third step is to demonstrate the existence of a widespread movement to dispense with causality in physics. Since, he claims, this movement cannot be related to any internal developments of physics, and particularly of quantum theory, it must be due (in a sociological sense) to the external influence of the milieu:

I am convinced, and . . . endeavour to demonstrate that the movement to dispense with causality in physics, which sprang up so suddenly and blossomed so luxuriantly in Germany after 1918, was primarily an effort by German physicists to adapt the content of their science to the values of their intellectual environment.[8]

Finally, if this is so for the rejection of causality in general, then it must be so for the particular case of the rejection of causality in quantum theory.[9]

The Weimar intellectual milieu
In analyzing the validity of Forman's contentions, we must look both at the individual stages of his argument and at its overall structure. With respect to his first claim, as to the nature of the Weimar milieu, there can be little doubt that his basic characterization of this as neo-romantic and existentialist is correct. The *Lebensphilosophie*, manifest already in pre-war literary, artistic and musical expressionism, was as he suggests transformed by Germany's defeat, and by the subsequent revolutions in Germany and Russia, into a dominant cultural force.[10] Insecure in the present, Germany sought to build a future out of her past, to return to the glory of Schiller and Goethe, and, rejecting the utilitarian standards of her conquerors, to re-establish herself on the cultural level as the leader of a new Europe. The previously lauded ideals of technology were now treated as suspect and tainted with Western (which is to say Anglo–French, Judaeo–Marxist) materialism.[11] Mathematics teaching in schools was reduced, and replaced by a return of the traditional emphasis on 'cultural' education.[12] The recent introduction of causal principles and mathematical procedures to the social sciences was condemned in favour of an intuitive approach.[13] In the life sciences, neo-

vitalist biology and Gestalt psychology reflected the return to romantic ideas of the individual.[14] And although it transcended the context of the Weimar milieu, Weyl's intuitionist attack on the traditional foundations of mathematics was also completely in tune with this milieu and shared, through the influence of Husserl's phenomenology, in its philosophical roots.[15]

The evaluation of Weyl's suggestion, however, entailed complex logical problems not easily related to the concepts of the milieu, and it is noteworthy that while both Weyl and his co-protagonist Brouwer were working outside Germany (in Switzerland and Holland respectively), the defence of classical mathematics, led by Hilbert, was conducted from Germany.[16] Thus, while a close connection between Weyl's ideas and the milieu cannot be doubted, a clear causal connection cannot easily be maintained either in respect of the origin of the ideas or in respect of their reception,[17] and this situation foreshadows the more complex one existing with respect to physics and causality. We are reminded that the history of ideas is rarely straightforward, and even in respect of the general attitude of the milieu to mathematics, physics and causality we should distinguish a subtlety that Forman has not. For while there were indeed many attacks upon mathematics and physics from outside these disciplines, they were in all cases attacks upon their *value*, rather than upon their *content*. Mathematics teaching in schools was reduced because it was not considered useful in the formation of the individual. A causal mathematical approach to the social sciences was attacked as being inapplicable to their particular subject matter. And both Troeltsch and Spengler attacked causality and physics as being symbolic of Western materialism.[18] But in none of these cases was there any conception whatsoever of physics and mathematics being able to adapt to the new ideals. Troeltsch restricted his criticisms to the context of the social sciences,[19] and if we look at Spengler's work we find that physics and causality were equated,[20] and that so long as physics survived at all in the present civilization it would continue to be characterized by the appropriate motif of the civilization, that is by causality. The civilization had now reached that stage where its causal aspect would have inevitably to resign dominance to that of its other leading motif, destiny, but this meant the death, not the transformation, of original physics.[21]

The existence of this equation, physics = causality, and of the related distinction between the value and content of the exact sciences does not affect the fact that both physics and causality were under strong attack from the milieu. There is no reason to suppose, moreover, that physicists did not react to the attack on the value of their discipline by adapting its content. But the distinction between value and content will be of some importance when we come to look at physicists' reactions, and it also draws attention to the important fact that physicists and mathematicians were to some extent *isolated from* rather than *attacked by* the forces of the milieu. Although recognizing the possibility of isolation from the milieu Forman in fact concentrates purely on adaptation or capitulation to it, and this inevitably produces a very one-sided picture.[22]

The general accommodation of physicists and mathematicians to the milieu

The distinction between value and content is most clearly apparent in the general reaction of physicists and mathematicians to the criticisms from the milieu. For while the many semi-popular addresses discussed by Forman do indeed establish a general accommodation to the milieu this was clearly, reflecting the nature of the criticisms, an accommodation of values. In these addresses physicists and mathematicians naturally used the language of the milieu, and justified the pursuit of their subject in terms that could be understood and appreciated by those who were questioning its cultural value. The discredited utilitarian motive for the pursuit of the exact sciences was replaced by the culture motive, Born's longing for community, Wien's idealism, von Mises's 'new intuition', and a general emotional need to strive after knowledge.[23] But while this indicates an awareness of the criticisms of the milieu, it implies no adaptation whatsoever of the content of the exact sciences, and even the accommodation of values is misleading. After all, the major motive of both mathematicians and physicists had never been a utilitarian one anyway.[24] The motives now cited publicly were just those that had always operated privately, and physicists were not so much forced into the new justification as released from the old. In this situation their accommodation can hardly be seen as a sacrifice of their ideals, or even as the first step on the road to such a sacrifice.[25]

If we are to seek a more meaningful adaptation of mathematicians and of physicists to the milieu, we must look to their work and to their private correspondence, but in these respects Forman has little to offer. Sommerfeld's use of the phrase 'number mysticism', which, incidentally, he was never allowed to forget, does represent a linguistic accommodation to the milieu, but the physics to which it was applied contains nothing that can be connected in any way with the milieu.[26] And the one piece of private correspondence that Forman uses in this context, a letter from Einstein to Born in January 1920, does not suggest to me the interpretation he puts upon it.[27]

The letter from Einstein, written in reply to one from Born that is now lost, deals mainly with social and political matters. Then, after these have been covered, Einstein declares that 'Spengler hasn't spared me either', and relates how one goes to bed convinced of Spengler's ideas, and wakes up seeing all the criticisms that can be made against them. In short, Spengler's book has been keeping Einstein awake and occupying his, and presumably also Born's, thoughts. Forman, however, interprets the quotation as meaning that Spengler's book included digs at both Einstein and Born; he admits in a footnote that he cannot find a dig against Einstein, but does not let this disturb his interpretations. Later on in the letter, Einstein declares that 'that business about causality causes me a lot of trouble too', that he does not quite have the courage of his convictions, but that he would be 'very unhappy to renounce *complete* causality'. From the context it is clear that Born has been discussing some of the internal arguments against causality in quantum theory, of which there were several, and that Einstein appreciated the force of these arguments, which troubled his firm conviction that strict causality would be maintained.[28] Einstein refers to 'Stern's interpretation', and we know that Stern had just started some experimental work with Gerlach on the phenomenon known as space quantization. If the quantum theoretical predictions were upheld (as they eventually were), this would impose a serious problem of interpretation, and, as Einstein and Ehrenfest later demonstrated, an internal threat to the causality ideal.[29] Omitting these considerations, however, Forman assumes that the topic is intimately related to the discussion of Spengler (no connection is made in the letter), and that Born was 'evidently ... willing to entertain the idea of acausality' for external reasons. Einstein's reluctance to give up complete causality is interpreted as a sop to Born, that he would be prepared to give up some of it.[30]

The causality issue
I have discussed Einstein's letter at some length because it indicates that when we come down to the content of physics, we must of necessity take into account internal as well as external considerations. Despite the criticisms that may be levelled against his analysis, Forman has succeeded in demonstrating that physicists and mathematicians were generally aware of the values of the milieu, and that this milieu did incorporate a marked hostility toward the causality principle. But when we come to the crucial claims, that there was a widespread rejection of causality in physics, and that there were no internal reasons for this rejection, then the weaknesses in his argument also become crucial. For there were strong internal reasons for the rejection of causality, and when these are taken into account, and Forman's supposed 'converts to acausality' critically re-examined, it would appear that the reaction of physicists to the causality challenge was far from being accommodation, that there may even have been a tendency to isolation.

Of Forman's eight supposed converts (a ninth, Sommerfeld, was added in the follow-up paper, but without any evidence for this being supplied),[31] two may be treated provisionally as irrelevant to his main argument.[32] Forman himself suggests that Exner's rejection of causality was independent of the milieu and of little contemporary relevance,[33] while Nernst, having himself got confused between the content of physics and its application, overcame this confusion by strongly defending the causality principle. In the original version of his paper, Nernst was faced with the common problem of physicists being attacked for asserting that the world was entirely causal, when they had in fact been concerned only with the causality of the mechanical, non-living world.[34] Trying to explain himself to a general audience he did begin, as Forman has shown, to accommodate to the anti-causal attitude of the milieu; but in the revised version a few months later he retracted fully from this accommodation, and insisted that causality was compatible with – and even necessary to – the ideals of the milieu.[35] Nernst's discussions show quite clearly that he was aware of the attitude of the milieu, and his confusion indicates the degree to which he felt pressurized by this attitude, but far from adapting to it he sought rather to adapt it to his own belief in causality. Of the remaining six of Forman's converts, Senftleben did not, as we shall see, reject causality either, while Schrödinger reconverted to it; Weyl, von Mises, Schottky and Reichenbach did all reject causality, but for reasons that were much more complex than Forman has indicated.

By the Weimar period, the concept of causality had long been a talking point in physics, and with the development of quantum theory it had already come under fairly strong internal pressure. The essential element of discreteness manifest in Planck's radiation law had promoted Jeans in 1910 and Poincaré in 1912 to ask whether differential equations were still the proper tool for physics[36]; and while Jammer's assertion in respect of Poincaré's question, that this was a mathematician's way of expressing a doubt in the causality principle, is in fact quite untenable, the fact that he could make it is indicative of the pressure in this direction.[37] Poincaré also suggested discrete units of time and the same year Planck, in presenting a new derivation of his law, felt obliged to stress explicitly that while he had been unable to stipulate a causal mechanism for the hypothesized discrete emission behaviour, such a mechanism was nevertheless supposed to exist.[38] In Bohr's atomic model of the following year the same situation was implicit.[39]

In discussing the light-quantum hypothesis in 1911, Einstein had argued that the only alternative to light-quanta was the rejection of strict energy conservation,[40] and although this was generally ignored at the time it could no longer be so after 1916, when he demonstrated that if energy and momentum conservation were to hold then light emitted from or absorbed by an atom would have to possess a directed momentum, i.e. to be particulate.[41] Since most physicists still felt the wave theory of light to be absolutely necessary, they had little option (unless, not actively working on the quantum theory, they could continue to turn a blind eye) but to consider seriously the possible rejection of energy conservation, and although this did not strictly imply a rejection of causality it clearly applied pressure in that direction.[42] By envisaging systematic energy changes or drawing a parallel with the statistical nature of the entropy law it would have been possible to accept a rational abandonment of strict energy conservation while refraining from too close an analysis of the implications for causality, but it would have been difficult to ignore altogether the prospects of such an analysis. And since it would have been difficult if not impossible to conceive of a workable system in which causality was retained but energy conservation abandoned, there must have been some pressure to link the two issues. Moreover, Einstein's 1916 paper was also notable for its introduction of probabilities governing the transitions between states of the Bohr atom, and although a causal mechanism behind the probabilities was again assumed, the explicit absence of such a mechanism was naturally provocative,[43] especially as there was no prospect whatsoever of finding one.[44]

Following these developments by Einstein there was, according to Heisenberg, a general acceptance among mainstream quantum physicists that energy conservation would have to be abandoned:[45] but this attitude was not expressed in print until 1921. Before that time we know of its certain existence only in the correspondence between Bohr and C. G. Darwin, where it was related to, but not equated with, the question of causality. Darwin, who seems to have been basing his exposition consciously on Bohr's ideas, wrote that

It may be that it will prove necessary to make fundamental changes in our ideas of space and time, or to abandon the conservation of matter and electricity, or even in the last resort to endow electrons with free will.[46]

The 'last resort' was never actively considered, as Darwin concentrated in the sequel solely upon the rejection of energy conservation[47]; Bohr, who agreed that 'conservation of energy seems to be quite out of the question', did not mention it at all.[48] But in a paper published the previous year he had already talked provocatively of 'spontaneous' transitions, and had emphasised repeatedly the breakdown of classical mechanics in the quantum context[49]; he was clearly aware of the pressures on causality, pressures that were as yet almost wholly internal, imposed by the problems and paradoxes of quantum theory.[50]

As Forman indicates, the first significant rejection of causality in physics, by Weyl in 1920, was not related explicitly to these problems.[51] But while here as in his mathematics the connection of Weyl's views with those of the milieu is clearly a close one, it is again far from straightforward. The primary external influence upon his physical ideas appears to have been an intuitionist philosophy that predated and transcended the Weimar milieu, and while his insistence, from 1917, on treating philosophy and physics as intimately related may perhaps be linked to the milieu, we should note that he himself appears to have treated the philosophy as secondary to physical, and especially mathematical, results.[52] His switch from the advocacy of a causal to that of an acausal field theory was moreover heavily conditioned by purely internal considerations.

In 1917–18, concurrent with his work on the foundations of mathematics, Weyl had developed and published a unified theory of gravitation and electricity, based on a further generalization of Einstein's 'general' theory of relativity.[53] Weyl's theory was regarded by his fellow relativists as being brilliant, but wrong, and in 1919 Pauli published some criticisms of it.[54] He objected that, contrary to experience, it was perfectly symmetric with respect to positive and negative charges,[55] and more fundamentally that it offered no reason why there should be only one mass corresponding to each type of electricity and, as a pure field theory, no explanation of the cohesive structure of matter. Since the only forces were those *produced by* the electrons and protons, the internal structure of these particles could not be accounted for: any attempt at a complete theory (such as Weyl's claimed to be) would have to incorporate complex properties of matter independent of the electromagnetic/gravitational field.[56] In May 1919, replying to Pauli's as yet unpublished views, Weyl wrote to him that he thought the non-equivalence of positive and negative electricity would reduce to that of past and future, or to the unidirectionality of time, but that neither problem could yet be solved.[57] As to his advocacy of a pure field theory, Pauli should not accuse him of being a dogmatist: he did not think he had found the philosophers' stone, and he was himself quite sure that there was something in matter independent of the field. In December, having seen Pauli's criticisms in print, Weyl wrote to him again, expanding on his views.[58] He held the past–future distinction to be more basic than that between the two electricities, and to completely determine the latter through the world geometry or field. Despite this determination, however, physics was fundamentally statistical and acausal, the acausality arising from something existing in matter, independent of the field, and to be thought of in terms of 'independent decisions'.[59] Later, he was to make it clear that, as his terminology suggests, the field was to be thought of as the prime *physical* reality.[60]

To fully comprehend Weyl's line of thought will require much greater research, but we may already see from the above that his rejection of causality, though fundamentally a metaphysical rather than an empirical decision, was nevertheless strongly rooted in a physical problem complex.[61] And although the quantum theory was not actually mentioned it was relevant, for Weyl's position stemmed from his defence of what was in effect a pure field theory against Pauli's insistence on the primacy of particulate matter. The problem was no other than the wave-particle (acausal–causal) problem of quantum theory, transferred from the context of light to that of matter. Moreover, Weyl's intuitionism was no more 'external' than was Pauli's epistemological operationalism.[62]

Although superficially Pauli and his co-believer Eddington may be placed with Einstein in the causal–particle camp, their positions were in fact less simple. While Einstein sought to derive particles from a pure field theory they, like Bohr, were actually searching for a far more fundamental change in our concepts. Although opposed on the field–particle issue, both Pauli and Bohr were agreed on the need for a change in our very concepts of space and time, a change that might well make the causality issue redundant; and as Pauli was able to convince both Born and Heisenberg at least partly of his position, it was one of some importance for German quantum physics – it was indeed to be the main factor in the birth of the new quantum mechanics.[63] Any attempt to polarise German quantum physicists in terms of causality is therefore inevitably somewhat artificial.

Outside this main stream of quantum physics the issue nevertheless remained a live one, and in 1921 Schottky and von Mises both published rejections of causality in the quantum theoretical context. As Forman has shown, both may be related to the milieu: von Mises expressed the popular version of his conclusions in a Spenglerian framework, and Schottky's presentation had a strongly existentialist tone.[64] But once again there were also – despite Forman's categorical assertion to the contrary – important internal factors.

During and immediately after the war there had been relatively little open debate on the fundamental wave–particle issue of quantum theory.[65] But at the Solvay Congress of 1921 the issue was re-opened by two papers arguing strongly for the necessity of the light–quantum concept, one by Maurice de Broglie,[66] and the other by Millikan, whose views represented a dramatic conversion after many years rejection of this concept.[67] In another paper to the same congress, Bohr suggested, for the first time in public, the abandonment of energy conservation as a preferable alternative to the light-quanta.[68] Von Mises's discussion of causality, which followed a few months later, was based on Bohr's pet theme, the failure of classical mechanics in quantum theory, and may well have been related to these developments. Schottky's analysis was explicitly related to Einstein's unsuccessful efforts at the development of a classical causal field theory in which the light-quanta and material particles appeared as singularities, and since Einstein himself seems to have put energy non-conservation and acausality on a similar footing, Bohr's paper could have been influential on Schottky also.[69] Einstein's position was quite untenable in the eyes of most of the mainstream quantum physicists, and Schottky's decision to reject the possibility of a causal pure field theory was a wholly reasonable one.[70] His choice of alternative, involving the rejection of causality rather than that of the fundamental space–time concepts, may be seen simply as a choice of Weyl's rather than Pauli's path, and need not necessarily have been related to the milieu.

In 1922–23 there was only one further challenge to the causality principle in physics, by Schrödinger; but there was in this period a widespread and open acceptance that energy conservation might have to be abandoned. Darwin and Bohr advocated this course repeatedly in their publications[71]; Sommerfeld indicated his sympathy with it in his influential textbook[72]; Born and Heisenberg adopted it as a provisional conclusion in some work on phenomena akin to that of the 'space quantization' experiments of Stern and Gerlach, and Einstein and Ehrenfest considered it, albeit reluctantly, in the context of these experiments.[73] Schrödinger's rejection of causality in the fall of 1922 was linked explicitly with the philosophical views of Exner, whose influence had already prompted him in 1918 to consider the general philosophical problem of causality;[74] but it was also tied in with the conservation issue, as well as with the response to Weyl's unified theory. On joining Weyl in Zurich in 1921 Schrödinger had shown himself sympathetic to this theory by attempting to apply it to quantum problems.[75] The next year he approached the problem of the Doppler effect in quantum theory from a purely wave theory of light,[76] and writing to Pauli he stressed his rejection of strict energy conservation and his conviction that there was as yet no other way out of the quantum paradox.[77] He thought that one might replace the quantum identity of energy and frequency by an equivalence between the effect produced by a material particle of given energy and that produced by a spherical wave of equivalent frequency, but without the energy necessarily being attributed to the wave, and thus without the wave having to be localized as a light-quantum. This would not be possible for a conservative mechanical system, but the devil knows, he argued, whether atomic systems are strictly causally determined, let alone conservative mechanical systems.[78] In his inaugural lecture a few weeks later, he rejected causality outright.[79]

Schrödinger's attitude to light-quanta at this time was quite close to Bohr's, and though he himself did not yet reject causality Bohr did stress in his papers of the period that a classical space–time description would have to be abandoned, and that there was, 'in the present state of science', no causal description of quantum phenomena.[80] In 1923 these two points were linked by H. A. Senftleben, who noted that whereas in the past one used to regard natural phenomena as arising statistically from causal microscopic changes, one now had to describe these changes themselves statistically.[81] If the causal requirement were defined as *Given a situation A in space–time we may determine a later situation B*, then, argued Senftleben, the quantum theory as it stood did not allow the premise, never mind the conclusion: 'Planck's constant "h" limits in principle the possibility of describing a process in space and time with arbitrary accuracy.' A precise situation A might exist, but it could not be described by quantum theory, and that theory could therefore be termed neither causal nor anticausal. Whether Senftleben's analysis was inspired by Bohr's work, or whether it was perhaps related to Planck's naive and unsound attempt to avoid discreteness in quantum theory by treating the quantum levels as averages over ranges of energy, the precise values within which were unknown, is not clear.[82] But it was neither a rejection of causality, nor, as Forman would also have it, the work of a 'quasi-crank'.[83]

In a manuscript that seems to date from the end of 1923 Bohr himself, apparently trying to break away from the classical mechanical framework through a phenomenological description of dispersion phenomena, suggested that the wave-particle problem was 'evidence of the unavoidable difficulties of giving a detailed description of atomic processes without departing essentially from the causal description in space and time that is characteristic of the classical mechanical description of nature'.[84] A little later, in the famous Bohr–Kramers–

Slater paper, he decided to 'abandon any attempt at a causal connexion between the transitions in distant atoms'.[85]

It is not clear, however, whether even in the Bohr–Kramers–Slater paper Bohr intended his rejection of causality to be anything other than provisional. It was certainly not emphasized, and his path appears to have been chosen on heuristic grounds as that which offered the most immediately promising opportunities for the developing theory of dispersion.[86] Moreover, only Einstein and Schrödinger, in responding to the new development, even mentioned the causality issue, Schrödinger welcoming Bohr's move,[87] and Einstein finding it 'quite intolerable that an electron should choose of its own free will, not only its moment to jump off, but also its direction' – a formulation that reflects his own prime concern with energy and momentum conservation.[88] Apart from Pauli, who thought the changes insufficiently fundamental, the opponents of the theory (including Slater, van Vleck, Compton, Stoner, Ehrenfest, Sommerfeld and, according to Pauli, 'many physicists, perhaps even the majority') argued either from the necessity of the light-quantum concept or from the tenuous and unproven nature of Bohr's assertions in general.[89] Kramers, Born, Ladenburg, Reiche and, after initial doubts,[90] Heisenberg, accepted the new technique, but in a spirit of pure positivism, as a new formula for saving the appearances.[91] The split was not therefore dictated explicitly by attitudes to causality, and these attitudes cannot even be brought in on an implicit level, for apart from Pauli the physicists divided simply as to whether they were or were not concerned with the technical problems to which the theory related.[92]

The Bohr–Kramers–Slater theory was not, then, the revolutionary rejection of causality that it is sometimes claimed to be, and the issue was not raised again until after the introduction of a new quantum mechanics when, following the experiments of Geiger and Bothe, the theory had been effectively and for the most part gladly dismissed.[93] The new quantum mechanics, in Schrödinger's as well as in Heisenberg's form, involved the rejection (primarily as a result of Pauli's influence) of any space–time description of the path of an electron inside the atom, and this changed the internal situation fundamentally.[94] Following Heisenberg's theory of 1925, though apparently unconnected with it, the causality principle was rejected by Reichenbach; and following the now determinist Schrödinger's theory of 1926 it was finally dismissed from the new quantum mechanics by Born.

Though Reichenbach's motivations are as yet obscure they do not appear to be directly related to physics, and they may well have reflected the influence of the milieu.[95] More important, however, as being the first rejection of causality of any real historical significance in quantum theory, was Born's decision. Having set himself to the interpretation of Schrödinger's new wave mechanics, Born had found Schrödinger's own classical interpretation to be untenable, as even for a free electron the wave-particle problem recurred and a space–time description proved impossible.[96] Announcing his own interpretation of Schrödinger's electron wave function in terms of a probability amplitude, he observed that:

> Here the whole problem of determinism presents itself. From the standpoint of our quantum mechanics there is no quantity which remains causal in the case of an individual collision effect; but in practice also we have no grounds to believe that there are inner eigenstates of the atom that stipulate a determined collision path. Should we hope to discover such eigenstates later (such as phases of internal atomic motions), and to determine them for the individual case? Or should we agree that the agreement of theory and experiment on the impossibility of giving a stipulation of the causal lapse is a pre-established harmony, which rests on the non-existence of such stipulations. My own inclination is that determinism is abandoned in the atomic world. But that is a philosophical question, for the physical arguments are not conclusive.[97]

It was with this inclination, which was soon confirmed by internal considerations, that the rejection of the classical causality principle became a serious possibility and, within only two years, a generally accepted feature of quantum theory. Ironically, though Forman excluded Born's acausality from his thesis as having been influenced by the new quantum mechanics, it may in fact have been related to the milieu. His wife was certainly influenced by this milieu, and had long mocked Born's scientific determinism.[98] while the context of a probabilistic interpretation suggests that Reichenbach's views on the primacy of probabilistic laws may well have been relevant.[99] Spengler's prediction that physics would degenerate into statistics may also have been in Born's mind,[100] for he later emphasized strongly the central role of statistics in his theory, referring to the 'close connection between mechanics and statistics',[101] and even to 'a fusion of mechanics and statistics'.[102] Finally, Weyl too could have been influential, for he had played an active role in the development of both matrix mechanics (with Born) and wave mechanics, and may well have been consulted by Born in the latter context.[103]

Apart from the possible external factors, however, there were also internal ones. Quantum theory had long involved an element of uncertainty, in the location of an orbit, the moment of a transition, etc.; but this had always been a case of uncertain conclusions following from uncertain data, as noted by Senftleben. In Born's analysis of the atom–electron collision, uncertain conclusions (the atom in a superposition of states and the electron spread throughout space) followed from apparently definite data on the state of the atom and motion of the electron. The existence of hidden microscopic parameters could have changed this, of course, but the theory did appear in some way to create uncertainty, and Born could well have deduced from this that physics was itself uncertain. Born's rejection of causality, equivalent as it was to the rejection of any relevant microscopic coordinates, was also closely tied to his insistence that his theory was final and complete, whether microscopic coordinates existed and were measurable or not, and this was the most remarkable feature of his presentation.[104] Heisenberg had built his theory upon quantities that were in principle observable, but Born restricted himself to those that were in practice observable, at the time of writing, and asserted that no future experiments could change the theory that he had evolved. This somewhat dangerous attitude appears to have stemmed directly from his assertion that the physical interpretation must follow uniquely from the mathematical formulation of the theory, and the assertion itself from a battle waged in Göttingen the previous winter between the advocates of a mathematical approach to the new quantum mechanics (Born, Hilbert, Weyl) and those of a physical approach (Heisenberg, Frank).[105]

Summary and discussion

If we accept the causality problem simply as Forman has expressed it, then the reaction of German physicists to their milieu would appear to have been closer to isolation than to adaptation, and the rejection of causality in quantum theory motivated 'primarily' by internal considerations. Despite very strong internal pressures on the causality principle this was rejected, before the arrival of the new quantum mechanics, by only a handful of physicists. Most of the rejections were closely linked with internal developments, and even if we allow the milieu to have been active in all cases the converts to acausality still represent only a miniscule proportion of German physicists as a whole. Whether the reaction of the vast majority took the form of isolation or neutrality is not clear, but they were certainly not ardently hoping or actively searching for an acausal quantum mechanics.

We may, then, dismiss Forman's thesis in his own terms, but only if we are prepared to accept the naive level of his own arguments; and this, it seems to me, we cannot do. For overwhelming the detailed evidence given by Forman and above are a range of serious problems relating to the causality issue in general, to the overall structure of Forman's demonstration, and to the whole question as to what is meant by an 'influence of the milieu', how this relates to the wider concept of sociological causation, and how this concept may be meaningfully applied to the history of science.

The most immediate problem to arise concerns the definition of causality and, harder still, of its rejection. Forman is quite correct to assume that causality was generally interpreted in terms of classical determinism or (equivalently for his purposes) the Machian concept of lawfulness[106]; but this is no justification for his consequent handling of the causality debate in simple black-and-white terms. To important contributors such as Weyl and Reichenbach, both of whom were well trained in the subtleties of philosophy of science, the issue was complex and the classical position naive.[107] And if we assume that those physicists who did not elaborate on their conceptions of causality were thinking simply in terms of the classical concept we are led into another difficulty. For in dealing with causality on this level we are dealing with a contentious and, in part at least, emotive issue; their reasoning conditioned by personal and unformalised philosophical or religious ideas, these physicists did not necessarily reach or accept what we might agree in retrospective detachment to be a logical conclusion. Two physicists might agree completely as to the definition of causality, but differ completely as to whether it had or had not been rejected in a theory. In particular the same situation might be seen by one as a rejection of causality and by the other as a temporary absence of a causal theory, and to make matters worse these two views might be expressed in exactly the same words.[108]

Most important still, the causality criterion was only one aspect of the prevailing scientific ideology, and though the most emotive aspect it was not the most basic. To most of those concerned with the fundamental problems raised by the quantum and relativity theories it was clearly secondary to the criterion of a consistent description of phenomena in ordinary space and time, and to that of an objective (observer-free) theory, and this was true in particular for those physicists most closely involved with the development of a new quantum mechanics, for Pauli, Heisenberg, and Bohr.[109] When this new mechanics came it was characterized most significantly by its rejection of the two latter criteria, and though the acausal interpretation applied by Born was carried into the popular conception of the new theory, the quantum physicists themselves were quick to emphasise its naivety. Both Heisenberg and Bohr, whose views formed the basis of the generally accepted Copenhagen interpretation that emerged in 1927, stressed that the causality concept was only applicable at all in the context of the traditional space–time concepts, and that these concepts themselves could not be consistently applied to the new theory.[110]

Subtleties such as those noted with respect to the causality concept make a mockery of Forman's attempt at a 'causal' analysis of the phenomena with which he is concerned. But even without these subtleties the attempt would anyway have been wrecked by sheer lack of information. With the information available, Forman has succeeded in demonstrating an influence of the milieu upon physicists' attitudes to causality, and were he to adopt a suitable concept of historical causation he could even assert quite reasonably that the attitudes were in some (weak) sense 'caused' by the milieu. But it is clear from the importance he attaches to the absence of internal motivations and from his insistence on the milieu as 'the primary' cause that his concept of historical causation is in fact a very strong one, and as such it must be supported by much more than the emotive value-laden discussion of examples that he offers.[111] In particular we are entitled to ask how German physicists' attitudes to causality compared with those of non-German physicists, whether the number of German physicists known to have rejected causality is significant in comparison with the total population – or whether, bearing in mind the non-uniform nature of this population, it represents only an atypical minority – and whether conclusions drawn for this large population may reasonably be carried over (as his argument requires) to the small and apparently un-representative group of quantum physicists. In setting out to demonstrate his thesis rather than to test it, Forman does not answer any of these questions; were he to try to, his argument would collapse.

Forman is also unforthcoming as to his concept of the influence of the milieu, and this raises problems of both a special and a general nature. To ask whether Schrödinger, for example, an Austrian working in Switzerland, may be included in the milieu is somewhat pedantic; but it does lead to the observation that the population with which Forman is concerned is far from being uniform, and that the supposition of a uniform reaction may therefore be misleading. We might for example expect the many Jewish physicists, who were already isolated from the milieu by their religion, to have reacted differently to this milieu from their Christian colleagues.[112] Another problem concerns individual influences such as those of Husserl on Weyl, Exner on Schrödinger, the nineteenth-century existentialist philosopher Kierkegaard on Bohr, or the Weimar existentialist theologian Karl Barth on Kramers: if a physicist was influenced by someone who shared the attitudes of the milieu, may he necessarily be said to have been influenced by the milieu? Even neglecting internal influences, the concept of social causation is far more complex than Forman would have us believe.[113]

When internal considerations are taken into account as well then it becomes immediately clear that no one set of influences – internal, social, philosophical, psychological, etc. – can be taken independently of the others, and that each physicist's reaction to a given problem will be determined by a complex of motivations, many of them immune to historical objectification. It may be that the development of scientific knowledge is ultimately anthropologically (sociologically or psycho-logically) determined, but if so the determination will probably operate at a level that is at present beyond objectifiable demonstration, through our very mode of thinking rather than through explicit conceptual-izations.[114] As the persistent failure of the advocates of a sociologically determined science to prove their case suggests, it will not be on the explicit level with which they, and Forman, have been concerned.[115] On this explicit level much can be interpreted in terms of social factors, and Forman's work has clearly demonstrated the paucity of a wholly internal treatment of issues such as that of causality. Physicists *were* influenced by the crisis-consciousness of post-war Europe and by the attitudes characteristic of the Weimar milieu. On the other hand, Forman's work has also demonstrated the dangers of a purely external treatment and the poverty of any naive social reductionism. Overall it has emphasized that if we are to talk meaningfully about the causal development of science we should not argue yet over which influences are the 'most important', but should first try to understand how they combine with, permeate, and act together with each other.

Source: History of Science (1980) forthcoming.

Notes

1. Paul Forman, 'Weimar culture, causality, and quantum theory, 1918–1927: adaptation by German physicists and mathematicians to a hostile intellectual environment', *Historical Studies in the Physical Sciences*, iii (1971): 1–116; 3. I shall refer to this paper as *Weimar*.

2. The only public response by a historian of science has been by Jon Dorling in a short address to the British Society for the History of Science in July 1976, and even this has not been published; but see 10 below.

3. Paul Forman, 'The reception of an acausal quantum mechanics in Germany and Britain', to be published in the proceedings of the symposium on 'The reception of unconventional science by the scientific community', American Academy for the Advancement of Science, February 1978. See note 115 below.

4. Forman, *Weimar*, 4.

5. Ibid., 33.

6. Ibid., 55.

7. Ibid., 7.

8. Ibid., 7.

9. For a clarification of this point of the argument I am indebted to a private communication from Paul Forman.

10. This picture of the milieu, which is in fundamental agreement with Forman's, is based largely on the historical studies he cites G. Lukács, *Die Zerstörung der Vernunft*. Berlin, 1954; K. Sontheimer, *Antidemokratisches Denken in der Weimarer Republik*. Munich, 1962; P. Gay, *Weimar Culture*. New York, 1968; and especially F. Ringer, *The Decline of the German Mandarins*. Cambridge, Mass., 1969. See also H. Lebovics, *Social Conservatism and the Middle Class in Germany, 1914–1933*. Princeton, 1969, and W. Laqueur, *Weimar, a Cultural History 1918–1933*. London, 1974; Laqueur rejects Forman's thesis on the grounds that the physicists did not split along clear ideological lines (Ch. 6).

11. In Spengler's *The Decline of the West*, the term 'Western' is ambiguous, relating on one hand to the overall decline of Western civilization, in which Germany is included, and on the other hand to the decline of Western, or non-German, ideals within that civilization. In general usage, however, it had simply the latter, derogatory meaning, and it was on this meaning that the polemic value of Spengler's work rested; indeed the rejection of Western in favour of German values was perhaps the central feature of the Weimar milieu.

12. See Ringer, op. cit. (note 10). Mathematics had only become established in the school curriculum toward the end of the nineteenth century.

13. See especially E. Troeltsch, 'Die Revolution in der Wissenschaft', *Schmoller's Jahrbuch*, xlv (1921): 1001–30.

14. There would appear to be some scope for a sociological study of the life sciences in Weimar Germany, but such a study has not, so far as I know, been undertaken.

15. Forman, *Weimar*, 60–1, 76–7; H. Spiegelberg, *The Phenomenological Movement*. The Hague, 1965.

16. C. Reid, *Hilbert*. New York, 1969.

17. Only the style of the debate may be put down with some confidence to the milieu: see Forman, *Weimar*, 61.

18. Troeltsch, op. cit. (note 13); O. Spengler, *Der Untergang des Abendlandes*.

Umrisse einer Morphologie der Weltgeschichte. Vol. 1: *Gestalt und Wirchlichkeit.* Munich 1918. Revised edition (33rd) 1923. Translation of revised edition, *The Decline of the West.* Vol. 1: *Form and Actuality.* New York, 1926.

19. See Forman, *Weimar*, 18.
20. See Forman, *Weimar*, 30–7; Spengler, op. cit. (note 18), especially section XI.
21. Ibid., section XI, especially pp. 380ff of translation.
22. It is noticeable that the historians of the Weimar milieu, op. cit. (note 10), all exclude science from their considerations; Gay (preface) and Laqueur (Ch. 6) treat it explicitly as isolated from the milieu.
23. Forman, *Weimar*, 47–58.
24. This was true, moreover, for the staunchest of determinists, Einstein and Planck among them, as much as for anyone else: it always had been true, and still is so.
25. The exception here would appear to be Doetsch: see Forman, *Weimar*, 52.
26. Ibid., 49–50. Concerning the repercussions see for example Pauli to Bohr, 31 Dec. 1924, where Sommerfeld's department is referred to jokingly as an institute for number mysticism: W. Pauli, *Briefwechsel, Bd. 1,* (eds) A. Hermann *et al.* New York, 1979, letter no. 79.
27. Forman, *Weimar*, 71–2; Einstein to Born, 27 Jan. 1920, English translation in M. Born and A. Einstein, *The Born–Einstein Letters.* New York, 1971, letter no. 13.
28. This conviction was apparently shared then by Born: see M. Born, *Physics in my Generation.* New York, 1956, and Einstein to Mrs Born, 1 Sept. 1919, letter no. 9 of op. cit. (note 27).
29. See Born's notes to the letter in op. cit. (note 27), and M. Jammer, *The Conceptual Development of Quantum Mechanics.* New York, 1966, pp. 133–4, for the technical background to this work.
30. The German original of Einstein's statement is however 'Ich verzichte aber sehr ungern auf die *vollständige* Kausalität', and this appears to allow for no such ambiguity.
31. Forman, op. cit. (note 3). Sommerfeld's name is simply included among those of the 'converts' in a summary of Forman's previous conclusions.
32. Though highly relevant to the general problem of the interaction between physicists and the milieu, they cannot be claimed to support his strong thesis, that physicists rejected causality in response to the pressures of that milieu.
33. F. Exner, *Vorlesungen über die physikalischen Grundlagen der Naturwissenschaften.* Vienna, 1919; Forman, *Weimar*, 74–6.
34. W. Nernst, *Zum Gültigkeitsbereich der Naturgesetze.* Berlin, 1921; Forman, *Weimar*, 84–5.
35. W. Nernst, 'Zum Gültigkeitsbereich der Naturgesetze', *die Naturwissenschaften,* x (1922); 489–95; 494–5, Forman, *Weimar*, 85–6.
36. J. H. Jeans, 'Non-Newtonian mechanical systems and Planck's theory of radiation', *Philosophical Magazine,* xx (1910): 943–54; H. Poincaré, 'L'hypothèse des quanta', in his *Dernières Pensées.* Paris, 1913, pp. 75–6.
37. Jammer, op. cit. (note 29), : 171. Poincaré had thought his suggestion would be amenable to the determinist Planck and there can be no doubt that he, like Jeans, intended no challenge to the causality principle. Jammer supports his assertion by citing Poincaré's discussion of the role of

chance in physics in 1904, but we should note that Poincaré then defined chance as a 'complex assemblage of causes': H. Poincaré, *La Valeur de Science*. Paris, 1904, : 110.

38. M. Planck, 'Ueber die Begründung des Gesetzes der schwarzen Strahlung', *Annalen der Physik*, xxxvii (1912): 642–56; 644.

39. N. Bohr, 'On the constitution of atoms and molecules', *Philosophical Magazine*, xxvi (1913): 1–25, 476–502, 857–75.

40. P. Langevin and M. de Broglie (eds.), *La Théorie du Rayonnement et les Quanta: Rapports et Discussions de la Réunion tenue à Bruxelles, du 30 octobre au 3 novembre 1911 sous les auspices de M. E. Solvay*. Paris, 1912, : 436.

41. A. Einstein, 'Zur Quantentheorie der Strahlung', *Physikalische Zeitschrift*, xviii (1917): 121–8.

42. Despite considerable evidence in support of the need for discrete and localised absorption of light – and hence for the existence of light-quanta – the evidence in support of the wave theory was overwhelming, and rather than incorporating a fundamental contradiction in their theories physicists naturally preferred to stick to the established theory: this may have been limited in application, but it was at least consistent. On this and the wave-particle issue in general see J. Hendry, 'Attitudes to the wave-particle duality of light and quantum theory, 1900–1920', *Annals of Science*, xxxvii (1980): 59–79.

43. While making it clear that causality was retained, Einstein did in fact adopt the terminology of 'chance'.

44. The situation was fundamentally different from that in statistical mechanics, for example, where the probabilities were derived from an assumed causal behaviour. In this case all attempts to specify an underlying causal behaviour had failed – and were to continue to do so.

45. Interview with W. Heisenberg, 1963, Sources for History of Quantum Physics archive: see T. Kuhn *et al.*, *Sources for History of Quantum Physics*. Philadelphia, American Philosophical Society, 1967. Heisenberg's reference was to the physicists at Munich (under Sommerfeld) and later at Göttingen (under Born and Frank).

46. C. G. Darwin, manuscript draft of a critique on the foundations of physics (July 1919), *Sources for History of Quantum Physics*. See also Darwin to Bohr, 20 July 1919, and Bohr to Darwin, July 1919, ibid.

47. Writing to Bohr, 20 July 1919, ibid., Darwin found the 'case against energy conservation quite overwhelming'.

48. Bohr to Darwin, July 1919, ibid.

49. N. Bohr, 'On the quantum theory of line spectra, part I', *Kongelige Danske Videnskabernes Selkabs Skrifter,* Series B, IV (1918–22): 1–118; 7. Jammer, op. cit. (note 29), : 114, found the terminology sufficiently provocative to interpret 'spontaneous' as 'acausal', but Bohr himself later defined it as 'without any assignable external stimulation', leaving plenty of room for causes internal or as yet unknown: N. Bohr, 'On the application of the quantum theory to atomic structure: Part I, The fundamental postulates', *Supplement to Proceedings of the Cambridge Philosophical Society* (1924): 24.

50. For a full discussion of the development of Bohr's ideas see K. M. Meyer-Abich, *Korrespondenz, Individualität, und Komplementarität*. Wiesbaden, 1965.

51. Forman, *Weimar*, 76–80; H. Weyl, 'Das Verhältnis der kausalen zur statistischen Betrachtungsweise in der Physik', *Schweizerische Medizinische Wochenschrift*, i (1920): 737–41, and *Raum-Zeit-Materie*, 4th edn: Berlin, 1921, : 283–4.

52. See Forman, *Weimar*, 76, footnote 176. Weyl was primarily a mathematician, using philosophical ideas for inspiration (on the foundations of mathematics) and for connecting mathematical results (of his unified theory) to the properties actually found in the physical world. See note 61 below.

53. H. Weyl, 'Reine Infinitessimalgeometrie', *Mathematische Zeitschrift*, ii (1918): 384–411; 'Gravitation und Elektrizität', *Sitz. Preuss. Akad. Wissenschaften*. (1918): 465–80; 'Eine neue Erweiterung der Relativitätstheorie', *Annalen der Physik*, lix (1919): 101–33; *Raum-Zeit-Materie*.Berlin, 1918. In progressing from the special to the general theory of relativity, Einstein had abandoned the assumption of Euclidean geometry that the directions of vectors at different points in space–time could be directly compared; in the Riemannian geometry that resulted, the relative direction of vectors at two points became dependent on the choice of paths joining the points, and the parameters defining this choice were associated with those of the gravitational field. In Weyl's theory, the same argument was applied also to the lengths of the vectors, and the parameters appropriate to this further degree of freedom were identified with those of the electromagnetic field.

54. W. Pauli, 'Zur Theorie der Gravitation und der Elektrizität von Hermann Weyl', *Physikalische Zeitschrift*, xx (1919): 457–67; see also W. Pauli, *Theory of Relativity*. London, 1958, : 206. For other responses to Weyl's theory, all of which combine criticism of his physical conclusions with extravagant praise of the underlying mathematical theory, see the letters from Mie to Weyl, 26 Oct. 1918, Einstein to Weyl, 8 Mar. 1918, Sommerfeld to Weyl, 3 July 1918, and Eddington to Weyl, 18 Aug. 1918, all in the archive of the Eigener Technische Hochschule, Zurich.

55. This criticism no longer held after the discovery of the positron, but it appeared to be valid at the time.

56. 'The continuum theories make direct use of the ordinary concept of electric field strength, even for fields in the interior of the electron. This field strength is however defined as the force acting on a test particle, and since there are no test particles smaller than an electron or a hydrogen nucleus, the field strength at a given point in the interior of such a particle would appear to be unobservable by definition, and thus be fictitious and without physical meaning.' This argument, cited from Pauli, op. cit. (note 54, 1958): 206, was included among Pauli's original criticisms and was referred to by Einstein, writing to Born on 27 Jan. 1920, op. cit. (note 27).

57. Weyl to Pauli, 10 May 1919, op. cit. (note 26), letter no. 1.

58. Weyl to Pauli, 9 Dec. 1919, ibid., letter no. 2.

59. Ibid. (my translation) 'That modern physics which finds no place in "lawful" or "field physics" can still be right. For I am quite convinced that statistics is something independent, opposed in principle to Causality, to "Law"; because it is in general paradoxical to introduce a field as some kind of prior existent. I imagine that the field physics really plays only the role of "world geometry"; in matter there is something else different, real, which is not causally comprehended, but which is perhaps to be thought of in terms

of independent decisions, and which we treat in physics through statistical calculation. It is quite possible that we must attribute to this the nature of the difference between past and future, between positive and negative electricity.'

60. H. Weyl, 'Was ist Materie', *die Naturwissenschaften,* xii (1924): 561–9, 505–93, 604–11; Forman, *Weimar,* 79.

61. In broad terms we may say that Weyl saw his mathematical field theory as being of prime importance. This, however, offered no reason as to why the real (Einstein) world should be preferred to any other world consistent with the theory, and it was this gap that Weyl's philosophical arguments were intended to fill.

62. For a full statement of Pauli's philosophical position, which was close to that later adapted by Eddington, see Pauli to Eddington, 20 Sept. 1923. op. cit. (note 26), letter no. 45.

63. See J. Hendry, 'The conceptual origins of quantum mechanics', typescript in circulation. A hint of Pauli's role is also to be found in D. Serwer, 'Unmechanischer Zwang: Pauli, Heisenberg, and the rejection of the mechanical atom, 1923–25', *Historical Studies in the Physical Sciences,* viii (1977): 189–256.

64. Forman, *Weimar,* 80–4; R. von Mises, 'Uber die gegenwärtige Krise der Mechanik', *Zeitschrift für angewandte Mathematik und Mechanik,* i(1921): 425–31, and *Naturwissenschaft und Technik der Gegenwart.* Leipzig, 1922; W. Schottky, 'Das Kausalproblem der Quantentheorie als eine Grundfrage der modernen Naturforschung überhaupt. Versuch einer gemeinverständlichen Darstellung', *die Naturwissenschaften,* ix (1921): 492–6, 506–11.

65. See Hendry, op. cit. (note 42).

66. Institut International de Physique Solvay, *Atoms et electrons: Rapports et Discussions du troisième Conseil de Physique tenue à Bruxelles du 1 au 6 avril 1921.* Paris 1921, : 80–100.

67. Ibid., : 120–30. For Millikan's earlier views see Hendry, op. cit. (note 42).

68. Ibid., : 228–47; the original English version of Bohr's paper is reproduced in N. Bohr, *Collected Works,* Vol. 3. Amsterdam, 1976, see especially : 373–5.

69. For Einstein's views on non-conservation and acausality see M. J. Klein, 'The first phase of the Bohr-Einstein dialogue', *Historical Studies in the Physical Sciences,* ii(1970): 1–39.

70. It was shared at this time by Pauli, Bohr, Born, Heisenberg, Schrödinger, Weyl and Sommerfeld, to name only the most important.

71. N. Bohr, 'The effect of electric and magnetic fields on spectral lines', *Proceedings of the Physical Society of London,* xxxv (1923): 275–302 (lecture delivered Mar. 1922), and op. cit. (note 49, 1924), completed in Nov. 1922 and originally published in *Zeitschrift für Physik,* xiii (1923): 117–65. C. G. Darwin, 'A quantum theory of optical dispersion', *Nature,* cx(1922): 841, and 'The wave theory and the quantum theory', *Nature* cxi(1923): 771–3.

72. A. Sommerfeld, *Atombau und Spektrallinien,* 2nd edn. Munich, 1922, translated as *Atomic Structure and Spectral Lines.* London, 1923, : 253. Although many physicists held fast to energy conversation (and although Sommerfeld himself was to vacillate on the issue), it should be noted that the writings of Bohr and Sommerfeld were the prime authorities for quantum theory during this period.

73. M. Born and W. Heisenberg, 'Die Elektronenbahnen im angeregten

Heliumatom', *Zeitschrift für Physik*, xvi(1923): 229–43; A. Einstein and P. Ehrenfest, 'Quantentheoretische Bemerkung zum Experiment von Stern und Gerlach', *Zeitschrift für Physik*, xi(1922): 31–4.

74. E. Schrödinger, notebook, 'Kausalität', 10 Sept. 1918, *Sources for History of Quantum Physics*.

75. E. Schrödinger, 'Ueber eine Bemarkenswerte Eigenschaft des Quantenbahnen eines einzelnen Elektrons', *Zeitschrift für Physik*, xii(1922): 13–23.

76. E. Schrödinger, 'Dopplerprinzip und Bohrsche Frequenzbedingung', *Physikalische Zeitschrift*, xxiii(1922): 301–3.

77. Schrödinger to Pauli, 8 Nov. 1922, op. cit. (note 26), letter no. 29.

78. The basic idea was to re-emerge later when, combined with Heisenberg's rejection of mechanical electron orbits in the atom, it was a fundamental feature of Schrödinger's wave mechanics.

79. E. Schrödinger, 'Was ist ein Naturgesetz', *die Naturwissenschaften*, xvii(1929): 9–11: inaugural lecture at Zürich, Dec. 1922; Forman, *Weimar*, 87–8.

80. N. Bohr, op. cit. (note 71), 279, and op. cit. (note 49, 1924), 20. In both cases the qualification, 'in the present state of science' was given in respect of the absence of causality.

81. H. A. Senftleben, 'Zur Grundlagen der Quantentheorie', *Zeitschrift für Physik*, xxii(1923): 127–56, quotation following from : 127.

82. For Planck's views see op. cit. (note 66), : 93–114.

83. What Forman means exactly by the phrase 'quasi-crank' (*Weimar*, 98) is unclear, but the crank seems to refer to a minor mental breakdown that he suffered in 1924 – a common enough occurrence that does not merit the nomenclature. As for the 'quasi', this may be related to the fact that a letter, cited by Forman as being written by Hans Albrecht Senftleben, is catalogued and indexed (incorrectly) in the *Sources for History of Quantum Physics* as being by someone completely different, namely Hermann Senftleben, an experimentalist. To make up for this unfairness, a paper by Hermann is indexed in van der Waerden's book, *Sources of Quantum Mechanics* (Amsterdam, 1967), as being by Hans Albrecht, while, to return to the catalogue, the letter catalogued as by Hermann and in fact by Hans Albrecht is cross-referenced as being by one H. R. Senftleben, who does not, mercifully, exist.

84. N. Bohr, 'Problems of the atomic theory', reproduced in op. cit. (note 68), quotation from : 571.

85. N. Bohr, H. A. Kramers and J. C. Slater, 'The quantum theory of radiation', *Philosophical Magazine*, xlvii(1924): 785–802; 791.

86. The abandonment of causality followed from the decision to apply the already existing virtual oscillator theory to a single atom rather than to a large number of atoms; this had strong heuristic advantages and was a natural response to Slater's suggestion of a virtual field guiding light-quanta, in the light of Bohr's rejection of light-quanta.

87. E. Schrödinger, 'Bohrs neue Strahlungshypothese und der Energiesatz', *die Naturwissenschaften*, xii(1924): 720–24, and Schrödinger to Bohr, 24 May 1924, *Sources for History of Quantum Physics*.

88. Einstein to Born, 29 Apr. 1924, op. cit. (note 27); see also Einstein to Ehrenfest, 31 May 1924, 12 June 1924, discussed by Klein, op. cit. (note 69), 33.

89. Pauli to Bohr, 2 Oct. 1924, and see also Pauli to Sommerfeld, Nov. 1924, 6

Dec. 1924; op. cit. (note 26), letters nos. 66, 70, 72. A. H. Compton, 'The scattering of X-rays', *Journal of the Franklin Institute*, cxcviii(1924): 61–71; 70. J. H. van Vleck, 'Quantum principles and line spectra', *Bulletin of the National Research Council*, liv(1926):70. E. C. Stoner, 'The structure of radiation', *Proceedings of the Cambridge Philosophical Society*, xxii (1925): 577–94; 582. Ehrenfest to Einstein, 9 Jan. 1925, quoted and translated by Klein, op. cit. (note 69), 31. J. C. Slater, 'The nature of radiation', *Nature*, cxvi(1925): 278. Sommerfeld's views are contained in the work cited by Compton.

90. Heisenberg first reported that 'I do not really see it as an essential progress': Heisenberg to Pauli, 4 Mar. 1924, op. cit. (note 26), letter no. 57. He only came round to it after Born had linked it up with a difference equation approach that they (Born and Heisenberg) had been pursuing since the previous fall, and had taken its technique 'independent of the critically important and still disputed framework of that theory': M. Born, 'Ueber Quantenmechanik', *Zeitschrift für Physik*, xxvi(1924): 379–95; 379. For the difference equation approach see Serwer, op. cit. (note 63).

91. H. A. Kramers, 'The law of dispersion and Bohr's theory of spectra', *Nature*, cxiii(1924): 673–4, and 'The quantum theory of dispersion', *Nature*, cxiv(1924): 310. Born, op. cit. (note 90). Heisenberg to Pauli, 8 June 1924, op. cit. (note 26), letter no 62. Ladenburg to Kramers, 31 May 1924, *Sources for History of Quantum Physics*, and see also Ladenburg to Kramers, 8 June 1924, ibid., where Einstein's reaction to the new theory is described as 'not unfavourable' (nicht ungünstig).

92. Kramers, Ladenburg and Reiche were between them responsible for the development of the Bohr dispersion theory prior to the introduction to the new technique, while Born and Heisenberg were able to incorporate this technique into their existing research programme, with considerable effect. Pauli had concerned himself, privately, with the dispersion theory, but had already hinted at innovations more fundamental than those suggested by Bohr: Paulo Sommerfeld, 6 June 1923, op. cit. (note 26), letter no. 37.

93. W. Bothe and H. Geiger, 'Experimentelles zur Theorie von Bohr, Kramers und Slater', *die Naturwissenschaften*, xiii(1925): 440–1. Born was already trying to develop de Broglie's theory before the experimental results were known, while even Bohr had prepared himself for an unfavourable result: Bohr to Heisenberg, 18 Apr. 1925, and Born to Bohr, 24 Apr. 1925, *Sources for History of Quantum Physics*.

94. See J. Hendry, 'The mathematical formulation of quantum theory and its physical interpretation, 1900–1927' (Ph.D. thesis, University of London, 1978), Chapter 3.

95. H. Reichenbach, 'Die Kausalstruktur der Welt und der Unterschied von Vergangenheit und Zukunft', *Bayerische Akademie der Wissenschaften, München, math.-naturwiss. Abteilung, Sitzungsberichte,* (1925): 133–75; Forman, *Weimar*, 88–91.

96. M. Born, 'Zur Quantenmechanik der Stossvorgänge', *Zeitschrift für Physik*, xxxvii(1926): 863–7.

97. Ibid., 865 (my translation).

98. Note 28.

99. H. Reichenbach, 'Wahrscheinlichkeitsgesetze und Kausalgesetze, *die Umschau*, ixxx(1925): 789–92.

100. Spengler, op. cit. (note 18), : 419 of translation.

101. M. Born, 'Das Adiabatenprinzip in der Quantenmechanik', *Zeitschrift für Physik*, xl(1927): 167–92; 192.

102. Ibid., 192.

103. Weyl's assistance was acknowledged by Schrödinger in his first paper on wave mechanics; for his involvement in the development of matrix mechanics see Weyl to Jordan, 13 Nov. 1925, 23 Nov. 1925, and 25 Nov. 1925, *Sources for History of Quantum Physics*, and Born to Weyl, 3 Oct. 1925, archive of the Eigener Technische Hochschule, Zurich. Despite their intellectual differences, Weyl was at this time virtually an honorary member of the mathematics department at Göttingen, and he may have seen Born there when the latter returned from America in the spring of 1926: see Courant to Weyl, 19 Feb. 1926 and 14 June 1926, ETH Zurich.

104. This attitude is clearest in his second paper on the subject, M. Born, 'Quantenmechanik und Stossvorgänge', *Zeitschrift für Physik*, xxxviii(1926): 803–27.

105. For a detailed discussion of Born's work and a discussion of this aspect of its background see J. Hendry, op. cit. (note 94): 125–31.

106. Forman, *Weimar*, 63–70.

107. For an idea as to the complexity of the causality issue see M. Bunge, *Causality*. Cambridge, Mass., 1959.

108. Hence the confusion over the Bohr–Kramers–Slater theory.

109. Note 63 above.

110. W. Heisenberg, 'Ueber den anschaulichen Inhalt der quantentheoretisch Kinematik und Mechanik', *Zeitschrift für Physik*, xliii(1927): 172–98; N. Bohr, *Nature* cxxi(1928), 580–90.

The same position was also taken early on by P. Jordan, 'Kausalität und Statistisch in der modernen Physik', *die Naturwissenschaften*, xv(1927): 105–7, and 'Philosophical foundation of quantum theory', *Nature*, cxix(1927): 566; it was also soon accepted by Born: M. Born, 'Ueber den Sinn der physikalischen Theorien', *die Naturwissenschaften*, xvii(1928): 109.

111. In a work that claims to be objective history, Forman's continued use of emotive language is most striking. One is even tempted to ask whether extrinsic influences led Paul Forman to ardently hope for, actively search for, and willingly embrace the theses he puts forward.

112. One would imagine that religious attitudes would constitute an important element of the milieu, but despite the importance generally attributed by historians to the role of anti-semitism in Weimar attitudes, Forman makes no mention of this, or of any other religious matter.

113. Yet another problem omitted by Forman concerns the isolation of intellectuals in general – even those adopting the views of the milieu – from the milieu. This is mentioned by Ringer and emphasised by Laqueur, op. cit. (note 10).

114. In this respect as in others the role of the observer in quantum theory would appear to be a far more promising subject of study than that of causality.

115. There does appear, on the other hand, to be a degree of social determination on the explicit level in respect of the *reception* (as opposed to the *creation*) of scientific theories. Forman's second paper on the causality issue, op. cit. (note 3), is concerned with the more promising

area of reception, and though it again runs into problems by leaving out of consideration differing degrees of familiarity with the internal background it does suggest a strong social element to the short-term reception of the new quantum mechanics.

Indexes

Subject index